Handbook of Poultry Science

Handbook of Poultry Science

Edited by Ralph Owens

SYRAWOOD
PUBLISHING HOUSE

New York

Published by Syrawood Publishing House,
750 Third Avenue, 9th Floor,
New York, NY 10017, USA
www.syrawoodpublishinghouse.com

Handbook of Poultry Science
Edited by Ralph Owens

International Standard Book Number: 978-1-68286-791-4 (Hardback)

Cataloging-in-Publication Data

Handbook of poultry science / edited by Ralph Owens.
      p. cm.
Includes bibliographical references and index.
ISBN 978-1-68286-791-4
1. Poultry. 2. Eggs--Production. 3. Aviculture. I. Owens, Ralph.
SF487 .H36 2019
636.5--dc23

TABLE OF CONTENTS

PREFACE

Every book is initially just a concept; it takes months of research and hard work to give it the final shape in which the readers receive it. In its early stages, this book also went through rigorous reviewing. The notable contributions made by experts from across the globe were first molded into patterned chapters and then arranged in a sensibly sequential manner to bring out the best results.

Poultry refers to the domesticated birds, which are reared for their eggs, feathers and meat. These birds belong to the Galliformes order, which consists of quails, turkeys and chickens. Poultry is widely eaten all over the globe. These birds provide nutritionally beneficial food, which contains low fat and high quality protein. Poultry are kept in coops and hatcheries, but they are more often reared in battery cages and furnished cages. They can suffer from diseases like avian influenza and also cause outbreaks of infectious agents like Salmonella, Campylobacter and Escherichia coli. Therefore, they are generally farmed intensively with routine use of antibiotics for preventing or managing disease outbreaks. All these aspects of poultry farming and disease management are studied under the domain of poultry science. This book provides comprehensive insights into the field of poultry science. The topics covered herein offer the readers new insights into this subject. This book is a vital tool for all researching or studying poultry science as it gives incredible insights into emerging trends and concepts.

It has been my immense pleasure to be a part of this project and to contribute my years of learning in such a meaningful form. I would like to take this opportunity to thank all the people who have been associated with the completion of this book at any step.

Editor

Energy Type and Amount in the Diets of Broiler Chickens: Effects on Performance and Duodenal Morphology

A. Ghahremani[1], A.A. Sadeghi[1*], S. Hesaraki[2], M. Chamani[1] and P. Shawrang[3]

[1] Department of Animal Science, Science and Research Branch, Islamic Azad University, Tehran, Iran
[2] Deptartment of Veterinary Medicine, Science and Research Branch, Islamic Azad University, Tehran, Iran
[3] Nuclear Agriculture Research School, Nuclear Science and Technology Research Institute, Atomic Energy Organization of Iran, Karaj, Iran

*Correspondence E-mail: a.sadeghi@srbiau.ac.ir

ABSTRACT

The present study was conducted to investigate the effects of dietary energy sources and levels on performance and small intestinal morphology in broiler chickens. A total of 600 one-day-old broiler chicks were randomly divided into five treatments with four replicates each. Chicks were fed diet based on corn as main energy source and energy level based on Cobb 500 manual instruction considered control group (C), basal diet with 3% lesser energy than control (T1), basal diet with 6% lesser energy than control (T2), basal diet based on corn and soy oil level according to Cobb 500 manual instruction (T3), basal diet based on corn and soy oil with 3% upper energy (T4) for 42 days. Results showed that chicks in T3 group had higher body weight, body weight gain and duodenum villus height compared to control group (C) and improved feed conversion ratio (FCR) at day 42 of age (P<0.05). Chicks in T2 group exhibited the lowest body weight (BW), body weight gain (BWG) and FCR but the highest feed intake (P<0.05). Feeding of diet T4 improved daily weight gain and duodenal villus height while caused concurrently increased FCR. Energy levels greater than Cobb recommendation significantly increased the villus height of the duodenum and decreased crypt depth compared to the control group (P<0.05). In order to achieve a higher weight more energy is needed than the recommended manual instruction for Cobb 500 but to have better feed conversion ratio the energy level recommended manual instruction is sufficient.

KEY WORDS broiler chickens, duodenum morphology, energy, performance, soy oil.

INTRODUCTION

The small intestine integrity in terms of both morphology and function is critical for animal development and growth (Ziegler *et al*. 2003). The intestinal epithelium of broiler chicken is responsible for the growth potential after hatching (Uni *et al*. 1998) and the development of intestinal morphology and function contributes to body weight increase of broiler chickens (Tarachai and Yamauchi, 2000). Small intestine of chickens develops rapidly during the first 5 days after birth (Noy and Sklan, 1998). Development of the intestinal villi in the chicken's early life could increase efficiency of nutrient utilization and enhance the growth performance. Furthermore, an increase in villi height may increase the intestinal surface area and nutrient absorption (Soltan, 2009). The small intestine is a metabolically active organ (Spratt *et al*. 1990) and many factors can affect its development (Thompson and Applegate, 2006). Energy is an important nutrient that constitutes the largest component of the diet and thus affects on potential growth performance, feed cost and profitability. The majority of the energy in poultry diets is obtained from carbohydrates while fat and protein can also yield energy (Leeson and Summers, 2005). Vegetable oils such as soy oil are frequently included in broiler diets to increase the energy density of the diet, improve efficiency and increase nutrient digestibility in broilers (Monfaredi et al. 2011). Soy oil also provide varying quantities of the essential fatty acid of linoleic acid (Leeson et al. 2001). Another important role of soy oil in diets is inhibition from de novo lipogenesis in broiler chickens (Wongsuthavas et al. 2011). Nutrient density is an important factor that may affect animal intestine development. High apparent metabolizable energy or high amino acid densities in the diets fed to broilers from 8 to 21 days of age improved their feed conversion ratio (Wang et al. 2014). Chickens fed a higher nutrient density diet grow faster throughout all growing phases (Nahashon et al. 2005; Zai et al. 2013). Intestinal structures may also be further modified to adapt to nutrition manipulation during growing phases and it seems to change energy levels can affect the small intestine structure. The appropriate structure of the small intestine can causes better nutrient absorption and ultimately improves growth performance in broiler chickens. In previous studies, experimental diets were provided according to broilers nutrients requirement suggested by NRC (1994). While nutritional needs of different strains of broiler chicken published by the companies of broiler breeder in recent years. In this experiment, energy and protein level in control diet were used according to Cobb 500 instruction manual, 2012. Genetic changes into improve performance that are provided by the companies producing strains of broiler chickens, because in this present study was decided , the effects of dietary energy level was used 3% lesser and upper based on Cobb 500 manual instruction on performance. Also, information on the effect of energy level in diets according to Cobb 500 manual instruction on duodenal morphology and performance of broiler chicken is lacking. Therefore, the aim of this study was to evaluate the effect of dietary energy sources and levels on growth performance and small intestinal morphology in broiler chickens.

MATERIALS AND METHODS

A total of 600 one-day-old Cobb 500 broiler chicks were obtained from a local hatchery randomly allocated to one of 5 dietary treatments (4 replicates per treatment and 30 chicks per replicate). Detailed descriptions of bird management (water, feed, light program and pen environment) and the arrangement of treatment groups in the broiler facility were presented in a companion study by Cobb 500 broiler chickens (Cobb 500, 2010). The lighting, tempera-

ture and air conditioning program used was consistent with the specifications in the Cobb lineage manual during the experimental period (Cobb 500, 2010). Feed intake and live body weight were recorded weekly and then feed conversion ratio (FCR) was calculated. Dead chicks were collected daily and weighed at the time of carcass removal; carcass weights were included in the FCR calculations.

In this experiment, chicks were fed by basal diet based on corn and energy level by Cobb 500 manual instruction as control group (C), basal diet with 3% lesser energy than control (T1), basal diet with 6% lesser energy than control (T2), basal diet based on corn and soy oil energy level according to Cobb 500 manual instruction (T3), basal diet based on corn and soy oil with 3% upper energy (T4). The experimental diets formulated by using (Cobb 500, 2012) (Table 1) and animal and poultry feed formulation (UF-FDA) software. Diets and fresh water were provided ad libitum during this experiment. Birds were fed the starter diet from day 1 to 10, the grower diet from day 11 to 28 and finisher diet from day 29 to 42. The experimental diets in mashed form were based on corn, soybean meal and in some treatments the gluten of corn and soy oil were used to adjust the levels of protein and energy contents.

The following growth performance variables were evaluated: body weight (BW), body weight gain (BW gain), feed intake (FI), daily feed intake (DFI) and feed conversion ratio (FCR). The birds were weighed on the first day of the experiment, then weighed weekly throughout the remaining experimental period (7 to 42 d of age). The feed was provided weekly and the leftover fed were weighed weekly to calculate the feed conversion ratio.

At 28 d, 4 birds per pen (20 chicks/dietary treatment) were dissected for determination of duodenum. Birds were weighed and killed by cervical dislocation. Duodenual samples (2 cm in length) were obtained from of the midpoint of each chick. The intestinal segments were flushed with cold PBS and fixed in 10% neutral buffered formalin phosphate for subsequent morphological examination.

The duodenum samples were processed by dehydration through a series of graded alcohol solutions (50, 70, 80, 90, 95 and 100%) cleared with xylene and embedded in paraffin. Paraffin sections (5-μm thickness) were mounted on glass slides. The slides were stained using routine procedures for Mayer's hematoxylin and eosin (Yaghobfar et al. 2006).

Villi were photographed under alight microscope (Olympus CX31RBSF attached cameraman) using the method presented by (Wang et al. 2015). Morphometric parameters of duodenum villi were performed at a magnification of 40X and goblet cells were performed at measurement of 400X all measurements were analyzed using image j software.

Table 1 Ingredients and chemical composition of the experimental diets for broiler chicks

Ingredients (%)	Starter (0-10 days old)					Grower (11-28 days old)					Finisher (29-42 days old)				
	C	T1	T2	T3	T4	C	T1	T2	T3	T4	C	T1	T2	T3	T4
Corn grain	63.3	63.2	62.1	58.2	54.2	69.2	68.8	67.7	65.3	59.3	70.2	71.47	71.5	65.1	62.4
Soybean meal	22.6	28.9	31.5	31.9	32.2	18.0	23.4	26.2	24.0	28.4	19.3	20.0	24.2	25.8	25.0
Soybean oil	-	-	-	2.5	4.14	-	-	-	2.00	5.00	-	-	-	3.50	5.00
Corn gluten meal	9.12	3.01	-	2.67	4.68	8.22	3.27	-	4.21	2.96	6.21	4.2	-	1.47	3.50
Dicalcium phosphate	2.07	2.06	2.06	2.05	2.05	1.90	1.90	1.90	1.90	1.90	1.70	1.70	1.70	1.70	1.70
Calcium carbonate[1]	1.06	1.03	1.01	1.01	1.01	1.05	1.05	1.05	1.05	1.05	0.92	0.92	0.920	0.900	0.900
Salt	0.380	0.380	0.380	0.380	0.380	0.370	0.370	0.370	0.370	0.370	0.32	0.32	0.320	0.320	0.320
DL-methionine	0.340	0.410	0.450	0.400	0.390	0.220	0.270	0.300	0.250	0.250	0.180	0.220	0.270	0.220	0.200
L-lysine	0.530	0.400	0.350	0.310	0.360	0.440	0.340	0.280	0.320	0.220	0.360	0.360	0.270	0.210	0.220
L-threonine	0.100	0.110	0.130	0.080	0.090	0.100	0.100	0.120	0.100	0.050	0.090	0.090	0.100	0.060	0.040
Vitamin and mineral per-mix[2]	0.500	0.500	0.500	0.500	0.500	0.500	0.500	0.500	0.500	0.500	0.500	0.500	0.500	0.500	0.500
Choline chloride	-	-	-	-	-	-	-	-	-	-	0.220	0.220	0.220	0.220	0.220
Filler[3]	-	-	1.52	-	-	-	-	1.58	-	-	-	-	-	-	-
Total	100	100	100	100	100	100	100	100	100	100	100	100	100	100	100
Calculated nutrient content															
ME (kcal/kg)	3035	2934	2853	3035	3120	3108	3014	2921	3108	3201	3185	3085	2990	3185	3275
CP (%)	21.0	20.3	19.7	21	21.5	19.0	18.4	17.8	19.0	19.5	18.0	17.4	16.8	18.0	18.5
Ca (%)	0.900	0.900	0.900	0.900	0.900	0.840	0.840	0.840	0.840	0.840	0.760	0.760	0.760	0.760	0.760
Available phosphorus (%)	0.450	0.450	0.450	0.450	0.450	0.420	0.420	0.420	0.420	0.420	0.380	0.380	0.380	0.380	0.380
Na (%)	0.170	0.170	0.170	0.170	0.170	0.170	0.170	0.170	0.170	0.170	0.160	0.160	0.160	0.160	0.160
Digestible methionine (%)	0.670	0.670	0.670	0.670	0.670	0.530	0.530	0.530	0.530	0.530	0.480	0.480	0.480	0.480	0.480
Digestible lysine (%)	1.18	1.18	1.18	1.18	1.18	1.05	1.05	1.05	1.05	1.05	0.950	0.950	0.950	0.950	0.950
Digestible threonine (%)	0.770	0.770	0.770	0.770	0.770	0.690	0.690	0.690	0.690	0.690	0.650	0.650	0.650	0.650	0.650

ME: metabolizable energy and CP: crude protein.
C: control; T1: basal diet with 3% lesser energy than control; T2: basal diet with 6% lesser energy than control; T3: basal diet based on corn and soy oil energy level according to Cobb 500 manual instruction and T4: basal diet based on corn and soy oil with 3% upper energy.
[1] Per kg contains: Ca: 23% and P: 18.5%.
[2] Supplied by Razak Co., Tehran, Iran, and provided per kilogram of diet: vitamin A: 11000 IU; vitamin D$_3$: 2000 IU; vitamin E: 18 IU; vitamin K: 4 mg; vitamin B$_{12}$: 0.015 mg; Thiamine: 1.8 mg; Riboflavin: 6.6 mg; Calcium pantothenic acid: 12.0 mg; Niacin: 30.0 mg; Pyridoxine: 2.9 mg; Folic acid: 1.0 mg; choline: 260.0 mg; Manganese: 64.5 mg; Zinc: 33.8 mg; Iron: 100.0 mg; Copper: 8.0 mg; Iodine: 1.9 mg and Selenium: 0.25 mg.
[3] Inert filler used to complete diet formulations to 100%.

Morphometric parameters recorded included total villus height (VH) (from the tip to the bottom of each villus), mid-point villus width (VW), crypt depth (CD) (from the base to its opening), villus:crypt ratio (V:C; calculated by villus height by crypt depth) and goblet cell count (GC count) according to the procedure of (Wang *et al.* 2015).

A randomized completely design with 4 replications was used to test for effects of dietary treatment on growth performance (BW, BW gain, FI and FCR) and duodenal morphology parameters. All parameters were analyzed using SPSS (SPSS, 2010). Data derived from experiment were analyzed by a one way ANOVA. Significant differences among means were determined by Duncan's multiple range test (P<0.05).

RESULTS AND DISCUSSION

The body weight, body weight gain, feed intake, and feed conversion ratio at the 28 and 42 days of age of the birds relative to the energy sources and levels in the diet are shown in Tables 2 and 3. The results showed that dietary treatments significantly affected on BW, BW gain, FI and FCR at 28 and 42 d (P<0.05). In this study it was shown that using of the T3 and T4 increased body weight and daily weight gain.

Table 2 Effects of dietary energy sources and levels on growth performance in d 28 of age

Treatment	Feed intake (g/day/bird)	Live weight (g)	Daily weight gain (g/day/bird)	FCR
C	58.4[b]	1279[b]	40.2[ab]	1.45[c]
T1	61.2[ab]	1199[d]	39.6[b]	1.54[b]
T2	66.8[a]	1086[e]	38.6[b]	1.72[a]
T3	59.9[b]	1243[c]	41.1[ab]	1.44[c]
T4	62.2[ab]	1337[a]	41.3[a]	1.50[bc]
P-value	0.002	0.000	0.158	0.000
SEM	0.811	19.825	0.384	0.026

C: control; T1: basal diet with 3% lesser energy than control; T2: basal diet with 6% lesser energy than control; T3: basal diet based on corn and soy oil energy level according to Cobb 500 manual instruction and T4: basal diet based on corn and soy oil with 3% upper energy.
The means within the same row with at least one common letter, do not have significant difference (P>0.05).
SEM: standard error of the means.
FCR: feed conversion ratio.

But feed conversion ratio was significantly higher for the birds that fed T4 diet compared to T3. Feed intake and feed conversion ratio were significantly increased, while body weight and daily weight gain were significantly reduced for the broilers fed diets with T1 and T2 compared to control diet. Birds fed high level of energy exhibited the best BW and daily weight gain (DWG). Using of the basal diet based on corn and soy oil energy level according to Cobb 500 manual instruction (T3) exhibited better growth parameters compared with basal diet based on corn and energy level by Cobb 500 manual instruction as control group (C).

Table 3 Effects of dietary energy sources and levels on growth performance in d 42 of age

Treatment	Feed intake (g/day/bird)	Live weight (g)	Daily weight gain (g/day/bird)	FCR
C	86.2[bc]	2160[b]	47.6[b]	1.81[c]
T1	88.4[ab]	2093[c]	45.3[c]	1.94[b]
T2	89.3[a]	2030[d]	43.3[d]	2.06[a]
T3	84.3[c]	2188[b]	48.5[b]	1.73[d]
T4	87.4[b]	2260[a]	50.0[a]	1.74[d]
P-value	0.002	0.000	0.000	0.000
SEM	0.490	18.690	0.579	0.029

C: control; T1: basal diet with 3% lesser energy than control; T2: basal diet with 6% lesser energy than control; T3: basal diet based on corn and soy oil energy level according to Cobb 500 manual instruction and T4: basal diet based on corn and soy oil with 3% upper energy.
The means within the same row with at least one common letter, do not have significant difference (P>0.05).
SEM: standard error of the means.
FCR: feed conversion ratio.

Morphological examination showed that dietary treatments significantly affected on duodenum villus height, crypt depth and villus height:crypt depth (P<0.05) (Table 4).

In this experiment, significantly higher villus height and villus height:crypt depth also lower crypt depth were observed in the duodenum of the birds fed T3 and T4 diets as compared with the control diet. The results showed that birds fed diets with low levels of energy (T1 and T2 diets) significantly exhibited lower villus height and villus height:crypt depth also higher crypt depth compared with the control. Birds fed dietary treatment with basal diet based on corn and soy oil with 3% upper energy (T4) significantly exhibited the highest villus height and villus height:crypt depth also lowest crypt depth versus the control.

No dietary effect was apparent for villus wide and goblet cell count at the duodenum (Table 4). But birds fed dietary treatment with low levels of energy (T1 and T2 diets) exhibited the highest villus wide and the lowest goblet cell count. Birds fed dietary treatment with basal diet based on corn and soy oil energy level according to Cobb 500 manual instruction (T3) exhibited the lowest villus wide and highest goblet cell count.

The results of this study indicating that different levels of energy significantly affected on BW, BW gain, FI and FCR were consistent with previous reports (Yang et al. 2007; Min et al. 2012). Both ME and crude protein (CP) significantly affected on BW and BW gain at 42 days of age (P<0.05), ME levels can significantly affected on FI (P<0.05) while CP did not (Min et al. 2012). Feed intake was reduced with energy increase to 3100 kcal/kg and a concurrent improvement on FCR was found when energy

increased to 3000 kcal/kg (Vieira et al. 2006). Our findings indicated that feeding broilers with soy oil can lead to significant improvements in growth. The results showed that using of basal diet based on corn and soy oil energy level according to Cobb 500 manual instruction (T3) exhibited the best FCR results, that was also consistent with previous observations (Houshmand et al. 2011). Soy oil is frequently included in broiler diets to increase the energy density of the diet, improve efficiency and increase nutrient digestibility in broilers (Monfaredi et al. 2011). The results showed that birds fed dietary treatment basal diet based on corn and soy oil energy level according to Cobb 500 manual instruction (T3) exhibited better BW, BW gain and FCR compared to control at 42 days of age. It seems, to add soy oil as source of energy was caused to better growth performance in broiler chicks. In previous researches, the positive effect of soy oil on growth performance of broilers is well documented (Griffiths et al. 1977; Scaife et al. 1994). The final live weight (FLW) was significantly highest in broiler chickens fed dietary treatment with normal energy and normal protein (NENP) and was lowest in broiler chickens fed dietary treatment with low energy and high protein (LEHP) (P<0.05). The BW gain also followed similar trend as FLW. Feed conversion ratio was significantly better in broiler chickens fed dietary treatment NENP, a trend also exhibited by the protein efficiency ratio (PER) (P<0.05) (Dairo et al. 2010). The results showed that birds fed dietary treatment with low levels of energy exhibited the lowest BW, BW gain and FCR and the highest feed intake. Lower energy diets containing insufficient energy for the protein synthesis have a consequence catabolism of AA to make up this deficiency; this would in turn result in reduced growth and poorer feed efficiency.

Birds fed dietary treatment with high level of energy significantly exhibited the highest BW and BW gain (P<0.05). Increased live weight was mostly due to higher ME consumption in same unit of diets by chickens, similarly supplementation of oil caused a positive trend in cumulative live weight gain (g/bird) of broilers at different ages (Das et al. 2014). In this study, observed limitation to expected FI increment was in part probably due to limitation in physical capacity of gastrointestinal tract. Our explanation is in agreement with that claimed by Griffiths et al. (1977) and Kamran et al. (2008).

In our study, birds fed basal diet based on corn and soy oil with 3% upper energy (T4) exhibited higher FI compared to control. This is in disagreement with Leeson et al. (1996). In this experiment, it seems that other reasons could affect on broiler chickens feed intake according to the same supply of nutrients in the diet and changing dietary energy levels. It seems, that energy density of rations had changed BW, BW gain, FI and FCR.

Table 4 Effects of energy sources and levels on duodenal morphology (µm) in 28 d

Treatment	Villus height (µm)	Villus width (µm)	Crypt depth (µm)	V/C[1]	Goblet cell (count/mm)[2]
C	375[b]	51.0	60.0[ab]	6.25[c]	140
T1	350[bc]	54.0	66.0[ab]	5.30[d]	138
T2	301[c]	62.0	74.0[a]	4.06[e]	137
T3	475[a]	50.0	54.2[b]	8.76[b]	147
T4	527[a]	52.0	50.5[b]	10.43[a]	146
P-value	0.000	0.626	0.027	0.000	0.606
SEM	20.338	2.784	2.852	8.132	2.201

[1] V/C: villus height to crypt depth ratio.
[2] Goblet cell densities were calculated as the number of goblet cells per unit of villus height.
C: control; T1: basal diet with 3% lesser energy than control; T2: basal diet with 6% lesser energy than control; T3: basal diet based on corn and soy oil energy level according to Cobb 500 manual instruction and T4: basal diet based on corn and soy oil with 3% upper energy.
The means within the same row with at least one common letter, do not have significant difference (P>0.05).
SEM: standard error of the means.
FCR: feed conversion ratio.

The results showed that birds fed dietary treatment with low levels of energy significantly exhibited the lowest duodenum villus height and VH:CD ratio also the highest crypt depth (P<0.05). A shortening of the villus and large crypt can lead to poor nutrient absorption, increased secretion in the gastrointestinal tract, and lower performance (XU et al. 2003).

Chicks exhibited a lower feed conversion ratio (FCR) and a higher BW gain when their crypts were shorter (Wang et al. 2015). In this study, birds fed dietary treatment with high level of energy significantly exhibited the highest duodenum villus height, villus height:crypt depth but the lowest crypt depth (P<0.05). Dietary treatment did not affect duodenum villus width and goblet cell count. But birds fed dietary treatment with low levels of energy exhibited the highest villus width and the lowest goblet cell count. The results showed that birds fed basal diet based on corn and soy oil energy level according to Cobb 500 manual instruction (T3) exhibited the lowest villus width and the highest goblet cell count. The results were in agreement with previous reports by Fan et al. (1997), Tarachai and Yamauchi (2000) and Xu et al. (2003).

Chickens fed a higher nutrient density diet grow faster throughout all growing phases (Nahashon et al. 2005; Zai et al. 2013). In contrast, increases in the villus height and villus height:crypt depth were directly correlated with increased epithelial cell turnover (Fan et al. 1997) and longer villi were associated with activated cell mitosis (Samanya and Yamauchi, 2002). The villi play a crucial role in the digestion and absorption processes of the small intestine, as is the first to make contact with nutrients in the lumen (Gartner and Hiatt, 2001). Dietary fat and probiotic supplementation significantly increased villus height of the duodenum. Villus height is known to correlate positively with nutrient absorption (Tarachai and Yamauchi, 2000) but improvement in broiler growth occurred among those fed soy oil diets only. Due to this reasons, using of the T3 and T4 diets exhibited the best duodenum villus height compared to control.

In this research, it seems that soy oil and nutrient density improved morphological parameters and growth performance.

Goblet cell secret mucin in the digestive tract to protect the intestinal membrane from digestive enzyme degradation and pathogen invasion (Wang et al. 2015). The results showed that birds fed basal diet based on corn and soy oil energy level according to Cobb 500 manual instruction (T3) exhibited numerous goblet cell counts. Basal diet based on corn and soy oil energy level according to Cobb 500 manual instruction (T3) improved performance and small intestine structure compared to control.

In this study, there was positive relationship between soy oil consumed in diets and small intestine structure with growth performance in Cobb 500 broiler chickens. Birds fed basal diet based on corn and soy oil energy level according to Cobb 500 manual instruction (T3) exhibited the best growth performance, probably resulting from improved morphological parameters.

CONCLUSION

The results showed that the higher energy level rather than nutritional needs based on Cobb 500 broiler chicken requirements as specified in the manual was affective on performance and morphological parameters. These effects indicate that the current nutritional recommendations are not sufficient for achieving the full genetic potential of current broiler strains for body weight gain but in terms of feed conversion ratio the most appropriate level of energy obtained from Cobb 500 broiler chicken requirements in instruction manual.

ACKNOWLEDGEMENT

The authors are grateful to the Islamic Azad University for research funding support. We also thank all staffs in the poultry unit, for the assistance in the care and feeding of chicks used in this research.

REFERENCES

Cobb 500. (2012). Cobb Broiler Performance and Nutrient Supplement Guide. Cobb-Vantress Inc., Siloam Springs, Arkansas.

Cobb 500. (2010). Cobb Broiler Management Guide. Cobb-Vantress Inc., Siloam Springs, Arkansas.

Das G.B., Hossain M.E. and Akbar M.A. (2014). Effects of different oils on productive performance of broiler. Iranian J. Appl. Anim. Sci. **4(1),** 111-116.

Dairo F., Adesehinwa A.S.A.O.K., Oluwasola T.A. and Oluyemi J.A. (2010). High and low dietary energy and protein levels for broiler chickens. *African J. Agric. Res.* **5(15),** 2030-2038.

Fan Y., Croom J., Christensen V., Black B., Bird A., Daniel B., Mcbride M. and Eisen E. (1997). Duodenal glucose uptake and oxygen consumption in turkey poulets selected for rapid growth. *Poult. Sci.* **76,** 1738-1745.

Gartner P. and Hiatt J.L. (2001). Color Textbook of Histology. Saunders, Baltimore, Maryland.

Griffiths L., Leeson S. and Summers J.D. (1977). Influence of energy system and level of various soy oil sources on performance and carcass composition of broiler. *Poult. Sci.* **56,** 1018-1026.

Houshmand M., Azhar K., Zulkifli I., Bejo M.H. and Kamyab A. (2011). Effects of non-antibiotic feed additives on performance, nutrient retention, gut ph, and intestinal morphology of broilers fed different levels of energy. *J. Appl. Poult. Res.* **20,** 121-128.

Kamran Z., Sarwar M., Nisa M., Nadeem M.A., Mahmood S., Barbar M.E. and Ahmed S. (2008). Effect of low protein diets having constant energy-to-protein ratio on performance and carcass characteristics of broiler chickens from one to thirty-five days of age. *Poult. Sci.* **87,** 468-474.

Leeson S., Caston L. and Summers J.D. (1996). Broiler response to diet energy. *Poult. Sci.* **75,** 529-535.

Leeson S., Scott L. and Summers J.D. (2001). Scotts Nutrition of the Chicken. University book, Guelp, Canada.

Leeson S. and Summers J.D. (2005). Feeding Programs for Laying Hens. Commercial poultry nutrition. University books, Guelph, Ontario.

Min Y.N., Shi J.S., Wei F.X., Wang H.Y., Hou X.F., Niu Z.Y. and Liu F.Z. (2012). Effects of dietary energy and protein on growth performance and carcass quality of broilers during finishing phase. *J. Anim. Vet. Adv.* **11(19),** 3652-3657.

Monfaredi A., Rezaei M. and Sayyahzadeh H. (2011). Effect of supplemental soy oil in low energy diets on some blood parameters and carcass characteristics of broiler chicks. *South African J. Anim. Sci.* **41,** 24-32.

Nahashon S., Adefope N.N., Amenyenu A. and Wright D. (2005). Effects of dietary metabolizable energy and crude protein concentrations on growth performance and carcass characteristics of french guinea broilers. *Poult. Sci.* **84,** 337-334.

Noy Y. and Sklan D. (1998). Yolk utilization in the newly hatched poultry. *Br. Poult. Sci.* **39,** 446-451.

NRC. (1994). Nutrient Requirements of Poultry, 9ᵗʰ Rev. Ed. National Academy Press, Washington, DC., USA.

Samanya M. and Yamauchi K. (2002). Histological alterations of intestinal villi in chickens fed dried *Bacillus subtilis var. natto. Comp. Biochem. Physiol.* **133,** 95-104.

Scaife J.R., Moyo J., Galbraith H., Michie W. and Carmpbell V. (1994). Effect of different dietary supplemental fats and oils on the tissue fatty acid composition and growth of female broilers. *Br. Poult. Sci.* **35,** 107-118.

Soltan M. (2009). Influence of dietary glutamine supplementation on growth performance, small intestinal morphology, immune response and some blood parameters of broiler chickens. *Int. J. Poult. Sci.* **8,** 60-68.

Spratt R.S., Mcbride B.W., Baylay H.S. and Leeson S. (1990). Energy metabolism of broiler breeder hens. 2. Contribution of tissues to total heat production in fed and fasted hens. *Poult. Sci.* **69,** 1348-1356.

SPSS Inc. (2010). Statistical Package for Social Sciences Study. SPSS for Windows, Version 11. Chicago SPSS Inc.

Tarachai P. and Yamauchi K. (2000). Effects of luminal nutrient absorption, intra-luminal physical stimulation, and intravenous parenteral alimentation on the recovery responses of duodenal villus morphology following feed withdrawal in chickens. *Poult. Sci.* **79,** 1578-1585.

Thompson K.L. and Applegate T.J. (2006). Feed withdraw alters small intestinal morphology and mucus of broilers. *Poult. Sci.* **85,** 1535-1540.

Uni Z., Ganot S. and Sklan D. (1998). Post-hatch development of mucosal function in the broiler small intestine. *Poult. Sci.* **77,** 75-82.

Vieira S.L., Viola E.S., Berres J., Olmos A.R., Conde O.R.A. and Almeida J.G. (2006). Performance of broilers fed increased levels energy in the pre-starter diet and on subsequent feeding programs having with acidulated soybean soap stock supplementation. *Brazilian J. Poult. Sci.* **8(1),** 55-61.

Wang X., Peebles E.D. and Zhai W. (2014). Effects of protein source and nutrient density in the diets of male broilers from 8 to 21 days of age on their subsequent growth, blood constituents, and carcass compositions. *Poult. Sci.* **93,** 1463-1474.

Wang X., Peebles E.D., Morgan T.W., Harkess R.L. and Zhai W. (2015). Protein source and nutrient density in the diets of male broilers from 8 to 21 d of age: effects on small intestine morphology. *Poult. Sci.* **94,** 61-67.

Xu Z.R., Hu C.H., Xia M.S., Zhan X.A. and Wang M.Q. (2003). Effects of dietary fructooligosaccharide on digestive enzyme activities, intestinal microflora and morphology of male broilers. *Poult. Sci.* **82,** 1030-1036.

Yaghobfar A., Rezaian M., Ashrafi-helan J., Barin H., Fazaeli S. and Sharifi D. (2006). The effect of hull-less barley dietary on the activity of gut microflora and morphology small intestinal of layer hens. *Pakistan J. Biol. Sci.* **9(4),** 659-666.

Yang J.P., Yao J.H. and Liu Y.R. (2007). Effect of feed restriction on growth performance and carcass characteristics of broilers chickens. *Acta Agric. Boreali-Occidentalis Sinica.* **16,** 51-56.

Zai W., Peebles E.D., Zumwalt C.D., Mejia L. and Corozo A. (2013). Effects of dietary amino acid density regimens on growth performance and meat yield of Cobb × Cobb 700 broilers. *J. Appl. Poult. Res.* **22,** 447-460.

2

Efficiency of Peppermint (*Mentha piperita*) Powder on Performance, Body Temperature and Carcass Characteristics of Broiler Chickens in Heat Stress Condition

S. Arab Ameri[1*], F. Samadi[1], B. Dastar[1] and S. Zarehdaran[1]

[1] Department of Animal Science, University of Agricultural Science and Natural Resources of Gorgan, Gorgan, Iran

*Correspondence E-mail: samiraarabameri@chmail.ir

ABSTRACT

This experiment was carried out to evaluate different levels of peppermint (*Mentha piperita*) plant powder usage on feed conversion ratio (FCR), body weight (BW), feed intake (FI), body temperature, carcass parts (breast and thigh) and internal organs (liver, heart, gizzard) weights in broiler chicken. A total of 192 broiler chicken were randomly divided into 4 experimental treatments with 4 replicates (12 birds per replicate) arranged in a completely randomized design. Experimental diets consisted of: (1) basal diet (control); (2) basal diet + 1% peppermint powder; (3) basal diet + 2% peppermint powder and (4) basal diet + 300 mg of vitamin E per kilogram. Heat stress performed by setting room temperature on 34 °C for 8 hours/day from 35 to 42 days of age. Results showed peppermint powder supplement in all levels significantly affected the FCR at 21 days of age and BW at 42 days of age (P<0.05). Birds treated by basal diet plus vitamin E and control diet showed the highest and lowest FCR values, respectively, at 21days of age. Body weight and feed consumption were significantly reduced in birds in the heat stress group. Peppermint powder supplementation at the level of 1% reduced body temperature compared with the control group during heat stress period (P<0.05). Significant differences were observed between dietary treatments for the relative weights of carcass, breast and thigh at 35 days of age and breast, gizzard and liver relative weights at 42 days of age (P<0.05). Birds fed basal diet plus vitamin E had higher carcass weight than the control groups on 35 days. In general, the results of this study revealed that peppermint powder as a natural antioxidant has beneficial effects on chicken growth performance, body temperature regulation and carcass and internal organ weights.

KEY WORDS broiler, carcass traits, heat stress, peppermint, performance.

INTRODUCTION

Medicinal plants have been used for many years in the treatment of various diseases of animals and humans. Much research has been done on the role of medicinal plants in the treatment of diseases of birds, which some of these reported beneficial effects of medicinal plants (Cross *et al.* 2007; Hernandez *et al.* 2004). Peppermint (Mentha) is a member of the Labiatae family and one of the oldest me-

dicinal plants in the world which is a cross between two species namely *M. aquatic* and *M. spicata*. These are endemic to Europe and America, India, China and the Soviet Union. In fact, they are used in tradition medicine of eastern and western nations (Aflatuni, 2005; Schuhmacher *et al.* 2003; Pavela, 2005; Aridogan *et al.* 2002). Different species of mint are of particular importance because of their abundant essential oils that include menthol, carvone, limonene, beta pinene, Mentone, alpha-pinene and geraniol, and effective pharmacological compounds. However, the main flavonoid component of mint is menthol (Schuhmacher *et al.* 2003; Dorman *et al.* 2003). Peppermint, in addition to the essential oils, also contains tannins, glycosides, saponins and other bioactive components (Edris *et al.* 2003). Studies show that medicinal plants due to its antioxidant and flavonoid compounds could play an important role in improving cardiovascular health and liver diseases (Pouramir *et al.* 2006). Peppermint is widely used as a herbal medicine, and it is believed to improve the response of the immune system, fight secondary infections, treat effectively irritable bowel syndrome and inflammation of the gallbladder. It also has antiseptic and antispasmodic effects. Its effect on endophytic fungi muucihas been well examined and its role has been proven (Mimica *et al.* 2003). It has been proved that essential oils of this plant inhibits some types of salmonella infection (Tssou *et al.* 1995) and act against *Candida ablicans* (Ezzat, 2001). As an aromatic plant, it has been traditionally used as medicine. It extends the shelf life of food, inhibit bacteria and fungi growth (Jamroz *et al.* 2003; Menezes *et al.* 2004).

Peppermint oil use as a sedative has expanded in the tropics. It has been applied on the affected area to reduce pain and to improve blood flow. Moderate antibacterial effect against both gram-positive and gram-negative has been reported for this plant (Schuhmacher *et al.* 2003; Jamroz *et al.* 2003). Peppermint oil or peppermint tea is often used to treat gas and indigestion, and it may also increase the flow of bile from the gallbladder (Mimica Dukic *et al.* 2003; Forster, 1996). Also Mentha species can be used in reducing gas and cramping, alcoholic beverages, rheumatism and toothache (Shah and Mello, 2004).

Heat stress refers to the high ambient temperature and the heat resulting from metabolism, which increases body temperature (Aengwanich and Chinrasri, 2002). This is one of the most challenging environmental conditions affecting commercial poultry. Compared to other species of domestic animals, broiler chickens are more sensitive to high ambient temperatures. In spite of a rapid metabolism and high body temperature, broiler chickens have no sweat glands. Furthermore, fast-growing lean broilers generate more heat than their free-living counterparts living in the wild (Geraert *et al.* 1993). These physiological characteristics, in combination with confined housing, make it difficult for broilers to regulate their body temperature. As environmental temperature rises, food consumption, growth rate, feeding efficiency, carcass quality, egg shell quality and survivability all decline (Hashemi *et al.* 2007; Mashaly *et al.* 2004; Borges *et al.* 2004). The most dangerous heat illness is heat shock which may cause mortality. When the environmental temperature is above the thermal comfort zone, birds will suffer from heat stress and changes in blood

acidity and metabolites occur. Cardiovascular system is a system that is involved in the regulation of body temperature. When the birds are housed under heat stress, cardiovascular system changes occur including the balance of acid-base, blood pH, respiratory alkalosis and decreased levels of blood viscosity, hematocrit and plasma protein concentration. And long-term heat stress may lead to damage to the lymphatic organs thus susceptible to bacterial diseases, viral and parasitic increases (Pardue and Thaxton, 1996). In order to reduce the harmful effects of heat stress, it is recommended that the birds be fed in the cool hours during the day and night, or to use food supplements, such as herbal medicine, along with the basal diet to reduce body temperature. These would result in increased feed consumption, feeding efficiency and bird's survival (Hayashi *et al.* 2004). Peppermint to have antimicrobial properties so prevent the growth of harmful bacteria in your digestive system thus improves digestion and absorption and increase the body weight (Aridogan *et al.* 2002). Also, Narimani-Rad *et al.* (2011) reported that dietary supplementation of medicinal plants mixture (1% Oregano, 0.5% Ziziphora and 0.5% Peppermint) caused performance and carcass quality improvement via more weight gain increase in carcass yield and then decreases abdominal fat deposition. And, they investigated 0.3 % ethanolic extract of peppermint to drinking water seem to have a positive influence on broiler performance productive via more carcass yield and decrease abdominal fat deposition. And Hernandez *et al.* (2004) reported supplementation of poultry diets with aromatic plants have a stimulating effects on digestive system of the animals through the increasing the production of digestive enzymes and by improving the utilization of digestive products through enhanced liver function. So that Lee *et al.* (2003) determined an increase in relative liver weight for birds given thymol, but this was seen only at the age of 21 d and not at 40 days that led increases of body weight. Also Galib and Al-Kassie (2010) showed the effect of peppermint on liver weight. They also reported that liver weight of control group was higher than those of the other groups. On the other hand Abbas (2010) presented that feeding of 3 g/kg of fenugreek, parsley and basil seeds had not significantly affected liver, carcass and abdominal fat. The aim of this study was to evaluate the effect of peppermint powder on growth performance of broilers, body temperature, body weight and the weight of internal organs.

MATERIALS AND METHODS

Plant preparation
Peppermint plant used in this experiment was collected in summer season when the plant was in vegetative stage, from the research farm (36°00'-16" north latitude and

59°00'-36" east longitude; altitude:985 m) of Mashhad University of Ferdowsi, Mashhad, Iran. Collected leaves were shadow dried and grounded with a laboratory hammer mill (Iran khodsaz gristmill, ELS 300 C, Iran).

The total values of phenolic compounds were measured colorimetrically, using Folin-Ciocalteu method (Guo et al. 2000).

Chickens, diets and experimental design
A total, 192 one-day-old broiler chickens (Ross, 308) were purchased from a local commercial hatchery and raised over a 42-day experimental period. The chicks were placed in an environmentally controlled poultry house with wood shavings as litter at the research farm of Animal Faculty, Gorgan University of Agricultural Science and Natural Resources, Gorgan, Golestan province, Iran. The temperature was set at 32 °C at 1 day of age and then decreased by 1 °C every 2 days until a permanent temperature of 24 °C was reached at 35 day of experiment. The heat stress was applied once daily (from 0800 to 1600 h=8 h/d) during the last experimental week by increasing room temperature to reach 34 °C. From 1600 to 0800 h, the environmental temperature was reduced to 21 °C. The lighting schedule provided 23 hrs of light per day.

A 2-phase feeding program was used, with a starter diet until 21 day and a finisher diet until 42 day of age. The compositions of the basal diets are shown in Table 1. Diets for each period were prepared with the same batch of ingredients, and all diets within a period had the same compositions. Diets were formulated to meet or exceed requirements by the NRC, (1994) for broilers of these ages.

The experiment was performed as completely randomized design in a 4 × 2 factorial arrangement, with 4 replicates of 12 chicks per replicate.

The dietary treatments included: basal diet (control), basal diet supplemented with 1 and 2 percent peppermint powder and basal diet supplemented with vitamin E. Broilers were provided with unlimited water and food *ad libitum*.

Measuring body temperature
For the measurement of body temperature on days 35 and 42 of the test, from each unit two chickens were randomly selected. The body temperature was measured using a digital thermometer in the time before and the peak of heat stress.

The relative weights of carcass and internal organs
In order to measure the relative weights of carcass and internal organs, at days 35 and 42 two chicks near the average weight of each experimental unit were selected and slaughtered. After slaughtering, carcass and internal organs weights were measured by a digital scale.

Table 1 Ingredients and calculated analyses of the basal diets[1]

Ingredient (%)	0 to 21 d	22 to 42 d
Corn grain	56.5	60.56
Soybean meal	37.27	32.33
Soybean oil	2.38	3.69
Dicalcium phosphate	1.44	1.09
Calcium carbonate	1.28	1.38
Vitamin premix[2]	0.25	0.25
Mineral premix[3]	0.25	0.25
DL-methionine	0.15	0.07
Salt	0.43	0.33
Calculated analysis		
Metabolizable energy (kcal/kg)	2950	3100
Crude protein (%)	21.2	19.38
Calcium (%)	0.92	0.87
P (available) (%)	0.41	0.34
Sodium (%)	0.18	0.15
Lysine (%)	1.15	0.03
Methionine (%)	0.48	0.37
Cysteine (%)	0.83	0.69
Threonine (%)	0.81	0.73

[1] Calculated composition was according to NRC (1994).
[2] Each kg of vitamin premix contained: vitamin A: 3600000 IU; vitamin D$_3$: 800000 IU; vitamin E: 9000 IU; vitamin K$_3$: 1600 mg; vitamin B$_1$: 720 mg; vitamin B$_2$: 3300 mg; vitamin B$_3$: 4000 mg; vitamin B$_5$: 15000 mg; vitamin B$_6$: 150 mg; vitamin B$_9$: 500 mg; vitamin B$_{12}$: 600 mg and Biotin: 2000 mg.
[3] Each kg of mineral premix contained: Mn: 50000 mg; Fe: 25000 mg; Zn: 50000 mg; Cu: 5000 mg; Iodine: 500 mg and Choline chloride: 134000 mg.

Statistical analysis
The experiment was performed as completely randomized design in a 4 × 2 factorial arrangement. A GLM procedure was performed (SAS, 2003) and the difference among the mean values was tested using the Duncan multiple range test at (P<0.05). Mean values and SEM are reported.

RESULTS AND DISCUSSION

Total phenol, flavonoids and antioxidants
The determined amounts of total phenol, flavonoids and antioxidants of alcoholic extract peppermint leaf are shown in Table 2.

Table 2 Total phenol, flavonoids and antioxidant of alcoholic extract peppermint (as dry weight)

Compounds	(mg/g)
Total phenol	2.7
Flavonoids	1.8
Antioxidants	72.0 %

Growth performance
The effects of peppermint powder on growth performance of broilers are shown in Table 3.

Table 3 The effects of peppermint on growth performance[1] of broiler chickens

Treatment	0 to 21			0 to 42		
	BWG	FI	FCR	BWG	FI	FCR
Control	643.36	956.65	1.48[b]	2005.02[ab]	4209.73	2.10
1% peppermint	619.48	951.57	1.53[ab]	1992.17[b]	4210.64	2.11
2% peppermint	650.32	975.12	1.49[ab]	2100.56[ab]	4291.60	2.03
300 mg vitamin E	626.15	968.76	1.54[a]	2121.36[a]	4361.59	2.05
SEM	15.24	17.72	0.02	38.60	51.78	0.03
P-value	0.469	0.774	0.097	0.079	0.169	0.427

The means within the same column with at least one common letter, do not have significant difference (P>0.05).
BWG: body weight gain (g); FI: feed intake (g) and FCR: feed conversion ratio (g of feed/g of BWG).
SEM: standard error of the means.

There were significant (P<0.05) differences among the treatments for feed conversion ratio (FCR) on day 21 and body weight (BW) gain on day 42 of experiment. Vitamin E supplemented group showed significantly (P<0.05) higher BW than the other treatments at 42 day of age. A significantly lower (P<0.05) FCR was observed in control group at 21 day of age.

Body temperature
The effect of dietary treatments on body temperature is shown in Table 4. Results showed significant differences between dietary treatments in body temperature in the peak time of heat stress on day 42. Body temperature was highest in the control group and lowest in 2 percent peppermint powder treated group (P<0.05).

The relative weights of carcass and internal organs
Data on the relative weights of carcass and internal organ are shown in Tables 5 and 6. Carcass, breast and thigh relative weights were significantly influenced by the dietary treatments at day 35 (P<0.05). The treatments did not show any significant effect on liver, heart and gizzard weights at day 35 (P>0.05). Peppermint powder did not affect on relative weights of carcass, breast and thigh on day 42 (P>0.05). The chicks fed diet supplemented with 1% peppermint powder had the highest gizzard and liver weights.
Our results showed that peppermint powder ameliorated the stress effects on BW gain and FCR. Medicinal plant supplements are used commonly as dietary additives for humans.

They are chosen for their non toxic chemical composition, relatively low cost and easy availability. Also, over the past decades, medicinal plants and their extracts have been used in animal diets as feed additives in order to improve their performance, health and the quality of their products .Consistent with our findings, Nobakht and Aghdam Shahriar (2011) report 2% mixture of medicinal herbs (mint, camel thorn and mallow) improved daily gain and FCR than the control group, probably because of the antibacterial and antifungal effects of botanicals used in the experiment.

Data reported by Galib and Al-Kassi, (2010) shows an improvement in BWG and FCR under dietary treatment with peppermint powder. Al-Ankari et al. (2004) observed the positive effect of wild mint on performance of broiler chicks. In contrast, Toghyani et al. (2010) reported no effect for mint leaves on broiler performance measures, which may be due to the amounts or combinations of the active ingredients. Amasaib et al. (2013) revealed no significant effect for Mentha spicata on feed intak. The insignificant effect of addition of spearmint to the basal diet may be due to the fact that, the diets were isocaloric and it is expected that the feed consumption could be similar or may be due to the similar environmental during this period. On the other hand Ocak et al. (2008) reported that adding 0.2 percent mint to basal diet help increase growth and reduce mortality in chickens. Medicinal plants with an impact on microbial population, enhance metabolizable energy. In addition, the mint with antimicrobial properties prevent the growth of harmful bacteria in the digestive tract, thereby improving the digestion and absorption as well as body weight gain (Cross et al. 2007). Results of this study are in agreement with some previous research that indicated herbs, plant extracts, essential oil and / or the main components of essential oil did affect body weight, feed intake and feed efficiency in broilers because they have appetizing and digestion-stimulating properties and antimicrobial effects (Abbasi and Samadi, 2014). Most studies about medicinal herbs have been carried out their essential oils. However, the vast majority of studies on dietary essential oil supplementation did not find any stimulating or depressive effect of oils on voluntary feed intake of broiler chickens (Hernandez et al. 2004). The findings about body temperature of this study are in agreement with some preceding studies. Heat stress results from a negative balance between the net amount of energy flowing from the animal's body to its surrounding environment and the amount of heat energy produced by the animal. This imbalance may be caused by variations of a combination of environmental factors (e.g., sunlight, thermal irradiation and air temperature, humidity and movement) and characteristics of the animal e.g., species, metabolism rate and thermoregulatory mechanisms) (Nienaber and Hahn, 2007; Renaudeau et al. 2012).

Table 4 The effects of peppermint on body temperature (°C) of broiler chickens

Treatment	35 d		42 d	
	Before heat stress	During peak heat stress	Before heat stress	During peak heat stress
Control	41.62	42.11	41.50	42.28[a]
1% peppermint	41.58	41.87	41.38	42.05[ab]
2% peppermint	41.60	41.90	41.45	42.07[ab]
300 mg vitamin E	41.59	41.93	41.43	42.07[ab]
SEM	0.97	0.16	0.96	0.14
P-value	0.08	0.07	0.07	0.07

The means within the same column with at least one common letter, do not have significant difference (P>0.05).
SEM: standard error of the means.

Table 5 The effects of peppermint on organs weight (% of body weight) of broiler chickens at 35 d

Treatment	Carcass	Thigh	Breast	Heart	Gizzard	Liver
Control	54.65[b]	16.70[b]	17.05[b]	0.58	1.87	2.71
1% peppermint	54.65[b]	16.58[b]	17.52[b]	0.60	1.97	2.86
2% peppermint	55.20[b]	17.06[b]	18.00[b]	0.60	1.96	2.76
300 mg vitamin E	57.67[a]	18.09[a]	20.19[a]	0.65	1.91	2.62
SEM	1.05	0.40	0.90	0.03	0.13	0.24
Gender						
Male	56.40	17.41	18.46	0.60	1.94	2.62
Female	54.90	16.94	18.04	0.62	1.92	2.83
SEM	1.02	0.32	0.93	0.05	0.14	0.22
P-value						
Diet	0.03	0.008	0.01	0.39	0.86	0.79
Gender	0.07	0.09	0.07	0.30	0.90	0.32

The means within the same column with at least one common letter, do not have significant difference (P>0.05).
SEM: standard error of the means.

Table 6 The effects of peppermint on organs weight (% of body weight) of broiler chickens at 42 d

Treatment	Carcass	Thigh	Breast	Heart	Gizzard	Liver
Control	58.42	18.73	19.57[ab]	0.61	1.62[ab]	2.25[b]
1% peppermint	57.30	18.31	18.69[b]	0.68	1.86[a]	2.71[a]
2% peppermint	58.70	19.09	19.68[ab]	0.61	1.71[ab]	2.41[ab]
300 mg vitamin E	59.29	18.86	21.27[a]	0.62	1.66[ab]	2.53[ab]
SEM	1.20	0.28	0.85	0.04	0.10	0.20
Gender						
Male	58.95	19.28[a]	20.14	0.62	1.60	2.39
Female	58.26	18.31[b]	19.96	0.64	1.71	2.46
SEM	0.90	0.36	0.96	0.06	0.09	0.17
P-value						
Diet	0.08	0.52	0.02	0.42	0.04	0.004
Gender	0.07	0.001	0.08	0.32	0.5	0.06

The means within the same column with at least one common letter, do not have significant difference (P>0.05).
SEM: standard error of the means.

Stresses such as heat stress leads to excessive production of free radicals, which reduces antioxidant capacity (Robert et al. 2003). In fact, heat stress affects the sympathetic nerves and causes catecholamine release thereby leading to an increase in free radicals in the blood and tissues of the body. Free radicals attack the structure of unsaturated fat hence damaging cell membranes (Curi et al. 2003). Free radicals give rise to peroxidation in cells and thereby increase lipoperoxide concentrations in the tissues.

Lipoperoxide surplus leads to reduced enzyme activity of glutathione peroxidase, superoxide dismutase and catalase (Du et al. 2000). In these conditions, the plasma levels of certain vitamins and minerals involved in the antioxidant system decreases and the amount of active oxygen radicals (ROS) increases. In order to prevent free radical damage to the body, the use of antioxidants such as herbs, vitamins etc and reduce heat stress are beneficial as these donate electrons to free radicals, converting them into harmless compounds and thus protects cells from oxidative damages (Aryaeian *et al.* 2011). Also, researchers have noted that heat stress causes an increase in body temperature, which consequently bring about alkalosis. Yet, heat stress, by increasing blood albumin, controls, up to some extent, the alkalosis (Aengwanich and Chinrasri, 2002). Mirsalimi *et al.* (1996) reported a male bird's body temperature is higher than female because ofhigher metabolism. Hashemi *et al.* (2007) argued that on day 38, the male bird's body temperature is significantly higher than the degree of female body temperature.

Under conditions of high temperatures, birds change their behavior and physiological homeostasis, or by taking some herbs, regulate body temperature and thus reduce their body temperature. This study has showed that peppermint can reduce body temperature under heat stress, which may be due to the presence of menthol in peppermint.

Wang *et al.* (1998) found that complex bioactive components in natural medicines, with their function not clearly identified, along with nutrients and bioactive compounds have antimicrobial activity, increase immunity and reduce stress such as heat stress.

The results of this study suggest the important role of peppermint in controlling the weight of carcass and internal organ. Such findings are consistent with the results Narimani Rad *et al.* (2011) who reported that the complementary mix of medicinal herbs (1% Oregano, 0.5% Ziziphora and 0.5% Mint) improves the performance and carcass quality. In fact, medicinal plants, by reducing harmful bacteria in the digestive tract, slow down the decomposition rate of amino acids and proteins in the digestive juices which results in more uptake, improved carcass quality and reduced protein conversion to fat (Szewczyk *et al.* 2006; Lee *et al.* 2004). In accordance with this Alçiçek *et al.* (2003) observed an improvement on carcass yield of broilers when supplemented with an essential oil combination in a broiler diet, in addition Chand *et al.* (2011) also used different levels of medicinal plants in broiler chicks and found significant improvement in leg weight. Toghyani *et al.* (2010) and Ocak *et al.* (2008) evaluted that, use of peppermint had not any significant effect on relative weight of organs and body parts. Also Hernandez *et al.* (2004) reported that, use of antibiotic or mixtures of plant extracts had not any significant effect on carcass traits of broilers. And, Gurbuz and Ismael (2015) said peppermint and basil as feed additive weren't significant effect on the carcass, carcass yield and abdominal fat.

Kusandi and Djulardi (2011) showed that high temperature reduces weight of the bursa of fabricius, liver and spleen. This reduction may be due to rising temperatures and loss of appetite. Khaligh *et al.* (2011) observed no significant differences in liver weight of chicks fed with herb

mixtures. In contrast, Guo *et al.* (2000) reported that the use of medicinal plants has led to the increased weight of the lymphoid organs such as thymus, spleen and bursa of fabricius in broiler chickens, this difference may be due to different levels of medicinal plants. Various research has been carried out on the use of medicinal plants on the weight of the carcass. The use of medicinal plants in the production of broiler chickens significantly improve the gizzard weight (Chand *et al.* 2011; Khan and Durrani, 2007).

CONCLUSION

In general, results indicated that supplementation of peppermint powder in the diet improve oxidative stability, performance, body temperature in heat stress, carcass and internal organ weights.

ACKNOWLEDGEMENT

We owe our debts of gratitude to the Ferdowsi University of Mashhad Administration that permitted us to sample from peppermint farms, the Animal Science Faculty of Gorgan University of Agricultural Science and Natural Resources and the Supervisor Dr Firouz Samadi for his invaluable helps to accomplish this project.

REFERENCES

Abbas R.J. (2010). Effect of using fenugreek, parsley and sweet basil seeds as feed additives on the performance of broiler chickens. *Int. J. Poult. Sci.* **9,** 278-282.

Abbasi F. and Samadi F. (2014). Artichoke (*Cynara scolymus*) powder as natural antioxidant source in Japanese quail diet and its effects on performance and meat quality. *J. Poult. Sci.* **2,** 95-111.

Aengwanich W. and Chinrasri O. (2002). Effect of heat stress on body temprature and hematological parameters in male layers. *Thai. J. Physiol. Sci.* **15,** 27-33.

Aflatuni A. (2005). The Yield and Essential Oil Content of Mint (*Mentha* ssp.) in Northern Ostrobothnia. Oulu University Press, Finland, Europe.

Al-Ankari A.S., Zaki M.M. and AI-Sultan S.I. (2004). Use of habek mint (*Mentha longifolio*) in broiler chicken diets. *Int. J. Poult. Sci.* **3,** 629-634.

Alçiçek A., Bozkurt M. and Çabuk M. (2003). The effects of an essential oil combination derived from selected herbs growing wild in Turkey on broiler performance. *African J. Anim. Sci.* **33,** 89-94.

Amasaib E.O., Elrahman B.H.A., Abdelhameed A.A., Elmnan B.A. and Mahala A.G. (2013). Effect of dietary levels of spearmint (*Mentha spicata*) on broiler chick's performance. *J. Anim. Feed Res.* **3,** 193-196.

Aridogan B.C., Baydar H., Kaya S., Demirci M.,Ozbasar D. and Mumcu E. (2002). Antimicrobial activity and chemical composition of some essential oils. *Arch. Pharmacal. Res.* **25**, 860-864.

Aryaeian N., Djalali M., Shahram F., Jaza-yeri S., Chamari M. and Nazari S.A. (2011). Betacarotene, vitamin E, MDA, glutathione reductase and arylesterase activity levels in patients with active rheumatoid arthritis. *Iranian J. Pub. Health.* **40**, 102-109.

Borges S.A., Fischer D., Silva A.V., Majorka A., Hooge D.M. and Cummings K.R. (2004). Physiological responses of broiler chickens to heat stress and dietary electrolyte balance (sodium plus potassium minus chloride, miliequivalents per kilogram). *Poult. Sci.* **83**, 1551-1558.

Chand N., Muhammad D., Durrani F., Qureshi M.S. and Ullah S.S. (2011). Protective effects of milk thistle (*Silybum marianum*) against aflatoxin B1 in broiler chicks. *J. Anim. Sci.* **24**, 1011-1018.

Cross D.E., McDevitt R.M., Hillman K. and Acamovic T. (2007). The effect of herbs and their associated essential oils on performance, dietary digestibility and gut microflora in chickens from 7 to 28 days of age. *J. Br. Poult. Sci.* **48**, 496-506.

Curi R., Newsholme P., Lima M.M.R., Pithon-curi T.C. and Procopio J. (2003). Glutamine and glutamate-their central role in cell metabolism and function. *Cell. Biochem. Funct.* **21**, 1-9.

Dorman H.J.D., Kosar M., Kahlos K., Holm Y. and Hiltunen R. (2003). Antioxidant properties and composition of aqueous extracts from Mentha species, hybrids, varieties, and cultivars. *J. Agric. Food Chem.* **51**, 4563-4569.

Du R., Lin H. and Zhang Z.Y. (2000). Peroxide status in tissues of heat stressed broilers. *Anim. Sci.* **10**, 1373-1376.

Edris A.E., Shalaby A.S., Fadel H.M. and Abdel-Wahab M.A. (2003). Evaluation of a chemotype of spearmint (*Mentha spicata*) grown in Siwa Oasis, Egypt. *European Food Res. Technol.* **218**, 74-78.

Ezzat S.M. (2001). *In vitro* inhibition of candida albicans growth by plant extract and essential oil. *J. Microbiol. Biotechnol.* **17(7)**, 757-759.

Forster S. (1996). Peppermint: mentha piperita. *Am. Botanical Council. Botanical. Series.* **306**, 3-8.

Galib M. and Al-Kassi M.W. (2010). A comparative study on diet supplementation with a mixture of herbal plants and dandelion as a source of probiotices on the performance of broilers. *Pakistani Int. J.* **9**, 67-71.

Geraert P.A., Guillaumin S. and Leclercq B. (1993). Are genetically lean broilers more resistant to hot climate? *Br. Poult. Sci.* **34**, 643-653.

Guo F.C., Savelkoul H.F.J., Kwakkel R.P., Williams B.A. and Verstegen M.W.A. (2000). Immunoactive medicinal properties of mushroom and herb polysaccharides and their potential use in chicken diets. *Poult. Sci.* **59**, 427-440.

Gurbuz Y. and Ismael I.A. (2015). Effect of peppermint and basil as feed additive on broiler performance and carcass characteristics. *Iranian J. Appl.* Sci. **6**, 149-156.

Hashemi S.R., Dastar B., Hassani S. and Ahangari Y.J. (2007). Growth performance, body temperature and blood proteins in broiler in response to betaine supplement and dietary protein level under heat stress. *J. Agric. Sci. Nat. Res.* **14**, 1-10.

Hayashi K., Yoshizaki R., Ohtsuka A., Toroda T. and Tuduki T. (2004). Effect of ascorbic acid on performance and antibody

production in broilers vaccinated against IBD under a hot environment. Pp. 106-107 in Proc. 22nd World's Poult. Cong. Istanbul, Turkey.

Hernandez F., Madrid J., Garcia V., Orengo J. and Megias M.D. (2004). Influence of two plant extracts on broilers performance, digestibility and digestive organ size. *Poult. Sci.* **83**, 169-174.

Jamroz D., Wertlecki J., Orda A., Wiliczkiewicz A. and Skorupinska J. (2003). Influence of phatogenic extracts on gut microbial status in chickens. Pp. 110-112 in Proc. 14th European Symp. Poult. Nutr. Lillehammer, Norway.

Khaligh F., Sadeghi G.H., Karimi A. and Vaziry A. (2011). Evaluation of different medicinal plants blends in diets for broiler chickens. *J. Med. Plant. Res.* **5**, 1971-1977.

Khan M. and Durrani F. (2007). Hematological hypolipidemic hypoglycemic anticoccidial hepatoprotective immunostimulant and growth promotant effect of *Withania somnifera* in broiler production. MS Thesis. KPK Agric Univ., Peshawar, Pakistan.

Kusandi E. and Djulardi A. (2011). Physiological dynamic of broiler at various environmental temperatures. *Int. J. Poult. Sci.* **10**, 19-22.

Lee K.W., Everts H., Kappert H.J., Frehner M., Losa R. and Beynen A.C. (2003). Effects of dietary essential oil components on growth performance, digestive enzymes and lipid metabolism in female broiler chickens. *Br. Poult. Sci.* **44**, 450-457.

Lee K.W., Everts H., Kappert H.J., Frehner M., Losa R. and Beynen A.C. (2004). Effects of dietary essential oil components on growth performance, digestive enzymes and lipid metabolism in female broiler chickens. *J. Br. Poult. Sci.* **44**, 450-457.

Mashaly M.M., Hendricks G.L., Kalama M.A., Gehad A.E., Abbas A.O. and Patterson P.H. (2004). Effect of heat stress on production parameters and immune responses of commercial laying hens. *Poult. Sci.* **83**, 889-894.

Menezes M.C., Souza M.M.S. and Bothelo R.P. (2004). *In vitro* evaluation of antimicrobial activity of Brazilian plants extracts on bacteria isolated from oral cavity of dogs. *Rev. Univ. Rural.* **24**, 141-144.

Mimica Dukic N., Bozin B., Sokovic M., Mihailovic B. and Matavulj M. (2003). Antimicrobial and antioxidant activities of three Mentha species essential oils. *Plan. Med.* **69**, 413-419.

Mirsalimi H., Panahidehghan M.R., Rasoolnejadfereidoni S., Modirsaneii M., Moafi M. and Niknafas F. (1996). Physiology of birds. Publications unit of the Education Department of Agricultural Economic Organization Kosar, Tehran, Iran.

Narimani-Rad M., Nobakht A., Aghdam Shahryar H., Kamani J. and Lotfi A. (2011). Influence of dietary supplemented medicinal plants mixture (Ziziphora, Oregano and Peppermint) on performance and carcass characterization of broiler chickeens. *J. Med. Plant. Res.* **5**, 5626-5629.

Nienaber J.A. and Hahn G.L. (2007). Livestock production system management responses to thermal challenges. *Int. J. Biometereol.* **52**, 149-157.

Nobakht A. and Aghdam Shahriar H. (2011). Effect of medicinal plants Mallow, camel thorn and mint on performance, carcass quality and blood metabolites in broilers. *J. Anim. Sci.* **3**, 51-63.

NRC. (1994). Nutrient Requirements of Poultry, 9th Rev. Ed. National Academy Press, Washington, DC., USA.

Ocak N., Erener G., Burak F., Sungu M., Altop A. and Ozmen A. (2008). Performance of broilers fed diets supplemented with dry peppermint (*Mentha piperita*) or thyme (*Thymus vulgaris*) leaves as growth promoter source. *Czech J. Anim. Sci.* **53,** 169-175.

Pardue S.L. and Thaxton J.P. (1996). Ascorbic acid in poultry: a review. *World's Poult. Sci.* **42,** 107-123.

Pavela R. (2005). Insecticidal activity of some essential oils against larvae of Spodoptera littoralis. *Fitoterapia.* **76,** 691-696.

Pouramir M., Sajadi P., Shahabi S., Rezaei S. and Samadi P. (2006). Effects of food diet of tomato and carrot juices on serum lipids in rats. *J. Med. Sci.* **13,** 55-59.

Renaudeau D., Collin A., Yahav S., Basilio V., Gourdine J.L. and Collier R.J. (2012). Adaptation to hot climate and strategies to alleviate heat stress in livestock production. *Anim. Sci.* **6,** 707-728.

Robert J., Edens F.W. and Ferket P.R. (2003). The effects of selenium supplementation on performance and antioxidant enzyme activity in broiler chicken. MS Thesis. North Carolina State Univ., USA.

SAS Institute. (2003). SAS®/STAT Software, Release 6.11. SAS Institute, Inc., Cary, NC. USA.

Schuhmacher A., Reichling J. and Schnitzler P. (2003). Virucidal effect of peppermint oil on the enveloped herpes simplex virus type 1 and type 2 *in vitro. Phytomedicine.* **10,** 504-510.

Shah P. and Mello P.M.D. (2004). A review of medicinal uses and pharmacohogical effects of *Mentha piperita. J. Nat. Prod. Rad.* **3,** 214-221.

Szewczyk A., Hanczakowska E. and FwiGtkiewicz M. (2006). The effect of nettle (*Urtica dioica*) extract on fattening performance and fatty acid profile in the meat and serum lipids of pigs. *J. Anim. Feed Sci.* **1,** 81-84.

Toghyani M., Toghyani M., Gheisari A.A., Ghalamkari G. and Mohammadrezaei M. (2010). Growth performance, serum biochemistry, and blood hematology of broiler chicks fed different levels of black seed (*Nigella sativa*) and peppermint (*Mentha piperita*). *Livest. Sci.* **129,** 173-178.

Tssou C.C., Drodinos E.H. and Nychas G.J.E. (1995). Effects of essential oil from mint on salmonella enteritidis and listeria monocytogenes in model food system at 4 and 10 °C. *J. Appl. Microbiol.* **78,** 593-600.

Wang R.J., Li D.F. and Bourne S. (1998). Can 2000 years of herbal medicine history help us solve problems in the year 2000? Pp. 273-269 in Proc. 14th Annual Symposium Nottingham, Nottingham, UK.

The Effects of Protexin Probiotic and Aquablend Avian on Performance and Immune System of Broiler Chickens

N. Jabbari[1], A. Fattah[1*] and F. Shirmohammad[1]

[1] Department of Animal Science, Shahr-e-Qods Branch, Islamic Azad University, Tehran, Iran

*Correspondence E-mail: amir1356fattah@qodsiau.ac.ir

ABSTRACT

This study was conducted to investigate the effects of protexin probiotic and aquablend avian antibody on the performance and immune system of broilers. In this experiment, 320 1-day-old male broiler chicks of the Ross 308, with 5 treatments and 4 replications in a completely randomized design were distributed in experimental units (cages). The treatments include: control (C), protexin (P1), protexin (P2) with double there commended dose, aquablend (A1) and aquablend (A2) aquablend twice there commended amount. The results showed that protexin probiotic and aquablend avian antibody in different doses had no significant effects on average weight gain and feed consumption of broilers at different week ($P>0.05$). A significant difference between treatments was observed in feed conversion ratio at the second and fifth weeks ($P<0.05$). Consumption protexin probiotic and aquablend avian antibody with different doses has significant effect on the relative weight of carcasses ($P<0.01$). Consumption protexin probiotic and aquablend avian antibody with different doses had no significant effect on antibody titers produced in primary and secondary challenge with sheep red blood cells ($P>0.05$). According to results of this research, it seems that the using of protexin probiotic and aquablend avian antibody has not importance effect on improving the performance of broilers except for carcass weight.

KEY WORDS aquablend avian, broilers, immune system, performance, protexin.

INTRODUCTION

Food safety and feed conversion ratio are very important in the production of foods of animal origin. The increased population sizes of farm animals, clustering of food animal production units and the intensive global transport of live animals and animal products facilitate the spread of zoonotic pathogens. Moreover, farm animal health is severely affected by gastro-intestinal infections that occur frequently under large sized farming conditions. Thus, strategies aimed to improve farm animal health may have impact on both animal and public health. In addition, improvement of intestinal health will lead to lower costs since animals with impaired intestinal health have a reduced appetite and / or diarrhea, resulting in a reduced nutrient uptake and, therefore, negatively affect the feed conversion ratio. Additionally, the following immune response may trigger muscle wasting by increasing catabolism to fulfill the excessive need for amino acids necessary to produce immune response effectors, such as cytokines and antibodies (Thomke and Elwinger, 1998). To promote growth, farm animal feed has been supplemented with sub therapeutical doses of antibiotics, so-called growth-promoting antibiotics (GPAs), since the mid-1940's (Dibner and Richards, 2005). During the first three decades of their use in feed, mean increases of body weight ≥ 8% for Penicillin and Tetracycline were reported (Graham *et al.* 2007). However, nowadays the magnitude of these effects is marginal, due to selective breeding, improved feed formulations and improved hygienic conditions in animal husbandry. Furthermore, the mechanisms involved in GPA-mediated enhanced growth are still under debate.

Foremost, GPAs are thought to inhibit subclinical infections (Gaskins et al. 2002), thereby preventing illness and thus maintaining the feed conversion ratio. Other proposed modes of action are suppression of carbohydrates and fat malabsorption, improved nutrient utilization by inhibiting the growth of normal GI tract flora, reduction of growth-depressing microbial metabolites, such as ammonia, aromatic phenols and bile degradation products, and enhanced nutrient uptake through the thinner intestinal wall in GPA-fed animals (Gaskins et al. 2002).

A more important point of concern was the relative ease at which microorganisms had demonstrated to be able to transmit antibiotic resistance genes via the exchange of transposons or plasmids and the possible transmission of this resistance to human pathogens. It became such a major concern that it resulted in a total European ban in 2006 on the use of antibiotics as a feed additive to promote growth (Phillips, 2007). A consequence of the ban is that pathogens, suppressed by the use of GPAs, can now reemerge. Because GPAs were almost entirely aimed at gram-positive bacteria (Witte, 2000), it can be expected that gram-positive bacteria, such as Clostridium perfringens will increasingly become a problem in the poultry sector. Regardless of a possible global ban on the use of GPAs, the many disadvantages involved with their use make it mandatory to search for alternative strategies to increase food safety and to promote growth in food animals. Not surprisingly, the body's natural defense mechanisms are now one of the focuses, in particular those of the digestive tract. The gastrointestinal (GI) tract comprises the largest mucosal surface in the body and is in direct contact with the external environment. A healthy GI tract harbors a wide variety of residential "non-pathogenic" and potential pathogenic microorganisms displaying complex symbiotic and competitive interactions (Verstegen and Williams, 2002). A disturbance in this balance could facilitate outgrowth of pathogenic microbiota, which will depress animal growth by competing with the host for nutrients and by producing toxic metabolites resulting in increased turnover of gut mucosa (Verstegen and Williams, 2002). Therefore, several strategies aim at shifting this delicate balance in the favor of beneficial microbiota by stimulation and/or activating growth of these subpopulations of intestinal microbiota via prebiotics, probiotics, organic acids, enzymes or herbs (Verstegen and Williams, 2002).

But in many cases due to the high cost of the introduced material or require high expertise for the production and consumption of these materials or absolute uncertainty of the effects of these products and their associated complications, products have not been able found to be commercially available (Rus et al. 2005). One of the cases that have been considered is the production of effective materials on the digestive system of animals. Stomach and intestinal mucosal surfaces are the body's major systems and are directly linked to the external environment, which is the balance and performance in natural state (Verstegen and Williams, 2002).

Any change in this area would facilitate and stimulate the growth of pathogens that causes loss and delay of the animal grow due to impaired absorption of nutrients (Verstegen and Williams, 2002). One of the main alternatives for antibiotics in poultry industry is probiotics. One of these probiotics is protexin (Ayasan, 2013). Also, Aquablend Avian® is a blend of beneficial bacteria and antibodies of natural origin extracted from dehydrated egg, both designed to provide the animal with adequate microflora accompanied with the presence of antibodies specific against predominant pathogens in the poultry. Animal needs a minimum amount of beneficial bacteria (microflora) in order to digest a normal digestion and also be protected against pathogens. Normally, when pathogens enter the intestinal tract, they tend to compete with the microflora in occupying inner intestinal surface (competitive exclusion). This creates an imbalance of the intestinal microflora reducing the dominant presence of lactic acid producing bacteria. In this situation, desired microbial in Aquablend Avian® provide the animal with both, beneficial desired microbial to recover the normal balance in the intestinal microflora and natural antibodies, specific against predominant pathogens in the animal. These antibodies attach themselves to their specific receptors in the inner intestinal surface. In this way, the antibodies inhibit the infectious action of the pathogens, which are then excreted, thus preventing animal infection. Because of the importance of poultry as an economic and nutritious form of animal protein and the fast growing characteristics of this animal, research workers have devoted studies to the use of probiotics in poultry. This study was conducted to investigate the effect of Protexin and aquablend avian antibody on the immune system, performance and carcass characteristics of broiler chickens.

MATERIALS AND METHODS

320 Ross broilers (308), 1 day-old male broiler chicks (mean weight 140.36±5.26 g) were equally allocated to five treatments containing 20 pens in each. In each pen were included 16 birds. Diets were formulated to meet the nutrient requirements for poultry (NRC, 1994); Tables 1 and 2. The birds were fed a starter diet from 1 to 10 d, grower diet from 11 to 22 d and finisher diet from 23 to 35 d.

Table 1 Nutrient content of the basal diet over different periods of production

Ingredients (%)	Starter (1-10 d)	Grower (11-22 d)	Finisher (23-35 d)
Corn	55.59	61.27	63.19
Soybean meal (44% CP)	36.74	31.04	28.31
Soybean oil	3.56	4.07	5.05
Common salt	0.30	0.30	0.35
Dicalcium phosphate	1.85	1.79	1.66
Limestone	1.25	1.22	1.15
DL-methionine	0.21	0.22	0.23
L-lysine HCl	0.00	0.09	0.06
Mineral and vitamin premix	0.50	0.50	0.50

Table 2 Nutrient content of the basal diet over different periods of production

Analysis results	Starter (1-10 d)	Grower (11-22 d)	Finisher (23-35 d)
Dry matter (%)	89.1	89.1	89.2
Crude protein (%)	21.4	19.3	17.4
Metabolizable energy (kcal/kg)	2915	2955	3009
Lysine (%)	1.16	1.03	0.91
L-methionine (%)	0.56	0.51	0.45
L-methionine + cystine (%)	0.85	0.77	0.70
Calcium (%)	0.9	0.88	0.89
Available phosphorus (%)	0.44	0.4	0.35

Birds received diets which were supplemented with 1 gram per liter (P1) and 2 gram per liter (P2) protexin in water, and supplemented with 5 gram per liter (A1) and 10 gram per liter (A2) Aquablend Avian® in water. The experiment lasted for 6 weeks. Feed intake was recorded by replicate every week. Feed efficiency was calculated as: feed consumption weight/body weight. During 42 days of experimental period, mean weight, feed consumption, feed conversion and carcass characteristics was measured as below (mortality was recorded as it occurred):

Feed consumption= (feed weight at beginning period-feed weight at the end of period) / age
Feed conversion= (feed consumption during period/total weight increases during period)

For preparation injectable suspension SRBC, blood from sheep jugular vein using a syringe containing EDTA anticoagulant was used. Sheep red blood cells were washed three times by phosphate buffered saline (Munns and Lamont, 1991).

Then 1% SRBC suspension at a rate of 2.0 mL at the age of 21 and 35 days to two chicks from each replicate through breast muscle injection, and seven days after the chicks via wing vein blood samples were taken. 16 hours after blood coagulation, serum samples were isolated at a temperature of 37 °C. Then total SRBC antibody titers were measured (Vander, 1980). All analyses were carried out with one-way analysis of variance of SAS (SAS, 1996). Means were compared by using Duncan's test.

RESULTS AND DISCUSSION

Result showed that protexin and aquablend avian antibody has not significant effects on mean weight of broilers in different ages ($P>0.05$); (Table 3). Result showed that protexin and aquablend avian antibody has not significant effects on feed consumption of broilers at different week ($P>0.05$); (Table 4). Result showed that protexin and aquablend avian antibody has significant effects on feed conversion ratio of broilers in 2 and 5 week ages at 95% confident level (Table 5), but in total has not significant effect. Result showed that protexin and aquablend avian antibody has significant effects on carcass weight of broilers at 95% confident level (Table 6), but in other factors not showed significant effect. Result showed that protexin probiotic and aquablend avian antibody with different doses had no significant effect on antibody titers produced in primary and secondary challenge with sheep red blood cells ($P>0.05$); (Table 7).

The results showed that the antibody and protexin had no significant effect on average weight and feed consumption of broilers at different week ($P>0.05$). Significant difference between treatments was observed in feed conversion ratioat the second and fifth weeks ($P<0.05$). Consumption of protexin probiotic and aquablend avian antibody with different doses has significant effect on relative weight of carcasses ($P<0.01$). Consumption protexin probiotic and aquablend avian antibody with different doses had no significant effect on antibody titers produced in primary and secondary challenge with sheep red blood cells ($P>0.05$).

Table 3 Protexin and aquablend avian antibody effects on mean weight of broilers (g)

Items	Week 1	Week 2	Week 3	Week 4	Week 5	Week 6	Total
Control	148.8	297.8	465.7	542.7	528	753.5	2736.7
P1	136.6	301.8	466.5	526	550	733.2	2714.2
P2	140.9	309.4	443.2	547.2	534.2	701.7	2676.9
A1	140.7	312.4	471.2	542.2	615.2	754.5	2836.4
A2	134.8	301.9	456.5	535.2	554.2	732.7	2715.4
Significant	0.18	0.32	0.71	0.91	0.20	0.94	0.36
SEM	5.26	2.43	6.49	7.05	12.78	17.59	25.84

P1: protexin; P2: protexin with double there commended; A1: aquablend and A2: aquablend twice there commended.
SEM: standard error of the means.

Table 4 Protexin and aquablend avian antibody effects on feed consumption of broilers (g)

Items	Week 1	Week 2	Week 3	Week 4	Week 5	Week 6	Total
Control	188.1	447.2	935	968.5	1047.7	1217.8	4804
P1	174.7	448.7	928.8	970	1099.1	1344.8	4966.7
P2	175.7	444	929.1	966	1091	1144.2	4750.5
A1	173.5	443.5	890.5	1005.4	1053.2	1301.4	4858
A2	170.0	441.0	907.5	965.3	1041.5	1366.6	4892.2
Significant	0.06	0.05	0.37	0.76	0.68	0.26	0.82
SEM	2.14	4.49	6.86	10.62	14.88	35.91	110.1

P1: protexin; P2: protexin with double there commended; A1: aquablend and A2: aquablend twice there commended.
SEM: standard error of the means.

Table 5 Protexin and aquablend avian antibody effects on feed conversion ratio of broilers

Items	Week 1	Week 2	Week 3	Week 4	Week 5	Week 6	Total
Control	1.26	1.5	2.0	1.78	1.98	1.62	1.76
P1	1.28	1.49	2.0	1.84	2.01	1.85	1.83
P2	1.25	1.44	2.11	1.76	2.06	1.63	1.77
A1	1.23	1.38	1.89	1.88	1.71	1.72	1.71
A2	1.26	1.46	1.98	1.8	1.87	1.89	1.80
Significant	0.43	0.05	0.33	0.77	0.04	0.42	0.25
SEM	0.17	0.03	0.03	0.02	0.04	0.05	0.03

P1: protexin; P2: protexin with double there commended; A1: aquablend and A2: aquablend twice there commended.
SEM: standard error of the means.

Table 6 Protexin and aquablend avian antibody effects on carcass characteristics of broilers

The main effects	The relative weight of the body organs								
	Live weight (g)	Carcass weight %	Digestive system %	Liver %	Bursa %	Thymus %	Spleen %	Heart %	Gut (m)
Treatment									
Control	2710	82.37[a]	3.85	1.85	0.04	0.17	0.05	0.33	1.84
P1	2980	79.08[b]	4.17	1.99	0.04	0.17	0.05	0.36	1.93
P2	2720	84.36[a]	3.12	1.73	0.04	0.16	0.06	0.32	1.67
A1	2820	83.79[a]	3.39	1.97	0.03	0.16	0.06	0.36	1.76
A2	2830	82.03[ab]	3.9	1.78	0.05	0.22	0.07	0.39	1.86
P-value	0.29	0.01	0.21	0.59	0.73	0.38	0.63	0.07	0.21
SEM	0.04	0.58	0.25	0.05	0.002	0.009	0.003	0.009	0.03

P1: protexin; P2: protexin with double there commended; A1: aquablend and A2: aquablend twice there commended.
SEM: standard error of the means.

Table 7 Protexin and aquablend avian antibody effects on immune system characteristics of broilers

Treatment	Immune system						
	Control	P1	P2	A1	A2	P-value	SEM
Main effects							
IgM 21 days	1.75	1.12	1.5	1.25	1.75	0.44	0.14
IgG 21 days	2.37	2.12	1.75	2	1.37	0.33	0.22
SRBC 21 days	4.12	4.25	3.25	3.25	2.87	0.39	0.26
IgM 35 days	2.25	2.37	2.33	2.37	2.62	0.22	0.36
IgG 35 days	2.25	2.87	3.5	2.75	3.62	0.31	0.26
Sheep red blood cells (SRBC) 35 days	4.5	5.25	6.16	5.12	6.25	0.38	0.32

P1: protexin; P2: protexin with double there commended; A1: aquablend and A2: aquablend twice there commended.
SEM: standard error of the means.

Kabir *et al*. (2003) showed that with antibiotic consumption, live weight gains obtained were significantly higher in experimental birds as compared to control ones at all levels during the period of 2nd, 4th, 5th and 6th weeks of age, both in vaccinated and non-vaccinated birds. A significantly (P<0.01) higher carcass yield occurred in broiler chicks fed with the probiotics on the 2nd, 4th and 6th week of age both in vaccinated and non-vaccinated birds. The weight of leg was found significantly (P<0.01) greater for experimental birds as compared to control ones on the 2nd, 4th and 6th week of age. A significantly (P<0.01) higher breast weight in broiler chicks fed with the probiotics was observed on the 4th and 6th week of age. Analogously a significantly (P<0.05) higher breast portion weight was found in experimental birds as compared to control ones during the 2nd week of age. The antibody production was found significantly (P<0.01) higher in experimental birds as compared to control ones. Significant differences were also observed in the weight of spleen and bursa due to probiotics supplementation. The results of the study thus revealed that probiotics supplementation promoted significant influence on live weight gain, high carcass yield, prominent cut up meat parts and immune response. Shahsavari (2006) reported that the probiotic (protexin) on the function (egg production, feed conversion and weight, egg mass) and quality characteristics of broiler breeder eggs were not affected. Balevi *et al*. (2000) investigated the effects of dietary supplementation of a commercial probiotic (protexin feed consumption, egg yield, egg weight, food conversion ratio and humoral immune response in layer hens. In 7 replicates, a total of 280 40-week-old layers were given diets containing either 0, 250, 500 or 750 ppm for 90 d 2. When compared with the controls, the food consumption, food conversion ratio and the proportions of damaged eggs were lower in the group consuming 500 ppm probiotic (P<0.05). There was no significant difference between the controls and the groups receiving 250 and 750 ppm probiotic in food consumption, food conversion ratio and proportion of damaged eggs. Similarly, the egg yield, egg weight, specific gravity, and peripheral immune response showed no statistically significant differences between the groups) on daily.

Ayasan *et al*. (2006) investigate the effects of grower diets, dietary three different levels of probiotic (protexin) in grower diet on Japanese quail (*Coturnix coturnix Japonica*). Results showed that age and body live weight of quails at the first laying was found significant different between groups. During the egg production period, probiotic supplementation to the diet did not affect feed intake and feed conversion. Probiotics and prebiotics alter the intestinal microbiota and immune system to reduce colonization by pathogens in certain conditions. Feeding birds in a report Protexin increased antibody titers against the Newcastle

disease vaccine (Zakeri and Kashefi, 2011). Also, chickens treated with protexin higher antibody titers against avian influenza viruses showed (Ghafoor *et al*. 2005). Some studies showed that parameters related to immune (antibody titer against SRBC, antibody titer against Newcastle disease and immunoglobulins IgG, IgM) were not affected by different levels of probiotics. However most SRBC and IgG antibody titers were obtained by applying the most probiotics, more immunoglobulin IgG were related to probiotic treatments were statistically significant difference between them and the other groups.

As with increase the use of antibiotics, environmental and stress status influence efficacy of prebiotics and probiotics, these products show promise as alternatives for antibiotics as pressure to eliminate growth promotant antibiotic use increases by improving the microbial balance of the intestine bird probiotics and digestive enzymes increase the activity of digestive enzymes and enabling increased nutrient availability indigestible and beneficial changes in the metabolism of food consumed, thus improving feed efficiency (Chen and Nakthong, 2005). Due to the fact that each of these additives and active ingredient are different compounds, dose and components used in the experiment can be obtained different results in the use of these substances are effective growth promoters (Lee *et al*. 2004). Defining conditions under which they show efficacy and determining mechanisms of action under these conditions is important for the effective use prebiotics and probiotics in the future.

CONCLUSION

The purpose of using protexin and antibodies at the same time is increasing the quality and quantity, while protexin and aquablend has not importance effect in improving the performance of broilers. Probiotics with increased feed intake and the efficiency of feed intake increase the weight. Overall increasing performance due to use of antibodies and probiotics may be due to many reasons including existence various chemical compounds and improve the efficiency of food consumption and eliminate the annoying factors including harmful microorganisms in the digestive tract and food. However, according to the survey results, it is recommended that more studies and promising for the study of the materials used in poultry diets.

REFERENCES

Ayasan T. (2013). Effects of dietary inclusion of protexin (probiotic) on hatchability of Japanese quails. Indian J. Anim. Sci. **83(1)**, 78-81.

Ayasan T., Ozcan B.D., Baylan M. and Canogullari S. (2006). The Effects of dietary inclusion of probiotic protexin on egg yield parameters of Japanese quails (*Coturnix coturnix Japonica*). *Int. J. Poult. Sci.* **5(8),** 776-779.

Balevi T., Ucan U.S., Coskun B., Kurtoglu V. and Cetingul S. (2000). Effect of a commercial probiotic in the diet on performance and humoral immune system in layers. *Hayvanc. Arast. Derg.* **10,** 25-30.

Chen Y.C. and Nakthong C. (2005). Improvement of laying hen performance by dietary prebiotic chicory oligofructose and inulin. *Int. J. Poult. Sci.* **4,** 103-108.

Dibner J.J. and Richards J.D. (2005). Antibiotic growth promoters in agriculture: history and mode of action. *Poult. Sci.* **84,** 634-643.

Gaskins H.R., Collier C.T. and Anderson D.B. (2002). Antibiotics as growth promotants: mode of action. *Anim. Biotechnol.* **13,** 29-42.

Ghafoor A., Naseem S., Younus M. and Nazir J. (2005). Immunomodulatory effects of multistrain probiotics on broiler chickens vaccinated against avian influenza virus. *Poult. Sci.* **4,** 777-780.

Graham J.P., Boland J.J. and Silbergeld E. (2007). Growth promoting antibiotics in food animal production: an economic analysis. *Public. Health Rep.* **122,** 79-87.

Kabir S.M.L., Rahman M.M., Rahman M.B., Rahman M.M. and Ahmed S.U. (2003). The dynamics of probiotics on growth performance and immune response in broilers. *Int. J. Poult. Sci.* **3(5),** 361-364.

Lee H.Y., Andalibi P., Webster S.K., Moon K., Teufert S.H., Kang J.D., Li M., Nagura T., Ganz D. and Lim J. (2004). Antimicrobial activity of innate immune molecules against *Streptococcus pneumoniae*, *Moraxella catarrhalis* and nontypeable *Haemophilus influenzae*. *BMC Infect. Dis.* **4,** 12.

Munns P.L. and Lamont S.J. (1991). Research note: effects age and immunization interval on the immunity response T-cell dependent and T-cell independent antigens in chickens. *Poult. Sci.* **70(11),** 2371-2374.

NRC. (1994). Nutrient Requirements of Poultry, 9th Rev. Ed. National Academy Press, Washington, DC., USA.

Phillips I. (2007). Withdrawal of growth-promoting antibiotics in Europe and its effects in relation to human health. *Int. J. Antimicrob. Agents.* **30(2),** 101-107.

Rus H., Cudrici C. and Niculescu F. (2005). The role of the complement system in innate immunity. *Immunol. Res.* **33,** 103-112.

SAS Institute. (1996). SAS®/STAT Software, Release 6.11. SAS Institute, Inc., Cary, NC. USA.

Shahsavari K. (2006). The effect of probiotics on the yield and quality eggs of breeder hens. MS Thesis. Tarbiat Modares Univ., Tehran, Iran.

Thomke S. and Elwinger K. (1998). Growth promotants in feeding pigs and poultry ii; mode of action of antibiotic growth promotants. *Ann. Zootech.* **47,** 153-167.

Verstegen M.W. and Williams B.A. (2002). Alternatives to the use of antibiotics as growth promoters for monogastric animals. *Anim. Biotechnol.* **13,** 113-127.

Witte W. (2000). Selective pressure by antibiotic use in livestock. *Int. J. Antimicrob. Agents.* **16(1),** 19-24.

Zakeri A. and Kashefi P. (2011). The comparative effects of five growth promoters on broiler chickens humoral immunity and performance. Anim. Vet. Adv. **10(9),** 1097-1101.

4

Evaluation of Laying Performance and Egg Qualitative Characteristics of Indigenous Hens Reared in Rural Areas of Isfahan Province

A.A. Gheisari[1*], G. Maghsoudinejad[2] and A. Azarbayejani[1]

[1] Department of Animal Science, Isfahan Research Center for Agriculture and Natural Resources, Isfahan, Iran
[2] Animal Science Research Institute, Karaj, Iran

*Correspondence E-mail: gheisari.ab@gmail.com

ABSTRACT

This study was carried out to investigate the egg production performance and egg quality characteristics of the indigenous chickens reared in rural areas of two different climates of Isfahan province. Totally, 2160 indigenous chickens were studied in this research. Two dominant climatic regions (cold and hot) were determined for Isfahan province. In each climate, two towns and three villages in each town were chosen. Chadegan and Khansar were considered as cold and Kashan as well as Varzane were considered as towns in hot climate. Furthermore, six families were determined as experimental units in each village (a total of 72 families). Primarily, thirty-six chicks and four roosters of 45 day-old ages were delivered to each family. Laying performance of the chicks was recorded during the laying period (21-72 weeks of age). The egg quality characteristics were recorded once every two months. The results showed that although climate affected egg production across 25-32 and 57-64 weeks of ages (P<0.05), it did not influence the average egg production during the entire laying period. The least egg production observed in Chadegan (25.9%) which was significantly lower than Varzane and Khansar (35.9% and 37.5%, respectively; P<0.05). Moreover, shell qualitative index in cold climate was statistically better than that in hot climate (P<0.05). Chadegan had the highest egg shell thickness and shell to egg weight ratio among four towns (P<0.05). Furthermore, yolk color index in hot climate was significantly improved compared to cold climate (P<0.05). In conclusion, the egg qualitative characteristics were affected by climate conditions.

KEY WORDS egg production, egg quality parameters, indigenous chicken, Isfahan rural regions.

INTRODUCTION

Rearing indigenous poultry is prevalent in many developing and underdeveloped countries that might improve the rural economy (Vali, 2008). The main reasons for rearing indigenous chickens are their resistance to suboptimal environmental conditions or diseases compared to exotic breeds. Furthermore, raising indigenous hens provide job opportunities, leading to extra income for civilians and consequently might reduce immigration from rural areas to big cities (Deljoisaraian *et al.* 2011). Moreover, these birds contribute considerably to protect rural poor communities from malnutrition. Therefore, rearing high potential birds, particularly indigenous chickens in developing countries meet the need of rural families to poultry productions and so research in this field seems necessary (Gheisari and Azarbayejani, 2013). On the other hand, low productivity of indigenous birds has hampered their potential to uplift the living standards of producers and decreased their contribution to rural development. In this respect, Khalafalla *et al.*

(2001) remarked that insufficient sanitary cares, inappropriate nests and low knowledge of rural people are the most important limitations to achieve optimum performance in chickens reared in rural regions. Furthermore, the low production of Kenya's chicken (40-100 egg per year) was related to factors such as genetics, poor diet, diseases and inadequate husbandry management (Kingori *et al.* 2010). Various factors including age of the chickens, diet, egg storage time and temperature were reported to affect the internal egg quality (Miles and Jacob, 2000).

For example, Elibol *et al.* (2002) announced that the egg size, height of albumen, and yolk weight were increased as they aged. Gheisari *et al.* (2008) reported that strength and Haugh unit of eggs in indigenous chickens tended to increase following the increasing in feed consumption. In another study, Nourollahi (2013) reported a significant effect of rural climates on laying production of indigenous hens. In this respect, chickens in hot climate had greater egg production than cold climate. Also, outbreak of disease such as Newcastle disease virus has been affected by climate variabilities (Nyaiyo, 2014) that can affect the performance of indigenous poultry. In general, data considering the production performance and egg qualitative characteristics of indigenous hens reared in rural areas of different climates is scarce.

Thus, the objective of present study was to evaluate laying performance and egg qualitative characteristics of indigenous hens reared in different climates and rural areas of Isfahan province.

MATERIALS AND METHODS

Birds, management, egg production and qualitative characteristics

A total of 2160 one-day-old chickens were provided from Isfahan indigenous chicken breeding center. The chickens fed a diet containing 2500 kcal metabolizable energy per kilogram and 18% crude protein, 0.9% Ca, 0.42% available phosphorous, 0.15% Na, 0.65% methionine + cystine and 1% lysine (Gheisari and Golian, 1996) up to 45 days of age. Two different hot and cold climates were determined based on reports obtained from Isfahan meteorological organization to evaluate the performance of the chickens in these two different climates.

Two towns were allocated to cold (Chadegan and Khansar) and two towns (Kashan and Varzane) to hot climate. Then, three villages were chosen from each town and six families were determined in each village as experimental units (72 families). Twenty 45-day-old chickens were delivered to each family (the hen to cock ratio was 7:1) and families were trained for rearing chickens and recording their weight and egg production during the experimental period. Due to the fact that we wanted to evaluate the effect of rural condition on the performance of indigenous chickens, their feeding program was under families' condition and decision after 45 days of age. The egg production performance was recorded daily but results were reported in percentage on month basis (Table 3). Produced eggs were weighed every two weeks using digital scales (0.1 g accuracy). Furthermore, eggs were used to determine eggshell breaking strength (kg/cm^2), eggshell weight (g), eggshell thickness (mm) and Haugh unit every two months. Eggshell breaking strength was measured using OSk13473 eggshell intensity meter (Ogawa Sheiki Co., Ltd., Tokyo Central and Tokyo, Japan).

Eggshell weight was measured after washing the interior egg membrane and after drying overnight at 80 °C. Eggshell thickness (mm) was measured using micrometer screw from Mitutoya (Japan) which was applied to central part of an egg in the area of maximum perimeter. The height of the albumen was measured off the Chalazae at a point mid-way between the inner and outer circumferences of the thick white with a Spherometer. Haugh unit was calculated using the following formula (Haugh, 1937).

$$HU= 100 \log (H+7.57-1.7 \times W \times 0.37)$$

Where:
HU: Haugh unit.
H: albumen height (mm).
W: egg weight (g).

The albumen and the yolk were separated and their weights were measured using digital scales (0.1 g accuracy). Yolk color was determined with the Hoffmann-La-Roche yolk color fan. The yolk and the shell weight ratio were calculated by dividing the weight of each part to the egg weight.

Statistical analysis

Data were analyzed as a complete randomized block design using the general linear procedure (GLM) of SAS 9.2 (SAS, 2002). Hot and cold climates were considered as blocks. Differences between the means were ascertained by LSD test and significance was declared at (P<0.05).

RESULTS AND DISCUSSION

Egg production

The average egg production performance in different rearing weeks and total period is presented in Table 1. The least egg production during 25-40 weeks of age was observed in Chadegan which was significantly lower than the other towns (P<0.05).

Table 1 Egg production percentage of indigenous chickens reared in different rural areas of Isfahan province

Main effects	Production period (week)													
	21-24	25-28	29-32	33-36	37-40	41-44	45-48	49-52	53-56	57-60	61-64	65-68	69-72	Total
Khansar	1.2	21.5[b]	46.9[a]	61.5[a]	57.4[a]	57.4[a]	39.9	33.9	41.3[a]	61.5[a]	32.5[a]	31.8[a]	25.8[a]	37.5[a]
Chadegan	0	2.4[c]	23.3[b]	34.4[b]	34.2[c]	34.2[c]	36.8	35	36.7[ab]	34.4[b]	27.5[b]	20.6[bc]	13[c]	25.9[b]
Varzane	4.4	42.4[a]	56[a]	51[a]	49[ab]	49[ab]	36.1	40.2	36.8[ab]	51[a]	30.3[ab]	25.8[ab]	22.5[ab]	35.9[a]
Kashan	1.4	41.8[a]	45.1[a]	51.8[a]	45.8[ab]	45.8[ab]	33.3	26.3	24.5[b]	51.8[a]	22.9[c]	17.9[c]	16.6[bc]	31.1[ab]
Cold	1.0	11.9[b]	35.1[b]	47.9	45.8	43.9	38.4	34.4	39	38.8[a]	30[a]	26.2	19.4	31.7
Hot	2.9	42.0[a]	50.6[a]	51.4	47.4	39.3	34.7	33.2	30.6	29.2[b]	26.6[b]	21.8	19.5	33
Total Mean	1.9	27	42.8	49.7	46.6	41.6	36.5	33.89	34.8	34	38.3	24	19.5	32.4
SEM	0.7	5.4	4.4	3.2	2.8	2.5	1.9	2.7	2.5	2.3	1.1	1.8	1.7	1.7

means within the same column with at least one common letter, do not have significant difference (P>0.05).
SEM: standard error of the means.

The egg production of Khansar was significantly greater than Kashan across 53-56 and 61-72 weeks of age (P<0.05). The least egg production from 60 weeks of age up to the end of the experimental period (72 weeks of age) was observed in Kashan and Chadegan. Furthermore, Chadegan had significantly lower egg production than the other towns across the entire production period (P<0.05). Indigenous chickens in hot climate outperformed the cold ones during 25–32 weeks of age (P<0.05). The greater egg production during 57 to 64 weeks of age belonged to hens kept in cold climate as compared with hot climate (P<0.05). The egg production trend of indigenous hens in various towns is demonstrated in Figure 1. As shown, indigenous hens kept in Khansar achieved the peak of egg production faster than hens reared in hot towns, especially Kashan. In addition, their egg production curve was steadily higher than the other egg production curves during the experimental period.

On the other hand, indigenous hens in Chadegan started egg production slightly later and had lower peak point as well as lower stability. Although egg production was different between climates in different weeks, the average egg production was not influenced by climate during the entire laying period. Regarding the laying performance between towns, the most and the least average egg productions were found in Khansar (37.5%) and Chadegan (25.9%), respectively. It has been reported that environmental factors are more important than genetics for traits such as egg production and egg weight of indigenous hens in rural conditions. In the current study, significant differences between the average egg production in towns located in cold climate (Chadegan and Khansar) could be due to different methods of management and family incomes. Furthermore, lower egg production of hens in Chadegan may be attributed in part to the suboptimal environmental condition during the rearing and laying periods that resulted in chicken malnutrition. Nourollahi (2013) declared that the highest egg production percentage was observed in native chickens kept in hot climate of Fars province.

Kingori et al. (2010) reported that lower egg production of native chickens was related to poor quality of their diet, diseases and improper husbandry management. Otherwise, the average egg production of Malaysian indigenous chickens was reported 17% and 48% in semi-intensive and intensive systems, respectively (Ramlah, 1996), suggesting lower laying performance compared with results in the extensive production system of this study. Accordingly, it seems that the laying potential of indigenous hens in our experiment is higher than that of indigenous hens in many above mentioned countries. It seems that, although indigenous birds are affected by environmental and management variations, using nutritional supplements, proper housing and control of disease are parameters by which the growth and production efficiency of indigenous chickens can be improved in rural regions.

Egg weight

The egg weight of indigenous hens at different ages and the average egg weight across entire experimental period (21-72 weeks of ages) are presented in Table 2. The egg weight of hens at 29-44 weeks of age was higher in Khansar than Kashan (P<0.05). Moreover, egg weight of the hens reared in Khansar was higher than the other towns across 41-44 and 49-52 weeks of ages (P<0.05). Additionally, the egg weight of hens reared in cold climate was significantly higher than that in the hot climate during 33-40 weeks of age (P<0.05). It should be possibly owing to the fact that high environmental temperature could be a reason for low feed intake and low egg weight in chickens (Etches, 1998). More research on the effect of climate variations on egg weight of indigenous hens is warranted. Total average egg weight in the studied indigenous chickens was 49.6 g during entire laying period. In this regard, Kalita et al. (2009) declared that the average egg weight of Indian indigenous chickens (Assam) was 36-40 g. Moreover, Gheisari and Golian (1996) determined that the average egg weight in Isfahan indigenous chickens was 53.3 g under an intensive production system.

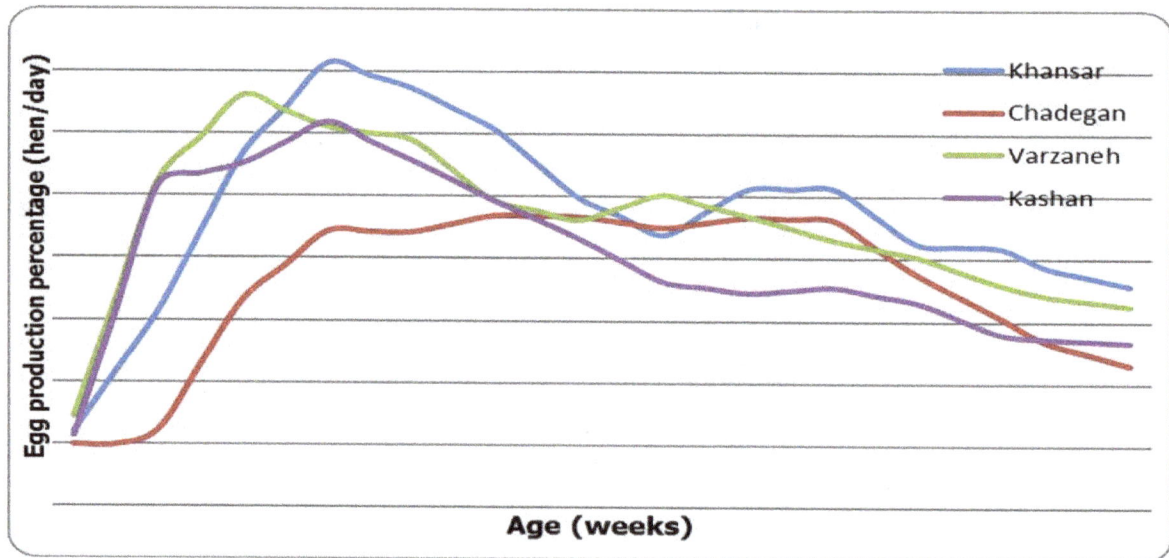

Figure 1 Egg production of studied hens in leaying period (21-72 weeks)

Table 2 Egg weight of indigenous chickens in different ages and the entire laying period reared in different rural areas of Isfahan province

Main effects	Production period (week)													
	21-24	25-28	29-32	33-36	37-40	41-44	45-48	49-52	53-56	57-60	61-64	65-68	69-72	Total
Khansar	48.1	48.5	50[a]	52[a]	53.3[a]	53.8[a]	51.9	53.2[a]	50.1	48.8	49	50.1	49.4	50.6
Chadegan	46	46.6	49.3[ab]	50.7[ab]	51.2[ab]	49.1[b]	50.3	50.4[b]	50.6	49.8	48.6	48.6	50.9	49.4
Varzane	48.8	49	48.7[ab]	50.1[ab]	48.9[b]	49.9[b]	50.3	50.6[b]	48.7	49	48.5	49	50	49.4
Kashan	45.8	45.9	46.4[b]	47.8[b]	48.7[b]	49.3[b]	51.2	50.4[b]	48.6	49.9	50.4	50.4	50.4	48.9
Cold	47.1	47.6	49.7	51.4[a]	52.2[a]	51.5	51.1	51.8	50.4	49.3	48.8	49.4	50.1	50.0
Hot	47.3	47.5	47.5	49[b]	48.8[b]	50	50.7	50.5	48.6	49.4	49.5	49.7	50.2	49.1
Total Mean	47.2	47.5	48.6	50.2	50.5	50.5	50.9	51.2	49.5	49.4	49.1	49.5	50.2	49.6
SEM	0.67	0.62	0.61	0.52	0.67	0.68	0.46	0.44	0.44	0.29	0.4	0.39	0.28	0.23

The means within the same column with at least one common letter, do not have significant difference (P>0.05).
SEM: standard error of the means.

Additionally, the average egg weight of Malaysian chicken eggs were reported between 39.7 g to 46 g in an intensive production system (Ramlah, 1996). In our study, nutritional situation was considerably affected by the life style of rural families in various regions which could be a main reason for the variations observed in the average egg weight. It has been shown that diet, body weight, age and genetics were effective factors on the egg weight (Afifian, 2006). On the other hand, Naroshin et al. (2002) reported that at least more than half of the egg weight was under genetic control. Pedroso et al. (2005) showed that the egg weight was increased by age promotion. Spratt and Leeson (1987) proposed that egg weight depends on the weight of its contents, although, changes in contents also associate with feed consumption and nutritional situation.

Egg qualitative characteristics
According to the results shown in Table 3, qualitative traits of eggshell (strength, thickness and shell weight) in the cold climate were higher than that in the hot climate (P<0.05).

Eggshell thickness in chickens reared in Chadegan was significantly greater than that of the other towns (P<0.05). The average eggshell weights of two hot climate towns (Kashan and Varzane) was lower than Chadegan (P<0.05). Eggshell strength in Kashan was significantly lower than towns in cold climate (P<0.05).

Yolk color index in hot climate was higher than the cold climate (P<0.05) but the effect of town was not significant. Hough unit of two climates were not different, however, between cold towns, egg of chickens in Chadegan had higher Haugh unit than Khansar (P<0.05).

Therefore, observed difference in Haugh unit of two cold towns can be due to the selection time and storage situation of the eggs. Moreover, the higher eggquality in cold than the hot climate in the present study might be attributed to the reduction in temperature that decrease the evaporations from the egg and consequently lead to save of albumen height and quality.

Otherwise, Nourollahi (2013) failed to show any effect of the climate on the egg qualitative traits.

Table 3 Average qualitative characteristics of the eggs produced by indigenous chickens reared in different rural areas of Isfahan province

Main effects	Shell				Yolk			Albumen	
	Weight (g)	Strength*	Thickness**	Shell to egg weight ratio (%)	Weight (g)	Color	Yolk to egg (%) weight ratio	Height (mm)	Haugh unit
...hansar	4[ab]	2.2[a]	33.5[b]	7.9[b]	14.6[a]	6.1[b]	28.4	5[b]	71.7[b]
...hadegan	4.2	2.4[a]	38.2[b]	8.9[b]	14.2[ab]	6.1[b]	28.5	5.9[a]	79.9[a]
...arzane	3.9[b]	2[ab]	33.3[b]	7.9[b]	14[a]	8.5[a]	28.6	5.2[ab]	75[ab]
...ashan	3.3[c]	1.7[b]	30[b]	7.3[b]	12.4[b]	8.9[a]	27.7	5[b]	76.5[ab]
...old	4.2[a]	2.3[a]	35.9[a]	8.4[a]	14.4	6.1[b]	28.4	5.5	75.8
...ot	3.6[b]	1.9[b]	31.6[b]	7.6[b]	13.2	8.7[a]	28.1	5.2	75.7
...otal Mean	3.92	2	33.8	8.02	13.8	7.4	28.3	5.3	75.7
...EM	0.114	0.081	0.855	0.187	0.361	0.369	0.684	0.164	1.150

*rength's unit is kg/cm².
*hickness's unit is hundreds of millimeters.
means within the same column with at least one common letter, do not have significant difference (P>0.05).
1: standard error of the means.

Gheisari *et al*. (2008) reported that average egg strength, thickness, shell weight, yolk weight, and Haugh unit produced by Isfahan indigenous chickens under intensive conditions were 2.7 (kg/cm²), 0.40 (mm), 4.93 (g), 16.04 (g) and 80.81, respectively. In another study on egg qualitative characteristics in Fars province, yolk weight percentage and Haugh unit of indigenous chickens' egg were 14.3 g, 33% and 80.5, respectively across 32-34 weeks of age (Pourreza, 1991). Yakuba *et al*. (2008) determined that shell weight, yolk weight, Haugh unit in Nigerian indigenous chickens were 4.7 g, 16.1 g and 71.4%, respectively.

CONCLUSION

In conclusion, results showed that indigenous chickens distributed in rural areas produced 130 eggs by the average weight of 45 g which had higher potential of egg production than studied indigenous chickens in some other countries. Environmental conditions such as climate can affect the egg qualitative traits in indigenous chickens. In addition, performance of these poultries in rural regions could be further promoted by improvement of environmental and nutritional conditions as well as amending management practices on indigenous chickens during rearing and laying periods.

ACKNOWLEDGEMENT

Authors are grateful to the Animal Science Department of Iran Ministry of Agriculture Jihad.

REFERENCES

Afifian A. (2006). Effect of use of different phase feeding methods on reproduction performance of native hens. MS Thesis. Department of Agriculture and Natural Resources, Islamic Azad University, Isfahan (Khorasgan) Branch, Isfahan, Iran.

Deljoisaraian J., Alipanah M. and Mohamadnia Koushki R. (2011). Factors affecting hatching and fertile eggs in indigenous chicken of Khazak. Pp. 51-56 in Proc. 1st Congr. Sci. New Technol. Agric. Zanjan, Iran.

Elibol O., Peak S.D. and Brake J. (2002). Effect of flock age, length of egg storage and frequency of turning during storage on hatchability of broilers hatching eggs. *Poult. Sci*. **81**, 945-950.

Etches R.J. (1998). Reproduction in Poultry. Acribia, Zaragoza, Spain.

Gheisari A.A. and Golian A. (1996). Effect of different dietary energy and protein levels on laying performance of Isfahan indigenous hens in their laying period. *Iranian J. Agric. Sci*. **2**, 29-35.

Gheisari A.A. and Azarbayejani A. (2013). Practical Principles to Keep Indigenous Hens. Arkan Danesh, Isfahan, Iran.

Gheisari A.A., Afifian A., Poureza J., Jahanfar H. and Sheikh Hadian H. (2008). Effect of use of different phase feeding methods on performance and egg quality characteristics in Esfahan native hens. *Pajouhesh and Sazandegi*. **78**, 65-73.

Haugh R.R. (1937). Haugh unit for measuring egg quality. US. *Egg Poult*. **43**, 552-555.

Kalita N., Gawande S.S. and Barua N. (2009). Production and reproduction performance of indigenous chicken in Assam under rural condition. *Indian J. Poult. Sci*. **44**, 5519-5529.

Khalafalla A.I., Awad S. and Hass W. (2001). Village Poultry Production in Sudan. University of Khartoum Publications, Khartoum, North Sudan.

Kingori A.M., Wachira A.M. and Tuitoek J.K. (2010). Indigenous chicken production in Kenya: a review. *Int. J. Poult. Sci*. **9**, 309-316.

Miles R.D. and Jacob P.J. (2000). Feeding the Commercial Egg Type Laying Hen. Department of Dairy and Poultry Science, University of Florida. Cooperative Extension Services, Institute of Food and Agricultural Science. Available at: http:/edis.ifas.ufi.edu.

Naroshin V.G., Romanow M.N. and Bogaryr V.P. (2002). Relation between pre-incubation egg parameters and chick weight after hatching in broiler breeders. *Anim. Prod. Technol*. **83**, 373-381.

Nourollahi H. (2013). Reviewing the Performance of Indigenous Chicken in Rural Areas of Fars Province. Agriculture and Natural Research Center of Fars, Fars, Iran.

Nyaiyo N.M. (2014). Effects of climate variability and other factors on the outbreak and spread of Newcastle disease in suneka division of Kisii county. MS Thesis. Kenyatta Univ., Kenya.

Pedroso A.A., Andrade M.A., Café M.B., Leandro N.S., Menten J.F. and Stringnini J.H. (2005). Fertility and hatchability of eggs lay in the pullet to breeder transition period and in the initial production period. *Anim. Rep. Sci.* **90**, 355-364.

Pourreza J. (1991). Scientific and Practical Principles of Poultry Production. Arkan Danesh, Isfahan, Iran.

Ramlah A.H. (1996). Performance of village chicken in Malasia. *World's Poult. Sci.* **52**, 75-79.

SAS Institute. (2002). SAS®/STAT Software, Release 9.2. SAS Institute, Inc., Cary, NC. USA.

Spratt R.S. and Leeson S. (1987). Broiler breeder performance in response to diet protein and energy. *Poult. Sci.* **66**, 683-693.

Vali N. (2008). Indigenous chicken production in Iran: a review. *Pakistan J. Biol. Sci.* **15**, 2525-2531.

Yakuba M., Ogah D.M. and Barde R.E. (2008). Productivity and egg quality characteristics for free range naked neck and normal feathered Nigerian indigenous chicken. *Poult. Sci.* **7**, 579-585.

5

The Effects of Different Levels of Canola Oil and Diet Mixing Time Length on Performance, Carcass Characteristics and Blood Lipids of Broilers

E. Ebdi[1] and A. Nobakht[1*]

[1] Department of Animal Science, Maragheh Branch, Islamic Azad University, Maragheh, Iran

*Correspondence E-mail: anobakht20@iau-maragheh.ac.ir

ABSTRACT

This experiment was conducted to investigate the effects of different levels of canola oil and diet mixing time length on performance, carcass traits and blood lipids in broilers. In this experiment 288 Ross-308 broilers were used from 11 up to 42 days as factorial arrangement (3×2) included three levels of canola oil (0, 3 and 6%) and two mixing time length (10 and 15 minute) in 6 treatments, 4 replicates and 12 birds in each replicate in a completely randomized design. Canola oil improved the performance of broilers ($P<0.01$). The highest values of daily weight gain, daily feed intake and final body weight were obtained using 6% of canola oil in diets. Mixing diet more than 10 minute, reduced the amounts of daily weight gain and final weight ($P<0.05$). In interaction between oil level and mixing time length, 6% canola oil × diet mixing for 10 minute, improved the performance of broilers ($P<0.05$). Dietary level of 6% canola oil increased the spleen percentage ($P<0.05$). Canola oil had reducing effect on the level of blood cholesterol ($P<0.05$). The lowest level of blood cholesterol was observed in group supplemented with 3% canola oil. Mixing diet more than 10 minute, increased the amount of low-density lipoprotein (LDL) in blood ($P<0.05$). In interaction effect, the lowest level of blood cholesterol was seen in 6% canola oil × mixed for 10 minute group. The overall conclusion is that using 6% canola oil and 10 minute mixing time for boiler diets could significantly improve the performance, and reduce their blood cholesterol.

KEY WORDS blood lipids, broiler chickens, canola oil, performance.

INTRODUCTION

The most practical method for increasing the energy density of diets in poultry feeding is through the addition of fats and oils (Peebles *et al.* 2000). It was reported that fat metabolism and deposition in poultry could be affected by different dietary fats and fatty acids (Snaz *et al.* 2000). Also they assist vitamin A and Ca absorption (Sklan, 1980; Leeson and Atteh, 1995). Some concerns that should be noted with fat utilization include: use of higher levels of fat may negate the effects of pelleting, measurement of metabolizable energy (ME) content can be difficult, there is the potential for rancidity, equipment needs relative to fat

additions must be adequate and potentially poor digestibility of saturated fats by the young bird (Chen and Chiang, 2005). Oils are the most important energy source of broiler rations. In order to get the optimum productivity from chickens, the protein and energy levels of ration should be high. By compensation of energy requirements of chickens with oils instead of carbohydrates, a better performance was attained (Leeson and summers, 2001). There are different sources of oils in the market for consuming, one of the most important of them is canola oil. Canola oil has been recognized as adequate mixture of essential unsaturated fatty acids such linolenic acid (C18:3) with its beneficial effect on broiler performance and linolenic acids that can be converted to longer chain omega-3 fatty acids (Sim *et al.* 1990; Yang *et al.* 2000) which are important factors in promoting the health of animals and human (Bezard *et al.* 1994). Adding 3% of canola oil and poultry fat resulted significant improvement in body weight and better feed conversion ratio than other groups. However, no significant differences were found in liver, breast, thigh weights between fat-supplemented groups in comparison with the control. Addition 6% poultry fat caused significant increasing on abdominal fat, gizzard weight was significantly higher in control group in comparison with supplemented groups (Shahryar *et al.* 2011). It has been accepted that dietary canola oil is excellent supplement for commercial fish such as salmon (Huang *et al.* 2008). However, canola oil contains less than 2% of erucic acid (docosenoic acid, C22:1, (-9) in relation to its total fatty acids content and less than 30 umoles of glucosinolates per gram of free oil on seed dry matter basis. In birds, the adverse effects of adding erucic acid to the diets are reflected on intake, growth performance and the apparent digestibilities of total lipid and individual fatty acids (Leeson and summers, 2001). Furthermore, chicks fed with diets containing erucic acid deposit less fat and utilize energy from this diet less frequently (Leeson and Summers, 2001).

Using 2% of canola oil in broiler diets positively improved their performance and carcass traits (Nobakht *et al.* 2012). In broilers using 5% canola oil increased their weight gain and blood triglyceride (Kiani *et al.* 2016). Adding canola oil up to 5% in diet of native turkeys had no significant effects on their blood lipids parameters (Salamatdoust Nobar *et al.* 2010). In laying hens incorporating 2% canola oil in diets improved their performance and reduced their egg and blood cholesterol level (Ismail *et al.* 2013).

Despite the fact that mixing is one of the most essential and critical operations in the process of feed manufacturing, yet it is frequently given little consideration. The main objective in mixing is to create a completely homogeneous blend. In other words, every sample taken should be identical in nutrient content. A functional definition of uniform mixing can be summarized in one sentence "All nutrients will be present in sufficient quantity in the daily feed intake of the target animal to meet its minimum growth requirements''.

A uniform and precise mixing of diets is very critical for providing energy and nutrients to meet the requirements for animals (Traylor *et al.* 1994). Poor mixing of feedstuffs may cause over supply or under supply of nutrients to animals, resulting in economic loss, poor performance, or both. Detrimental effects of insufficient mixing time on average daily gain and gain to feed ratio have been reported (Groesbeck *et al.* 2007).

Horizontal and vertical feed mixers are widely used for mixing the animal feeds at feed mills. These types of feed mixers have been tested for the degree of uniformity in diet mixing (McCoy, 2005). Cone-bottom type vertical mixers have been tested for mixing performance (Groesbeck *et al.* 2007).

In a newly experiment, Hyunwoong *et al.* (2015) indicated that the most suitable mixing time of broiler diets is 202 seconds. On the base of their recommendation, the mixing length should not be least than 200 seconds.

As fats have beneficial role in diet mixing and diet mixing time is very critical for having a uniform diet, in the current study the effects of different levels of canola oil as a fat source and different mixing time length of diets on broiler performance and their blood lipids profile have been investigated.

MATERIALS AND METHODS

In this experiment 288 Ross-308 broilers were used from 11 up to 42 days as factorial arrangement (3×2) included three levels of canola oil (0, 3 and 6%) and two mixing time of diets (10 and 15 minutes) in 6 treatments, 4 replicates and 12 birds in each replicate in a completely randomized design. The diets were formulated to meet the requirements of birds established by the Ross Company (2014) for broilers in grower (11-24 days) and finisher (25-42 days) periods (Tables 1 and 2).

In all experiment periods the diets and water were provided *ad-libitum* for birds. The lighting program during the experimental periods consisted of a period of 23 hours light and 1 hour of darkness. Environmental temperature was gradually decreased from 33 °C to 25 °C on day 21 and was then kept constant. Body weight, feed intake and feed conversion ratio were determined at the end of each experimental period. Mortality was also recorded if it occurred. At the end of the experiment, two birds from each replicate were randomly chosen for blood collection and approximately 5 ml blood samples were collected from the brachial vein of randomly chosen birds. The blood was centrifuged to obtain serum for determining the blood lipids which included cholesterol, triglyceride, low-density lipoprotein (LDL) and high-density lipoprotein (HDL). Kit packages (Pars Azmoon Company; Tehran, Iran) were used for determining the blood biochemical parameters using Anision-300 autoanalyzer system (Nazifi, 1997). Also, at 42 day of age, two birds from each replicate randomly chosen based on the average weight of the group and sacrificed. Dressing yield was calculated by dividing eviscerated weight by live weight. Abdominal fat, gizzard, liver, spleen, breast and thigh were collected, weighed and calculated as a percentage of carcass weight.

Table 1 Ingredients and chemical composition of broiler diets in grower period (11-24 days)

Feeds ingredients	Canola oil level		
	0%	3%	6%
Corn grain	62.83	56.13	47.43
Soybean meal (42% CP)	3317	35.28	37.39
Canola oil	0.00	3.00	6.00
Inert (sand)	0.27	1.90	5.52
Oyster shell	0.25	0.21	0.17
Bone meal	2.11	2.14	2.17
Salt	0.40	0.40	0.41
Vitamin premix[1]	0.25	0.25	0.25
Mineral premix[2]	0.25	0.25	0.25
DL-methionine	0.28	0.29	0.30
L-lysine hydrochloride	0.19	0.15	0.11
Calculated composition			
Metabolizable energy (kcal/kg)	2900	2900	2900
Crude protein (%)	19.80	19.80	19.80
Ca (%)	0.83	0.83	0.83
Available phosphorus (%)	0.42	0.42	0.42
Sodium (%)	0.18	0.18	0.18
Lysine (%)	1.14	1.14	1.14
Methionine + cysteine (%)	0.88	0.88	0.88
Tryptophan (%)	0.24	0.24	0.24

[1] Vitamin premix per kg of diet: vitamin A (retinol): 2.7 mg; vitamin D_3 (cholecalciferol): 0.05 mg; vitamin E (tocopheryl acetate): 18 mg; vitamin K_3: 2 mg; Thiamine: 1.8 mg; Riboflavin: 6.6 mg; Panthothenic acid: 10 mg; Pyridoxine: 3 mg; Cyanocobalamin: 0.015 mg; Niacin: 30 mg; Biotin: 0.1 mg; Folic acid: 1 mg; Choline chloride: 250 mg and Antioxidant: 100 mg.
[2] Mineral premix per kg of diet: Fe ($FeSO_4.7H_2O$, 20.09% Fe): 50 mg; Mn ($MnSO_4.H_2O$, 32.49% Mn): 100 mg; Zn (ZnO, 80.35% Zn): 100 mg; Cu ($CuSO_4.5H_2O$): 10 mg; I (K_1, 58% I): 1 mg and Se ($NaSeO_3$, 45.56% Se): 0.2 mg.

Table 2 Ingredients and chemical composition of broiler diets in finishing period (25-42 days)

Feeds ingredients	Canola oil level		
	0%	3%	6%
Corn grain	65.30	57.59	50.91
Soybean meal (42%CP)	30.22	32.34	34.35
Canola oil	0	3.00	6.00
Inert (sand)	1.18	3.81	5.47
Oyster shell	0.14	0.11	0.07
Bone meal	2.07	2.10	2.14
Salt	0.37	0.37	0.37
Vitamin premix[1]	0.25	0.25	0.25
Mineral premix[2]	0.25	0.25	0.25
DL-methionine	0.18	0.18	0.19
L-lysine hydrochloride	0.04	0	0
Calculated composition			
Metabolizable energy (kcal/kg)	2900	2900	2900
Crude protein (%)	18.58	18.58	18.58
Ca (%)	0.77	0.77	0.77
Available phosphorus (%)	0.40	0.40	0.40
Sodium (%)	0.17	0.17	0.17
Lysine (%)	0.96	0.96	0.96
Methionine + cysteine (%)	0.75	0.75	0.75
Tryptophan (%)	0.22	0.22	0.22

[1] Vitamin premix per kg of diet: vitamin A (retinol): 2.7 mg; vitamin D_3 (cholecalciferol): 0.05 mg; vitamin E (tocopheryl acetate): 18 mg; vitamin K_3: 2 mg; Thiamine: 1.8 mg; Riboflavin: 6.6 mg; Panthothenic acid: 10 mg; Pyridoxine: 3 mg; Cyanocobalamin: 0.015 mg; Niacin: 30 mg; Biotin: 0.1 mg; Folic acid: 1 mg; Choline chloride: 250 mg and Antioxidant: 100 mg.
[2] Mineral premix per kg of diet: Fe ($FeSO_4.7H_2O$, 20.09% Fe): 50 mg; Mn ($MnSO_4.H_2O$, 32.49% Mn): 100 mg; Zn (ZnO, 80.35% Zn): 100 mg; Cu ($CuSO_4.5H_2O$): 10 mg; I (K_1, 58% I): 1 mg and Se ($NaSeO_3$, 45.56% Se): 0.2 mg.

Statistical analyses

The data were subjected to one-way analysis of variance procedures appropriate for a factorial arrangement as completely randomized design using the general linear model procedures of SAS, (2005).

Means were compared using the Toky test (Valizadeh and Moghaddam, 1994). Statements of statistical significance were based on ($P < 0.05$).

RESULTS AND DISCUSSION

Performance

The effects of different levels of canola oil and diet mixing time length on performance of broilers are shown in Table 3. Adding canola oil into diets without having any effects on feed conversion ratio, increased the amounts of daily weight gain, daily feed intake and final live weight (P<0.01). Diet mixing for 15 minutes significantly reduced the amounts of final live weight (P<0.05). In interaction between canola oil levels and mixing time length, the best performance was obtained with 6% canola oil × 10 minute mixing time length. Therefore the highest amounts of daily weight gain, daily feed intake and final body weight belonged to this experimental group.

Carcass traits

The effects of different levels of canola oil and diet mixing time length on carcass traits of broiler chicks are shown in Table 4.

Except for spleen percentage, canola oil had no significant effects on carcass traits of the birds (P>0.05). Adding 6% canola oil to the diet increased the spleen percentage (P<0.05).

Using canola oil at 3% did not affect the size of spleen compared to control group (group without oil). Diet mixing time and interaction between diet mixing time and canola oil levels did not have any significant effects on carcass traits of chicks (P>0.05).

Blood lipids

The effects of different levels of canola oil and diet mixing time length on blood lipids of broiler chicks are provided in Table 5. The incorporation of canola oil in diets significantly reduced the amount of blood cholesterol (P<0.05). Mixing diet for 15 minute increased the blood level of LDL (P<0.05). In interaction between canola oil and mixing time, using 6% canola oil × 10 minute mixing diet, reduced the amount of blood cholesterol (P<0.05).

Canola oil more than energy is a rich source of essential fatty acids and fat soluble vitamins. So, using it in broilers diet support their performance by supplying the essential nutrient requirements. One of the most important roles of fats in diets is to improve the diets palatability (Leeson and summers, 2001). As, the palatability increase the amount of feed intake, for this reason, the amount of broilers feed intake, increased by increasing to canola oil level in diets. If we accept that the amount of weight gain in this experiment was in line with the amount of feed intake, then using canola oil could not significantly change the feed conversion ratio. There is strange positive relation between daily weight gain and final live weight (Leeson and summers, 2001). As the highest amount of daily weight gain was obtained by using 6% canola oil in diets, so, the highest final live weight was seen in group contained 6% canola oil in it diet. Our findings in the present study about the effects of incorporating of canola oil in broiler diets on their performance is in agree to previous research (Nobakht et al. 2012; Ismail et al. 2013; Kiani et al. 2016).

Table 3 The effects of canola oil and diet mixing time length on performance of broilers (11-42 days)

Treatments	Weight gain (g/d/h)	Feed intake (g/d/h)	FCR	Final weight (g)	Livability (%)	Production index
Canola oil level (%)						
0	66.77[c]	109.32[c]	1.64	2217.00[c]	97.92	318.29
3	68.93[b]	112.72[b]	1.64	2290.42[b]	97.22	323.79
6	70.38[a]	116.90[a]	1.66	2327.92[a]	96.53	322.80
SEM	0.50	1.03	0.01	11.88	1.27	3.56
P-value	0.0001	0.0001	0.2349	0.0001	0.7436	0.5172
Diet mixing time length (minute)						
10	69.35[a]	113.99	1.64	2299.16[a]	97.22	325.40
15	67.97[b]	111.97	1.65	2257.83[b]	97.20	317.90
SEM	0.40	0.84	0.01	9.70	1.04	2.91
P-value	0.0232	0.1024	0.8302	0.0056	.9999	0.0791
Canola oil levels (%) × diet mixing time (minute)						
0 × 10	67.21[ab]	110.40[b]	1.65	2240.84[bc]	97.23	319.32
0 × 15	66.32[b]	108.25[b]	1.63	2194.67[c]	98.62	317.27
3 × 10	69.56[a]	112.99[ab]	1.63	2300.84[a]	97.23	328.29
3 × 15	68.11[a]	112.46[ab]	1.65	2280.00[a]	97.23	319.30
6 × 10	71.29[a]	118.60[a]	1.66	2357.00[a]	97.23	328.58
6 × 15	69.48[a]	115.21[a]	1.67	2298.84[ab]	95.84	317.02
SEM	0.69	1.46	0.02	16.80	1.79	5.03
P-value	0.0411	0.0236	0.3194	0.0344	0.7436	0.6260

The means within the same column with at least one common letter, do not have significant difference (P>0.05).
FCR: feed conversion ratio.
SEM: standard error of the means.

Table 4 The effects of canola oil levels and diet mixing time length on carcass traits (carcass %) of broilers (42 days)

Treatments	Carcass	Abdominal fat	Gizzard	Liver	Spleen	Breast	Thigh
Canola oil level (%)							
0	70.75	3.72	3.95	3.83	0.21[b]	35.37	28.78
3	69.67	3.92	4.19	2.80	0.21[b]	34.50	29.00
6	70.60	3.76	4.05	2.92	0.023[a]	34.96	28.46
SEM	0.85	0.09	0.1	0.08	0.01	0.76	0.79
P-value	0.6308	0.3097	0.2503	0.5896	0.0364	0.7235	0.9200
Diet mixing time							
10	70.61	3.80	4.12	2.86	0.22	34.93	28.55
15	70.08	3.75	4.02	2.84	0.23	34.96	28.87
SEM	0.70	0.07	0.08	0.07	0.01	0.63	0.64
P-value	0.5895	0.9532	0.3934	0.7908	0.7916	0.9718	0.7315
Canola oil levels × diet mixing time (minute)							
0 × 10	70.79	3.71	4.07	2.88	0.21	36.21	29.48
0 × 15	70.72	3.74	3.84	2.78	0.21	34.54	28.07
3 × 10	71.28	3.86	4.15	2.72	0.22	33.87	27.93
3 × 15	68.07	3.97	4.24	2.88	0.20	35.13	29.88
6 × 10	69.78	3.83	4.13	2.99	0.23	35.72	28.26
6 × 15	71.43	3.71	3.98	2.85	0.24	35.21	28.67
SEM	1.21	0.12		0.14	0.01	1.08	1.11
P-value	0.8825	0.6444	0.4926	0.4218	0.4886	0.3901	0.3333

The means within the same column with at least one common letter, do not have significant difference (P>0.05).
SEM: standard error of the means.

Table 5 The effects of canola oil and diet mixing time length on blood lipids (mg/dL) of broilers (42 days)

Treatments	Cholesterol	Triglyceride	HDL	LDL
Canola oil level (%)				
0	129.12[a]	84.92	60.71	51.36
3	113.30[b]	78.88	55.30	47.94
6	122.30[ab]	68.58	51.08	54.00
SEM	3.85	12.68	5.48	5.51
P-value	0.0258	0.3566	0.4712	0.7403
Diet mixing time (minute)				
10	120.70	63.00	65.26	42.64[b]
15	122.74	78.58	46.83	59.55[a]
SEM	3.14	10.36	4.47	4.50
P-value	0.6625	78.58	0.0060	0.0138
Canola oil levels × diet mixing time (minute)				
0 × 10	120.46[ab]	78.77	59.27	45.39
0 × 15	137.78[a]	91.07	62.15	57.33
3 × 10	122.40[ab]	54.48	73.44	42.04
3 × 15	124.21[a]	63.28	37.17	53.84
6 × 10	119.45[b]	55.77	63.09	40.50
6 × 15	126.21[ab]	81.39	39.07	67.50
SEM	5.45	17.94	7.75	7.79
P-value	0.0100	0.8853	0.0624	0.5416

LDL: low-density lipoprotein and HDL: high-density lipoprotein.
The means within the same column with at least one common letter, do not have significant difference (P>0.05).
SEM: standard error of the means.

Homogenous mixing of diets has been shown can improve the production performance of broilers (Traylor *et al.* 1994; Groesbeck *et al.* 2007). Therefore, the confirmation of mixing efficiency of a mixer is very important before using the mixer for animal feed production. The uniformity of mixing diet is affected many factors including particle size, ingredient density, sequence of ingredient addition, amount of ingredients mixed, mixer design, mixing time and cleanliness of the mixer (McCoy, 2005).

In the present study, reducing the amounts of daily weight gain and final live weight in mixing diets more than 10 minutes may be related to de mixing and separating of diets particles. For having optimum performance, the poultry diets uniformity is critical (McCoy, 2005).

In an experiment, 202 seconds recommended as the best mixing time of broiler diet, but mixing for 300 and 420 seconds had adverse effect in this respect (Hyunwoong *et al.* 2015).

Improving of the performance by using 6% canola oil and diet mixing for 10 minutes, clearly showed the efficacy of high levels of fat and moderate diet mixing time on performance of broiler, whereas omitting of canola oil and prolonged of mixing time, had adverse effects on performance of broilers. These results may be related to the beneficial effects of canola oil (Nobakht *et al.* 2012; Ismail *et al.* 2013; Kiani *et al.* 2016) and optimum mixing time (McCoy, 2005) on performance of broilers. Using canola oil without enlargement in spleen size had no effect on carcass traits of broilers. Spleen is an immune organ and enlarge in the spleen size may be related to immune upgrading (Nazifi, 1997).

Canola oil such other fat sources, contain highly amount of fat soluble vitamins. Fat soluble vitamins such as A and E have important role on increasing of body immunity and disease prevention (Leeson and summers, 2001). Increase in the size of spleen by using canola oil in the present research may be related to fat soluble vitamins content of this oil. Our findings in the present study not in line with Shahryar *et al.* (2011) and Nobakht *et al.* (2012) reported results that those indicated using 2% and 3% of canola oil in broiler diets have beneficial effects on their carcass traits improvement.

Oils those originated from plants, such other plant derivatives do not have cholesterol in their compounds, so, in contrast to animal originated fats, using them, not only do not increase the level of blood cholesterol, but also can reduce it.

As canola oil usually is free of cholesterol, so, using it in broiler diets in contrast to control group, without having any effects on other blood lipids, significantly reduced the amount of blood cholesterol. This result is in agree with findings of Ismail *et al.* (2013), how indicated that using 2% canola oil in laying hens diets reduced their egg and blood cholesterol content.

Whereas is not agree with Salamatdoust Nobar *et al.* (2010) report that using canola oil up to 5% of native turkeys diets had no effects on their blood lipids.

Increase in the amount of blood LDL in mixing diets for 15 minutes may be related to de mixing and separating of diets particles from each other and lowering the feed quality. Significantly reducing the blood level of cholesterol in group contained 6% canola oil with 10 minutes of mixing time, show the efficacy of using high level of canola oil and diet optimum mixing time in this respect.

CONCLUSION

The overall conclusion is that using 6% canola oil and 10 minute mixing time for boiler diets could significantly improve the performance, and reduce their blood cholesterol.

ACKNOWLEDGEMENT

This study has been supported by Islamic Azad University of Maragheh Branch. The author likes to appreciate Dr Mehmannvaz for his supporting during experimental period in poultry farm.

REFERENCES

Bezard J., Blond J.P., Bernard A. and Clouet P. (1994). The metabolism and availability of essential fatty acids in animal and human tissues. *Reprod. Nutr. Dev.* **34,** 539-568.

Chen H.Y. and Chiang S.H. (2005). Effect of dietary polyunsaturated/saturated fatty acid ratio on heat production and growth performance of chicks under different ambient temperature. *Anim. Feed Sci. Technol.* **120,** 299-308.

Groesbeck C.N., Goodband R.D., Tokach M.D., Dritz S.S., Nelssen J.L. and DeRouchey J.M. (2007). Diet mixing time affects nursery pig performance. *J. Anim. Sci.* **85,** 1793-1798.

Huang S.S.Y., Chl F.U., Higgs D.A., Balfry S.K., Schulte P.M. and Brauner C.J. (2008). Effects of dietary canola oil level on growth performance, fatty acid composition and ionoregulatory development of spring Chinook salmon parr, *Oncorhynchus tshawytscha. Aquaculture.* **274(1),** 109-117.

Hyunwoong J.o., Changsu K., Doo Seok N. and Beob Gyun K. (2015). Mixing performance of a novel flat-bottom vertical feed mixer. *Int. J. Poult. Sci.* **14(11),** 625-627.

Ismail I.B., Al-Busadah K.A. and El-Bahr S.M. (2013). Effect of dietary supplementation of canola oil on egg production, quality and biochemistry of egg yolk and plasma of laying hens. *Int. J. Biol. Chem.* **7,** 27-37.

Kiani A., Sharifi S.D. and Ghazanfari S. (2016). Effect of graded levels of canola oil and lysine on performance, fatty acid profile of breast meat and blood lipids parameters of broilers. *Anim. Sci. Res. J.* **26(2),** 109-121.

Leeson S. and Atteh J.O. (1995). Utilization of fats and fatty acids by turkey poults. *Poult. Sci.* **74,** 2003-2010.

Leeson S. and Summers J.D. (2001). Nutrition of the Chicken. Published by University Books, Ontario, Canada.

McCoy R.A. (2005). Mixer testing. Pp. 35-46 in Feed Manufacturing Technology. V.S.K. Schofield, Ed. American Feed Industry Association, Arlington, Virginia.

Nazifi S. (1997). Hematology and Clinical Biochemistry of Birds. Shiraz University Publication, Shiraz, Iran.

Nobakht A., Ariyana A. and Mazlum F. (2012). Effect of different levels of canola oil with vitamin E on performance and carcass traits of broilers. *Int. Res. J. Appl. Basic. Sci.* **3(5),** 1059-1064.

Peebles E.D., Zumwalt C.D., Doyle S.M., Gerard P.D., Latour M.A., Boyle C.R. and Smith T.W. (2000). Effects of dietary fat type and level on broiler breeder performance. *Poult. Sci.* **79,** 629- 639.

Salamatdoust Nobar N., Gorbani A., Nazeradl K., Ayazi A., Hamidiyan A., Fani A., Aghdam Shahryar H., Giyasi ghaleh kandi J. and Ebrahim Zadeh Attari V. (2010). Beneficial effects of canola oil on breast fatty acids profile and some of serum biochemical parameters of Iranian native turkeys. *J. Cell.*

Anim. Biol. **4(8),** 125-130.

SAS Institute. (2005). SAS®/STAT Software, Release 9.1. SAS Institute, Inc., Cary, NC. USA.

Shahryar H.A., Salamatdoustnobar R., Lak A. and Lotfi A.R. (2011). Effect of dietary supplemented canola oil and poultry fat on the performance and carcass characterizes of broiler chickens. *Curr. Res. J. Biol. Sci.* **3,** 388-392.

Sim J.S. (1990). Flax seed as a high energy/protein/ omega-3 fatty acid feed ingredient for poultry. Pp: 65-72 in Proc. 53rd Flax Inst. United States, Fargo, North Dakota, USA.

Sklan D. (1980). Site of digestion and absorption of lipids and bile acids in the rat and turkey. *Comp. Biochm. Physiol.* **65(1),** 91-95.

Snaz M., Lopez-Bote C.J., Menoyo D. and Bautista J.M. (2000). Abdominal fat deposition and fatty acid synthesis are lower and β-oxidation is higher in broiler chickens fed diets containing unsaturated rather than saturated fat. *J. Nutr.* **130,** 3034-3037.

Traylor S.L., Hancock J.D., Behnke K.C., Stark C.R. and Hines R.H. (1994). Uniformity of mixed diets affects growth performance in nursery and finishing pigs. *J. Anim. Sci.* **72(2),** 59-67.

Valizadeh M. and Moghaddam M. (1994). Experimental Designs in Agriculture. Pishtaz Elem Production, Tehran, Iran.

Yang C.X., Ding L.M. and Rong Y. (2000). N-3 fatty acid metabolism and effects of alpha- linolenic acid on enriching n-3 FA eggs. *J. Chi. Agric. Univ.* **95,** 117-122.

6

Comparison of Different Selenium Sources and Vitamin E in Laying Hen Diet and Their Influences on Egg Selenium and Cholesterol Content, Quality and Oxidative Stability

F. Asadi[1], F. Shariatmadari[1*], M.A. Karimi-Torshizi[1], M. Mohiti-Asli[2] and M. Ghanaatparast-Rashti[2]

[1] Department of Poultry Science, Faculty of Agriculture, Tarbiat Modares University, Tehran, Iran
[2] Department of Animal Science, Faculty of Agricultural Science, University of Guilan, Rasht, Iran

*Correspondence E-mail: shariatf@modares.ac.ir

ABSTRACT

An experiment was carried out to compare the effects of laying hen's diet supplemented with inorganic and different organic sources of selenium (Se) on quality and oxidative stability of eggs during storage. A total of 81, (35-week old) laying hens of Lohmann LSL-White were assigned to cages in a completely randomized design with 9 groups of treatment and 3 replicates of 3 birds. Hens in each group were fed their corresponded diet included the basal diet supplemented with sodium selenite, Se-enriched yeast, Cytoplex-selenium and Selenomax at two different levels of 0.3 and 0.6 mg/kg, or 200 mg/kg vitamin E. To prevent brand judgment challenge, A, B and C letters were applied for different organic source of Se. After 56 days of feeding experimental diets, eggs were collected from the hens to analysis. Egg weight loss during storage at 4 °C was lower (P<0.05) in the group fed 0.3 mg of B source Se/kg of feed. Vitamin E and Se supplemented groups had lower malondialdehyde values than those from the non-supplemented (P<0.01). The C Source of organic Se resulted in lower malondialdehyde compared with the other sources of Se or control. The supplementation of Se in diet increased (P<0.01) yolk Se concentration, with the effect being more significant by C source of Se. Selenium and vitamin E supplementation decreased serum and yolk cholesterol content (P<0.01). The results demonstrate the better efficacy of the C source of organic Se to increase Se deposition in egg and improved egg quality compared with the other sources of Se.

KEY WORDS egg quality, laying hen, oxidative stability, selenium, vitamin E.

INTRODUCTION

Selenium (Se) is an important natural antioxidant which is essential in many metabolic processes in living organisms. It is found in nature in two forms of inorganic and organic. Inorganic Se refers to different minerals such as selenite, selenate and selenide, and organic Se is related to seleno-amino acids such as selenomethionine (SeMet) and seleno-cysteine (SeCys). Selenium deficiency in birds, especially if combined with the lack of vitamin E, causes the occurrence of exudative diathesis and encephalomalacia (Leeson S. and Summers, 2001). In human, Se intake is often lower than the recommended daily allowance (Tanguy et al. 2012); therefore, there is a need to increase Se consumption in the common human foods. Se-enriched eggs can be produced by adding inorganic or organic Se compounds to the hen diet. In the recent years, different sources of organic Se such as Se-enriched yeast, Se-proteinate and Se-amino acids, have been introduced to the animal feed industries. A Se-enriched yeast product has been produced by cultivation of *Saccharomyces cerevisi*ae in Se-enriched media, thus yeast can accumulate large amounts of Se and incorporate them into organic Se-containing compounds, mainly SeMet (Demirci, 1999; Schrauzer, 2001). Another organic Se source, Se-proteinate, is one of these organic Se sources which is produced with enzymatically hydrolyzed soy protein. Organic Se supplements have been reported to increase egg Se content more than inorganic Se (Payne *et al.* 2005; Mohiti-Asli *et al.* 2008). In addition, it has been shown that addition of organic Se to laying hens diet can improve the quality of stored eggs (Mohiti-Asli *et al.* 2008; Jlali *et al.* 2013).

Mohiti-Asli *et al.* (2008) reported that the diets supplemented with sodium selenite, seleno-yeast or vitamin E, improved egg quality, oxidative stability and fatty acid composition during the storage without any negative effect on hen performance. However, the efficacies of numerous organic Se products have yet to be evaluated (Jang *et al.* 2010; Tufarelli *et al.* 2015). Hence, having a lot of varieties of Se in market and getting confused about choosing the products, the objective of this experiment was to evaluate the effect of three organic Se products compared with inorganic sodium selenite and vitamin E in the diet of laying hens on serum and yolk cholesterol content, egg quality, Se content and oxidative stability.

MATERIALS AND METHODS

Egg quality measurements and laboratory analysis

Egg quality characteristics including shell thickness, shell resistance, shell weight, albumen quality (Haugh unit score; HU) and yolk weight, were measured in two eggs collected from each replicate at the end of 8-wk experimental period. Eggshell strength was tested on the Egg Shell Force Gauge (Robot mation Co. Ltd., Tokyo, Japan; Mohiti-Asli *et al.* (2008)), and egg weight, albumen height, and HU, were measured with the Egg Multi Tester EMT-5200 (Robot mation Co. Ltd.; Mohiti-Asli *et al.* (2008)). The yolk and albumen were separated and weighed to determine the yolk weight and albumen weight. Eggshell was weighed with eggshell membranes giving eggshell weight. Eggshell thickness was measured at the blunt, equatorial and sharp regions to get the average value using an Ultrasonic Thickness Gauge (Echometer 1062, Robot mation Co. Ltd., Japan) as indicated by Mohiti-Asli *et al.* (2008). From each replicate, two eggs were collected at the last week of the experiment in order to measure the yolk cholesterol content by method of Pasin *et al.* (1998). Eggs collected at the last week of experiment were stored in two different temperatures (4 °C and 25 °C) for 14 days to determine egg quality and oxidative stability. To study egg oxidative stability, malondialdehyde (MDA), was measured as a secondary

oxidation product according to the thiobarbituric acid (TBA) method described by Botsoglou *et al.* (1994) using third derivative spectrophotometry (UV-visible S2100, Scinco, Korea). Tetraethoxy propane (1, 1, 3, 3- Tetraethoxy propane, T9889, 97%, Sigma, St. Louis, MO 63103) was used as MDA precursor in the standard curve. At the end of the experiment, 3.0 mL of blood was drawn from the brachial vein of two hens in each cage. The sera were separated and serum cholesterol was determined using a commercial diagnostic kit enzyme method (Pars Azmoon Co., Iran).

Table 1 Composition and calculated analysis of the basal diet

Ingredients	% in diet
Yellow corn grain	62.20
Soybean meal (44%)	23.87
Soybean oil	2.20
Oyster shell	9.20
Dicalcium phosphate	1.54
Common salt	0.34
DL-Methionine	0.15
Vitamin premix[1]	0.25
Mineral premix[2]	0.25
Calculated analysis	
Metabolizable energy (kcal/kg)	2800
Crude protein (%)	16.02
Lysine (%)	0.79
Methionine (%)	0.42
TSAA (%)	0.67
Calcium (%)	3.73
Available phosphorous (%)	0.40
Linoleic acid (%)	2.58
Se (mg)	0.21

[1] Vitamin premix provided per kilogram of diet: vitamin A: 10000 IU; vitamin D_3: 2500 IU; vitamin E: 20 mg; vitamin K_3: 3 mg; vitamin B_1: 1 mg; vitamin B_2: 4 mg; Pantothenic acid: 10 mg; vitamin B_6: 3 mg; Niacin: 30 mg; vitamin B_{12}: 0.025 mg; Folic acid: 0.5 mg; Biotin: 0.05 mg and Cholin chloride: 400 mg.
[2] Mineral premix provided per kilogram of diet: Manganese: 100 mg; Iron: 25 mg; Copper: 5 mg; Iodine: 0.5 mg; Selenium: 0.16 mg and Zinc: 60 mg.

Statistical analysis

All data were analyzed as a completely randomized design with eight treatments using the General Linear Model procedure of SAS (SAS, 2002). The following model was fitted:

$$Y_{ij} = \mu + T_i + e_{ij}$$

Where:

Y_{ij}: trait of interest for hens.

μ: overall mean.

T_i: treatment effect.

e_{ij}: residual error.

Normal distribution of residuals and variance homogeneity of the data were tested by UNIVARIATE procedure and the Levene's test, respectively. The experimental unit was the collected eggs for egg quality measurements, constituting of two observations in each replicate.

For all the remaining studied traits including egg production, egg weight and feed conversion ratio (FCR), the cage was used as the experimental unit. Differences were considered significant at (P<0.05). Significant differences between means were separated by Duncan's multiple range test.

RESULTS AND DISCUSSION

Performance

To prevent brand judgment challenge, A, B and C letters were applied for different organic source of Se (see the Tables). Hen performance parameters, including egg production, egg weight, egg mass and FCR values, were not significantly affected by the vitamin E or any of the supplement sources of in the diets. However, feed consumption was higher in Se-supplemented groups than that of the control group (Table 2).

Egg quality

Neither vitamin E nor organic and inorganic sources of Se affected any internal or external quality characteristics of eggs stored at 4 °C (Table 3; P>0.05). However, egg and shell weights were significantly higher in hens fed B source of Se (0.3 mg/kg of Se) than the control. Quality traits of eggs stored for 14 days at 25 °C (including yolk weight, shell thickness, shell resistance, shell and egg weights) were not influenced by any treatments (Table 4; P>0.05). On contrary, the Haugh unit (HU) value was significantly higher in hens fed vitamin E supplemented diet. The other experimental treatments also showed higher HU values than the control group (P<0.05).

Dietary supplementation of vitamin E, organic sources and inorganic Se decreased MDA content of the egg yolks in the fresh and stored eggs, which were more pronounced with vitamin E and the C source of organic Se (Table 5). Eggs from hens fed C source of Se in diet had lower MDA in yolk in both levels of 0.3 and 0.6 mg/kg compared with the other sources of Se and the control. All sources of selenium supplements increased yolk Se content (Table 6; P<0.01). Organic Se from B and C sources (in the level of 0.6 mg/kg of the diet) increased yolk Se content more than inorganic Se and source A of Se (P<0.01). Yolk Se was higher in eggs from hens fed higher dietary Se level, however 0.6 mg/kg Se from C source did not increase yolk Se content than 0.3 mg/kg. Selenium and vitamin E decreased yolk cholesterol content with the effect being more significant by vitamin E (Table 5; P<0.01). Also, serum cholesterol was lower in hens fed vitamin E in diet than those fed the control diet (P<0.01).

Many previous reports suggested that the dietary Se supplementation has no effect on feed intake, egg production

and egg weight (Patton et al. 2000; Puthpongsiriporn et al. 2001; Dvorska et al. 2003; Jiakui and Xialong, 2004; Payne et al. 2005; Mohiti-Asli et al. 2008; Jlali et al. 2013). It could be supposed that Se dependent enzymes with a key role in the antioxidant systems maybe not involved in egg production (Jiakui and Xialong, 2004). However, some reports indicated that the effect of Se on egg production was only significant when Se content in the diet was below the requirement (Cantor and Scott, 1994). In the current study, adding organic source of Se to diet influenced weight of the stored eggs at 4 °C with the effect being more significant for B source of organic Se. Several factors including shell porosity, albumen quality and egg size, could affect loss of egg weight during storage. Therefore, a probable reason for minor weight loss in stored eggs from hens fed organic source of Se might be related to their better shell quality parameters such as higher shell thickness, shell resistance and shell weight. In the current study no significant difference was detected in shell thickness and resistance, although B source of organic Se had the highest shell weight among the treatments.

Reis et al. (2009) reported that egg weight was not different in broiler breeder hens fed different levels of sodium selenite or Zn-SeMet. Payne et al. (2005) found a linear increase in egg weight by supplementation of Se yeast from 0 to 3 mg/kg. They proposed that higher Se concentrations in diet could affect protein synthesis and consequently egg weight.

Organic Se incorporation into the egg has been suggested to be in a large fraction as SeMet (Surai, 2002). It has been shown that methionine supplementation per se affected egg weight positively (Calderon and Jensen, 1990). A similar response for SeMet, which leads to increased egg weight, could be expected if SeMe methionine were fully incorporated into egg proteins.

Eggs from hens fed vitamin E or 0.6 mg/kg B source of organic Se supplemented-diets had higher HU after 14 days of storage at 25 °C compared to the other groups. It has been reported that vitamin E (Mohiti-Asli et al. 2008) and organic Se (Pappas et al. 2005) reduced deterioration of egg albumen. However, the effect of Se on albumen quality was not consistent in the previous studies. Arnold et al. (1973) reported that HU of eggs were improved by sodium selenite. Payne et al. (2005) reported that HU values of eggs stored at 23-27 °C was similar in either dietary levels of sodium selenite or seleno-yeast. The concentration of MDA was increased in eggs stored for 14 days at 4 °C or 25 °C. The least concentration of MDA was found in hens fed diet supplemented with vitamin E. Several reports indicated that dietary administration of supplemental levels of α-tocopherol improved the stability of egg yolk lipids (Galobart et al. 2001; Mohiti-Asli et al. 2008).

Table 2 Effect of Se sources and vitamin E on laying hen performance[1]

Treatment	Egg production (% hen day)	Egg weight (g)	Egg mass (g/day)	Feed intake (g/hen/day)	FCR
Control	95.6±0.15	59.5±0.97	56.8±1.56	106.2±1.33[c]	1.87±0.05
Selenite, 0.3 mg/kg	93.6±0.61	57.8±0.32	53.9±1.01	114.7±1.75[a]	2.13±0.09
A, 0.3 mg/kg	95.5±0.95	59.0±0.89	56.4±0.91	115.1±1.03[a]	2.05±0.06
B, 0.3 mg/kg	95.0±0.98	60.1±0.59	57.2±1.91	111.8±1.01[ab]	1.96±0.05
C, 0.3 mg/kg	93.4±0.43	58.5±0.83	54.6±1.30	112.5±1.85[ab]	2.07±0.14
A, 0.6 mg/kg	92.7±0.63	58.6±0.28	54.3±1.80	112.4±1.14[ab]	2.08±0.10
B, 0.6 mg/kg	94.0±0.77	59.3±0.92	55.7±1.11	111.4±1.54[ab]	2.01±0.03
C, 0.6 mg/kg	94.7±0.17	58.8±0.28	55.8±0.86	115.2±1.81[a]	2.07±0.05
Vitamin E	91.9±0.19	57.6±0.88	52.4±1.41	108.4±1.74[bc]	2.08±0.07
Effect	NS	NS	NS	*	NS

[1] All data reported in Table are Means ± SE.
The means within the same column with at least one common letter, do not have significant difference (P>0.05).
* (P<0.05).
NS: non significant.
FCR: feed conversion ratio.

Table 3 Effect of Se source and vitamin E on egg quality characteristics (Means±SE) stored at 4 °C for 14 days

Treatment	Egg weight (g)	Shell weight (g)	Albumen weight (g)	Yolk weight (g)	Haugh unit	Shell thickness (mm×10²)	Shell resistance (kg/cm²)
Control	56.4±1.63[bc]	5.28±0.14[cd]	34.06±1.62	17.11±0.68	66.3±1.97	31.39±0.29	3.21±0.40
Selenite, 0.3 mg/kg	56.2±1.02[bc]	5.31±0.11[cd]	32.73±1.72	18.19±0.87	71.4±1.89	31.29±0.61	3.40±0.18
A, 0.3 mg/kg	57.8±1.21[bc]	5.63±0.14[abc]	34.28±1.27	17.92±0.47	69.1±1.72	31.53±0.91	3.32±0.08
B, 0.3 mg/kg	61.6±1.78[a]	5.86±0.25[a]	36.27±1.56	19.52±0.28	72.8±1.01	31.63±0.95	3.57±0.10
C, 0.3 mg/kg	54.6±1.80[c]	4.90±0.43[e]	32.23±1.11	17.45±0.26	67.8±1.86	29.28±0.62	3.59±0.20
A, 0.6 mg/kg	59.1±1.48[ab]	5.65±0.13[abc]	35.11±.76	18.36±0.48	71.8±1.59	31.68±0.29	3.35±0.26
B, 0.6 mg/kg	58.2±1.21[abc]	5.46±0.06[bcd]	33.93±1.80	18.81±0.24	71.8±1.89	31.81±0.45	3.78±0.25
C, 0.6 mg/kg	56.7±1.41[bc]	5.23±0.15[de]	34.65±1.58	16.86±0.60	68.7±1.50	29.83±0.54	3.53±0.58
Vitamin E	57.8±1.46[bc]	5.71±0.12[ab]	33.75±1.49	18.37±0.68	66.9±1.80	31.39±0.48	2.96±0.28
Effect	*	**	NS	NS	NS	NS	NS

The means within the same column with at least one common letter, do not have significant difference (P>0.01 and P>0.05).
* (P<0.05) and ** (P<0.01).
NS: non significant.

Table 4 Effect of Se source and vitamin E on egg quality characteristics (Means±SE) stored at 25 °C for 14 days

Treatment	Egg weight (g)	Shell weight (g)	Albumen weight (g)	Yolk weight (g)	Haugh unit	Shell resistance (kg/cm²)	Shell thickness (mm×10²)
Control	53.73±1.94	5.40±0.21	29.00±0.91	19.94±0.33	40.63±0.96[d]	3.21±0.40	29.33±0.51
Selenite, 0.3 mg/kg	54.68±1.84	5.51±0.05	28.66±1.42	20.50±0.12	43.20±1.86[bcd]	3.40±0.18	31.60±0.47
A, 0.3 mg/kg	55.25±1.26	5.72±0.17	30.19±1.62	19.31±0.61	42.72±1.21[bcd]	3.32±0.08	32.65±0.50
B, 0.3 mg/kg	56.63±1.62	5.73±0.12	31.28±1.96	19.61±0.38	41.49±1.42[cd]	3.57±0.10	31.02±0.10
C, 0.3 mg/kg	52.51±1.72	5.45±0.35	28.14±1.68	18.92±0.52	44.19±1.48[bcd]	3.59±0.20	30.76±0.49
A, 0.6 mg/kg	53.00±1.78	5.33±0.52	28.98±.1.65	18.66±0.94	47.08±1.02[abc]	3.35±0.26	30.81±0.21
B, 0.6 mg/kg	53.79±1.42	5.57±0.08	29.30±1.29	18.92±0.25	45.15±1.89[abcd]	3.78±0.25	31.56±0.11
C, 0.6 mg/kg	55.71±1.62	5.61±0.24	31.00±0.92	19.01±0.48	47.91±1.52[ab]	3.53±0.58	30.55±0.77
Vitamin E	55.16±1.00	5.57±0.12	29.41±1.28	20.17±0.95	50.21±1.91[a]	2.96±0.28	31.32±0.28
Effect	NS	NS	NS	NS	*	NS	NS

The means within the same column with at least one common letter, do not have significant difference (P>0.05).
* (P<0.05).
NS: non significant.

This result is due to the antioxidant property of vitamin E and the role of Se in glutathione peroxidase activity. Galobart et al. (2001) reported that adding 100 or 200 mg α-tocopherol per kg diet of laying hens decreased TBA in eggs stored for 6 months and concluded that dietary vitamin E in this level has antioxidant role. Eggs from hens fed C source of Se had lower concentration of MDA in yolk compared with the other sources of Se and the control.

Therefore, the C source of organic Se in the current study had more potential to increase oxidative stability of eggs. Dvorska et al. (2003) reported increased glutathione peroxidase activity in eggs after laying hens were fed diets containing Se and vitamin E. This increase in glutathione peroxidase activity would protect the egg from damage by free radicals, resulting in decreased potential of cellular damage to the shell or fluid egg.

Table 5 Effect of Se sources and vitamin E on MDA of fresh eggs and eggs stored at 4 °C and 25 °C for 14 days[1]

Treatment	Fresh egg	Stored at 4 °C	Stored at 25 °C
Control	1.31±0.05[a]	1.74±0.05[a]	3.35±0.05[a]
Selenite, 0.3 mg/kg	1.15±0.03[b]	0.93±0.02[f]	2.18±0.03[bc]
A, 0.3 mg/kg	1.08±0.03[bc]	1.53±0.05[c]	2.06±0.07[c]
B, 0.3 mg/kg	1.13±0.05[b]	1.05±0.05[e]	2.07±0.05[c]
C, 0.3 mg/kg	0.97±0.05[cd]	1.49±0.03[c]	1.90±0.03[d]
A, 0.6 mg/kg	1.16±0.05[b]	1.65±0.05[b]	2.21±0.05[bc]
B, 0.6 mg/kg	1.03±0.05[c]	1.16±0.03[d]	2.28±0.05[b]
C, 0.6 mg/kg	0.71±0.05[d]	0.88±0.03[fg]	1.78±0.03[d]
Vitamin E	0.71±0.05[d]	0.81±0.03[g]	1.51±0.02[e]
Effect	**	**	**

The means within the same column with at least one common letter, do not have significant difference (P>0.01).
** (P<0.01).
NS: non significant.

Table 6 Effect of Se sources and vitamin E on eggs Se and cholesterol content (Means±SE)

Treatment	Yolk Se content (ng/g)	Serum cholesterol (mg/dL)	Yolk cholesterol (mg/g)
Control	1.12±0.02[c]	172.1±1.17[a]	12.64±0.56[a]
Selenite, 0.3 mg/kg	1.37±0.03[b]	142.0±1.22[b]	10.70±0.47[dc]
A, 0.3 mg/kg	1.21±0.03[b]	148.9±1.23[b]	10.59±0.60[dc]
B, 0.3 mg/kg	1.43±0.01[b]	175.5±1.04[a]	11.18±0.44[bc]
C, 0.3 mg/kg	1.80±0.01[a]	140.2±1.07[b]	10.61±0.59[dc]
A, 0.6 mg/kg	1.41±0.09[b]	178.2±1.12[a]	11.16±0.29[bc]
B, 0.6 mg/kg	1.88±0.06[a]	141.9±1.34[b]	10.54±0.35[dc]
C, 0.6 mg/kg	1.90±0.03[a]	143.6±1.15[b]	10.67±0.56[dc]
Vitamin E	-	143.6±1.18[b]	9.24±0.55[e]
Effect	**	**	**

The means within the same column with at least one common letter, do not have significant difference (P>0.01).
** (P<0.01).
NS: non significant.

They reported that storing the eggs for 14 days at 20 °C caused yolk lipid peroxidation and increased MDA content in the yolk. However, lipid peroxidation significantly decreased in eggs by Se supplementation in diet.

Dietary supplementation of Se increased yolk Se content, with the effect being more for the C source of organic Se. However, no difference was generated in yolk Se content either by 0.3 or 0.6 mg of C source or by 0.6 mg of B source of organic Se per kg of diet. Many of researchers have reported that increasing Se content in diet increased Se content of eggs (Ort and Latshaw, 1978; Mohiti-Asli et al. 2008; Cobanova et al. 2011). Payne et al. (2005) reported that a Se-enriched yeast diet was more effective than a sodium selenite diet for increasing the Se content of eggs. Utterback et al. (2005) reported that adding seleno-yeast to the laying-hen diet yielded a 4.8-fold increase in eggs Se content compared with a 2.8-fold increase for the sodium selenite diet over the un-supplemented diet. Se content in serum and yolk were higher in some organic Se sources than the others, probably due to various bioavailability of different sources of Se.

Dietary supplementation of selenium and vitamin E decreased serum and yolk cholesterol content, with the effect being more significant for vitamin E.

Sahin et al. (2002) reported that higher vitamin E in diet resulted in a decrease in serum cholesterol concentrations of Japanese quails. El-Demerdash (2004) reported that vitamin E or Se alone significantly decreased the levels of cholesterol in rats. Mohiti-Asli and Zaghari (2010) reported that dietary supplementation of vitamin E reduced egg yolk cholesterol content. Hens could eliminate considerable amounts of cholesterol in the egg (Andrews et al. 1968). Moreover, Andrews et al. (1968) showed that egg cholesterol originates from serum cholesterol. But, other reporters (Washburn and Nix, 1974; Shivaprasad and Jaap, 1979) indicated that the plasma cholesterol concentration is not closely associated with the concentration of yolk cholesterol. Laying hen VLDLy resists the lipolytic activity of LPL (Bacon et al. 1978) and provides triacylglycerol and cholesterol for egg yolk. Therefore, yolk cholesterol is not related to serum low density lipoprotein (LDL) and high density lipoprotein (HDL).

Vitamin E affected hepatic synthesis and catabolism of cholesterol in the chicken (Sklan, 1983). The activity of hepatic cholesterol 7α-hydroxylase was reduced in vitamin E-deficient rabbits (Chupukcharoen et al. 1985) and paradoxically, the activity of this enzyme was also reduced in rats fed high levels of vitamin E (Kritchevsky et al. 1980).

CONCLUSION

It can be concluded that Se supplementation, especially in organic forms, to laying hens diet can increase Se contents of eggs. The source of organic Se supplemented to the diets is also important for egg enrichment and its physiological and antioxidant properties.

REFERENCES

Andrews J.W., Wagstaff R.K. and Edwards H.M. (1968). Cholesterol metabolism in the laying fowl. *Am. J. Physiol. Legacy Content.* **214**, 1078-1083.

Arnold R.L., Olson O.E. and Carlson C.W. (1973). Dietary selenium and arsenic additions and their effects on tissue and egg selenium. *Poult. Sci.* **52**, 847-854.

Bacon W.L., Leclercq B. and Blum J.C. (1978). Difference in metabolism of very low density lipoprotein from laying chicken hens in comparison to immature chicken hens. *Poult. Sci.* **57**, 1675-1686.

Botsoglou N.A., Fletouris D.J., Papageorgiou G.E., Vassilopoulos V.N., Mantis A.J. and Trakatellis A.G. (1994). Rapid, sensitive, and specific thiobarbituric acid method for measuring lipid peroxidation in animal tissue, food and feedstuff samples. *J. Agric. Food Chem.* **42**, 1931-1937.

Calderon V.M. and Jensen L.S. (1990). The requirement for sulfur amino acids by laying hens influenced by protein concentration. *Poult. Sci.* **69**, 934-944.

Cantor A.H. and Scott M.L. (1994). The effect of Se in the hen's diet on egg production, hatchability, performance of progeny and selenium concentration in eggs. *Poult. Sci.* **53**, 1870-1880.

Chupukcharoen N., Komaratat P. and Wilairat P. (1985). Effects of vitamin E deficiency on the distribution of cholesterol in plasma lipoproteins and the activity of cholesterol 7α-hydroxylase in rabbit liver. *J. Nutr.* **115**, 468-472.

Cobanova K., Petrovic V., Mellen M., Arpasova H., Gresaková L. and Faix S. (2011). Effects of dietary form of selenium on its distribution in eggs. *Biol. Trace Elem. Res.* **144**, 736-46.

Demirci A. (1999). Enhanced organically bound selenium yeast production by feed-batch fermeAntation. *J. Agric. Food Chem.* **47**, 2496-2500.

Dvorska J.E., Yaroshenko F.A., Surai P.F. and Sparks N.H.C. (2003). Selenium-enriched eggs: quality evaluation. Pp. 23-24. in Proc. 14th European Symp. Poult. Nutr. Lillehammer, Norway.

El-Demerdash F.M. (2004). Antioxidant effect of vitamin E and selenium on lipid peroxidation, enzyme activities and biochemical parameters in rats exposed to aluminium. *J. Trace Elem. Med. Bio.* **18**, 113-121.

Galobart J., Barroeta A.C., Baucells M.D., Codony R. and Ternes W. (2001). Effect of dietary supplementation with rosemary extract and α-tocopheryl acetate on lipid oxidation in eggs enriched with ω3-fatty acids. *Poult. Sci.* **80**, 460-467

Jang Y.D., Choi H.B., Durosoy S., Schlegel P., Choibr B.R. and Kim Y.Y. (2010). Comparison of bioavailability of organic Se sources in finishing pigs. *Asian-Australas J. Anim.* **23**, 931-936.

Jiakui L. and Xiaolong W. (2004). Effect of dietary organic versus inorganic selenium in laying hens on the productivity, seleniumdistribution in egg and selenium content in blood, liver and kidney. *J. Trace Elem. Med. Bio.* **18**, 65-68.

Jlali M., Briens M., Rouffineau F., Mercerand F., Geraert P.A. and Mercier Y. (2013). Effect of 2-hydroxy-4-methyl selenobutanoic acid as a dietary selenium supplement to improve the selenium concentration of table eggs. *J. Anim. Sci.* **91**, 1745-1752.

Kritchevsky D., Nitzche C., Czarnecki S.K. and Story J.A. (1980). Influence of vitamin E supplementation on cholesterol metabolism in rats. *Nutr. Rep. Int.* **22**, 339-342.

Leeson S. and Summers J.D. (2001). Scott's Nutrition of the Chicken. Published by University Books, Guelph, Ontario, Canada.

Mohiti-Asli M., Shariatmadari F., Lotfollahian H. and Mazuji M.T. (2008). Effects of supplementing layer hen diets with selenium and vitamin E on egg quality, lipid oxidation and fatty acids composition during storage. *Canadian J. Anim. Sci.* **88**, 475-483.

Mohiti-Asli M. and Zaghari M. (2010). Does dietary vitamin E or C decrease egg yolk cholesterol? *Biol. Trace Elem. Res.* **38**, 60-68.

Ort J.F. and Latshaw J.D. (1978). Thetoxic level of sodium selenite in the diet of laying chickens. *J. Nutr.* **108**, 1114-1120.

Pappas A.C., Acamovic T., Sparks N.H.C., Surai P.F. and McDevitt R.M. (2005). Effects of supplementing broiler breeder diets with organic selenium and polyunsaturated fatty acids on egg quality during storage. *Poult. Sci.* **84**, 865-874.

Pasin G., Smith G.M. and O'Mahony M. (1998). Rapid determination of total cholesterol in egg yolk using commercial diagnostic cholesterol reagent. *Food Chem.* **61**, 255-259.

Patton N.D., Cantor A.H., Pescatore A.J. and Ford M.J. (2000). Effect of dietary selenium source, level of inclusion and length of storage on internal quality and shell strength of eggs. *Poult. Sci.* **79**, 75-116.

Payne R.L., Lavergne T.K. and Southern L.L. (2005). Effect of inorganic versus organic selenium on hen production and egg Se concentration. *Poult. Sci.* **84**, 232-237.

Puthpongsiriporn U., Scheideler S.E., Shell J.L. and Beck M.M. (2001). Effect of vitamin E and C supplementation on performance, *in vitro* lymphocyte proliferation, and antioxidant status of laying hens during heat stress. *Poult. Sci.* **80**, 1190-1200.

Reis R.N., Vieira S.L., Nascimento P.C., Peña J.E. and Barros R. (2009). Torres CA. selenium contents of eggs from broiler breeders supplemented with sodium selenite or zinc-L-semethionine. *J. Appl. Poult. Res.* **18**, 151-157.

Sahin K., Sahin N. and Yaralioglu S. (2002). Effects of vitamin C and vitamin E on lipid peroxidation, blood serum metabolites, and mineral concentrations of laying hens reared at high ambient temperature. *Biol. Trace Elem. Res.* **85**, 35-45.

SAS Institute. (2002). SAS®/STAT Software, Release 9.2. SAS Institute, Inc., Cary, NC. USA.

Schrauzer G.N. (2001). Nutritional selenium supplements: product types, quality and safety. *J. Am. Coll. Nutr.* **20**, 1-4.

Shivaprasad H.L. and Jaap R.G. (1979). Egg and yolk production as influenced by liver weight, liver lipid and plasma lipid in three strains of small bodied chickens. *Poult. Sci.* **56,** 1384-1390.

Sklan D. (1983). Effect of high vitamin A or tocopherol intake on hepatic lipid metabolism and intestinal absorption and secretion of lipids and bile acids in the chick. *British J. Nutr.* **50,** 409-416.

Surai P.F. (2002). Selenium in poultry nutrition: 1. Antioxidant properties, deficiency and toxicity. *World Poult. Sci.* **58,** 333-347.

Tanguy S., Grauzam S., de Leiris J. and Boucher F. (2012). Impact of dietary selenium intake on cardiac health: experimental approaches and human studies. *Molecul. Nutr. Food Res.* **56,** 1106-1121.

Tufarelli V., Ceci E. and Laudadio V. (2015). 2-hydroxy-4-methylselenobutanoic acid as new organic selenium dietary supplement to produce selenium-enriched eggs. *Biol. Trace Elem. Res.* **171,** 453-458.

Utterback P.L., Parsons C.M., Yoon I. and Butler J. (2005). Effect of supplementing selenium yeast in diets of laying hens on egg selenium content. *Poult. Sci.* **84,** 1900-1901.

Washburn K.W. and Nix D.F. (1974). A rapid technique for extraction of yolk cholesterol. *Poult. Sci.* **53,** 1118-1122.

7

Evaluation of Thyroid Hormones, Blood Gases, Body Antioxidant Status, the Activity of Blood Enzymes and Bone Characteristics in Broiler Chickens with Cold Induced Ascites

R. Abdulkarimi[1]*, M.H. Shahir[1] and M. Daneshyar[2]

[1] Department of Animal Science, Faculty of Agriculture, University of Zanjan, Zanjan, Iran
[2] Department of Animal Science, Faculty of Agriculture, Urmia University, Urmia, Iran

*Correspondence E-mail: rahim.abdulkarimi@yahoo.com

ABSTRACT

A total of 150 day old female chickens (Ross 308) were randomly allocated to 2 groups with 5 replicate and 15 chicks in each replicate to determine the effects of cold induced ascites on performance, antioxidant status, blood enzyme activities and bone metabolism. The two experimental treatments were: 1) chicks that reared under normal temperature (NT) and 2) chicks that reared under cold temperature (CT). The experiment was terminated at 42 day of chicken age. Feed intake was reduced significantly by ascites during the starter period. Weight gain of NT birds was higher than CT birds during the starter, grower, finisher and the whole period (P<0.05). Feed conversion ratio was greater for CT birds during the grower, finisher and the whole period (P<0.05). Total mortality was greater in CT birds than NT ones during the whole period (P<0.05). Cold-induced ascites increased the right ventricular, total ventricular and ventricular septum weights and right ventricular (RV)/total ventricular (TV) ratio at day 42 (P<0.05). Blood pO_2, O_2 saturation, pH and T4 level were lower and blood pCO_2, T3 and calcium level was higher in CT birds than NT birds (P<0.05). The birds of both treatments had the same tibia length but femur length was shorter in CT birds (P<0.05). The diameter of both tibia and femur was smaller in CT birds (P<0.05). CT birds had a higher incidence of leg problems than NT ones during the whole period of the experimental phase (P<0.05). In conclusion, cold induced ascites reduced the performance, increased mortality and caused leg problems in broiler chickens.

KEY WORDS ascites, broiler, leg problem, performance.

INTRODUCTION

Pulmonary Hypertension Syndrome (PHS) or ascites is one of the considerable causes of high mortality in modern broilers. The economic losses due to ascites are about 1 billion US $ annually around the world (Maxwell and Robertson, 1997). It is estimated that 5% of broilers and 20% of roaster birds are dying due to ascites (Balog, 2003). Nowadays, the fast growing broiler chickens do not always have enough lung capacity to provide their oxygen de-

mands. This inadequacy results in impaired ability to regulate the energy balance under extreme conditions, such as low ambient temperature or high altitude (Luger *et al.* 2001) that leads to increase the cardiac output, pulmonary hypertension, right ventricle and ventricular hypertrophy, cardiopulmonary dysfunction and finally ascites and death (Julian, 1993; Julian, 1998; Wideman and Bottje, 1993; Maxwell *et al.* 1995). Many changes in heart tissue (Daneshyar *et al.* 2009), blood gases (Olkowski *et al.* 1999; Daneshyar *et al.* 2007), antioxidant status (Han *et al.* 2005;

Wang *et al.* 2012), blood enzymes (Fathi *et al.* 2011) and thyroid hormones (Scheele *et al.* 1992; Moayyedian *et al.* 2011) can be occurred during the ascites process. Tekeli (2014) revealed that the level of oxygen saturation decreased and inversely blood hemoglobin, hematocrit and right ventricular diameter significantly increased in ascitic broilers under high altitude.

Moayyedian *et al.* (2011) detected that broilers under cold temperature had a higher mortality, RV/TV ratio and venous pCO_2 and lower pO_2 than control birds. In other studies, Han *et al.* (2005) and Fathi *et al.* (2011) indicated that lung and liver ROS, plasma nitric oxide (NO) level, super oxide dismutase (SOD) activity and total antioxidant capacity (TAC) were lower and plasma malondialdehyde (MDA) were higher in cold-induced ascitic birds. Ipek and Sahan, (2006) detected that ascites mortality, right ventricular (RV) and RV/TV ratio was higher in cold stress birds. Daneshyar *et al.* (2009) did not detect any changes of lactate dehydrogenase (LDH), aspartate aminotransferase (AST), and alanine aminotransferase (ALT) in ascetic birds but mortality and RV/TV value was higher. In an experiment, Camacho-Escobar *et al.* (2011) indicated that the treatments with high ascites mortality accompanied with higher incidence of leg disorders level during the experiment. Leg problems is a major economic and welfare problem (Bradshaw *et al.* 2002) with incidence of typically less than 2 or 3% in broilers flocks and can affects the diet accessibility, feed conversion ratio and growth rate. Leg problems and developmental disorders of long bones can be affected by genetics, breeder nutrition, incubation, infectious diseases, and environmental stressors (Bradshaw *et al.* 2002; Oviedo Rondón *et al.* 2006a; Oviedo Rondón *et al.* 2006b).

It has been distinguished that environmental conditions such as low ambient temperature might negatively change the bone deformities or leg problems in modern strain with fast growing rate.

Some information is available about performance, mortality, heart tissue, blood indices, blood gases, thyroid hormones, blood enzyme activity of LDH, AST, alkaline phosphatase (ALP), NO and antioxidant status, but more information about leg disorder with ascites syndrome are needed.

The first aim of this study was to evaluate the possible effects of ascites on performance, mortality, heart tissue, blood indices, blood gases, thyroid hormones, blood enzyme activity and antioxidant status in female broilers. Since, there is no report regarding the effects of cold induced ascites on bone characteristics and related mineral metabolism in broiler chickens, the possible changes of bone structure and leg problem were investigated as the second aim of the current experiment.

MATERIALS AND METHODS

A total of one hundred and fifty day old female chicks (Ross 308, initial weight: 45±3 g) were obtained from a commercial hatchery randomly allocated to 2 groups; a control and ascitic groups with 5 replicates of 15 chicks each. Water and feed were provided *ad libitum*. All the chickens were fed the similar starter (day 1-10 of age), grower (day 11-24 of age) and finisher (day 25-41 of age) diets in mash form (Table 1).

Rations were formulated by amino feed[1] software according to the Ross requirement recommendation (Aviagen, 2014). The ascitic birds were reared in a house with 1670 m altitude under cold temperature condition. The temperature was 33 °C during the first week of the experiment and reduced to 26.0 ± 1 °C, 20.0 ± 1 °C and 15.0 ± 1 °C at 7, 14 and 21 day of age, respectively, and was maintained at 15.0 ± 1 °C until the end of the experiment (Luger *et al.* 2001). The control group were raised under normal temperature (NT birds) with 33 °C during the first week of the experiment. The temperature for NT birds was regularly reduced by 3 °C per week up to 3 wk of age and then was kept constant at 24.0 ± 1 °C until the end of the experiment. The average feed intake (FI, g/bird/day), average body weight gain (BWG, g/bird/day) and feed conversion ratio (FCR) were determined during the starter, grower, finisher and the whole periods. During the experiment, mortality was recorded daily and all the dead chickens were diagnosed for ascites according to ratio of right ventricle to total ventricle weight, amber-colored fluid in the abdominal cavity and pericardium. Leg problems were estimated visually on chicks with arthritis, tibial dyschondroplasia, rickets, osteochondrosis, prosis, prostrate status, or difficulty to walk (Williams *et al.* 2000). At day 42 of age, 2 birds/pen (ten for each treatment) were selected, weighed and killed by decapitation to obtain the heart tissue, blood and bone samples. The heart was dissected and proportional weight of total ventricular (TV) and right ventricular (RV) was determined and ventricular septum thickens (VS) measured by digital caliper. The heart ratio was calculated by the weight of the right ventricle as a percentage of the total ventricle weight (RV/TV). Blood samples were collected in heparinized syringes (1 mL) and kept on ice and instantly moved to laboratory less than two hours for blood gas measurements by pH/blood gas Analyzer (ABL50, ABL995, France; Daneshyar *et al.* 2012). Some blood samples collected in non heparinized syringes (2 mL) and allowed to clot for 2 h at 37 °C. The serum was then decanted and stored at 20 °C for later analyses (Daneshyar *et al.* 2012). Serum samples were thawed and serum malondialdehyde (MDA) content were determined by thiobarbituric acid method.

Table 1 Composition of experimental diets

Ingredients (%)	Starter (1-10 d)	Grower (11-24 d)	Finisher (25-42 d)
Corn grain	44.029	52.05	58.445
Corn gluten	9.409	-	-
Soybean meal	39.369	40.078	34.319
Soybean oil	2.392	3.815	3.397
Dicalcium phosphate	2.223	1.956	1.811
Carbonate calcium	1.219	0.969	0.955
L-lysine HCl	0.242	0.029	0.02
DL-methionine	0.214	0.245	0.207
Vitamin and mineral premix[1]	0.5	0.5	0.5
Salt	0.255	0.326	0.326
Bicarbonate sodium	0.149	0.031	0.02
Total	100	100	100
Calculated analysis			
Dry matter (%)	89.801	89.535	89.17
ME (kcal/kg)	2950	3000	3050
Crude fat (%)	4.294	5.799	5.581
Calcium (%)	1.027	0.88	0.831
Available phosphorus (%)	0.491	0.44	0.411
Chloride (%)	0.23	0.23	0.23
Sodium (%)	0.167	0.162	0.158
Methionine (%)	0.643	0.569	0.506
Lysine (%)	1.478	1.256	1.104
Arginine (%)	1.632	1.512	1.347
Methionine + cystine (%)	1.083	0.924	0.835
Threonine (%)	1.004	0.856	0.771
Tryptophan (%)	0.298	0.278	0.246

[1] Supplied per kilogram of diet: Retinol: 9000 IU; Alpha tochopherol acetate: 36 IU; Cholecaciferol: 2000 IU; Cyanocobalamin: 15 mg; Riboflavin: 6.6 mg; Calcium pantothenate: 9.8 mg; Niacin: 30 mg; Choline chloride: 625 mg; Biotin: 0.1 mg; Thiamine: 1.75 mg; Pyridoxine: 3 mg; Folic acid: 1 mg; Menadione: 2 mg; Antioxidant (ethoxy queen): 100 mg; Manganese: 248 mg; Zinc: 211 mg; Copper: 25 mg; Iron: 125 mg; Iodine: 2.5 mg and Selenium: 0.5 mg.

The plasma activities of glutathione peroxidase (GPX), superoxide dismutase (SOD) and total antioxidant capacity (TAC) were measured by Randox Kits (Randox Labs, Crumlin, UK). Uric acid were determined colorimetrically (Pars Azmon kit, Tehran, Iran) and plasma thyroid hormones (T3, T4) were measured by ELISA method (Pishtaz Tab Zaman Company Kite, Iran). Plasma LDH, AST and ALP enzyme activities and blood Ca and P levels were determined by spectrophotometric device (Alcyon 300, USA). Plasma nitric oxide (NO) assayed according to Katrina *et al.* (2001) method (UV-visa array, spectrophotometer photonix, AR 2015). After carcass separation, tibia bone for each bird was removed and length and diameter measured by digital caliper. The bone-dry matter and ash contents were determined by AOAC method (1990) and percent of calcium (Ca) and phosphorus (P) measured spectrophometrically.

The data were analyzed based on a completely randomized design using the GLM procedure of SAS (SAS, 2002). When the overall model was statistically different ($P<0.05$), the Tukey-Kramer multiple comparison test was used to compare the mean values of healthy and ascitic birds ($P<0.05$).

RESULTS AND DISCUSSION

The results of performance traits are shown in Table 2. Feed intake (FI) was affected significantly by ascites during the starter period and NT birds had a higher FI than CT birds ($P<0.05$). There was a significant difference between the treatments for WG, NT birds had higher WG than CT ones during the whole rearing periods ($P<0.05$). Ascetics increased the FCR during all of the periods except starter. Total rate of mortality due to ascites was significantly greater in CT treated birds than that of NT ones (46.6% *vs.* 4%) during the whole period (Figure 1; $P<0.05$). Cold environment significantly increased the weights of heart parts such as right ventricular (RV), total ventricular (TV), RV/TV ratio and ventricular septum thickness (VS). (Table 3; $P<0.05$).

The RV (0.1 *vs.* 0.04), TV (0.43 *vs.* 0.29), RV/TV ratio (27.9% *vs.* 15.3%) and VS (3.08 *vs.* 1.86) of CT exposed birds was greater than those of NT exposed ones. The results of thyroid hormones, blood gases and blood minerals are shown in Table 4. The blood pO2, O_2 saturation and pH were lower and pCO_2 was higher in CT exposed birds as compared to NT exposed ones ($P<0.05$).

Table 2 Feed intake (g/d/bird), weight gain (g/d/birds) and feed conservation ratio of female broiler chicks reared under normal (NT) and cold (CT) environmental temperatures

Treatments	NT	CT	P-value	Pooled SEM
Feed intake				
Starter	26.31[a]	23.98[b]	0.04	0.6
Grower	64.04	69.8	0.08	1.68
Finisher	142.8	140.2	0.82	5.5
Total	77.7	77.9	0.95	2.07
Weight gain				
Starter	18.68[a]	15.96[b]	0.003	0.55
Grower	36.9[a]	31.1[b]	0.009	1.25
Finisher	75.9[a]	52.9[b]	0.001	4.4
Total	43.8[a]	33.3[b]	0.004	1.94
Feed conversion ratio (FCR)				
Starter	1.38	1.5	0.054	0.03
Grower	1.73[b]	2.25[a]	0.004	0.11
Finisher	1.89[b]	2.65[a]	0.0001	0.13
Total	1.66[a]	2.13[b]	0.001	0.08

The means within the same column with at least one common letter, do not have significant difference (P>0.05).
SEM: standard error of the means.

Table 3 Proportional weights of total ventricular (TV), right ventricular (RV) and ventricular septum (VS) and RV/TV ratio of broiler chicks reared under normal (NT) and cold (CT) environmental temperatures

Treatments	TV	RV	VS	RV/TV
NT	0.29[b]	0.04[b]	1.86[b]	0.15.3[b]
CT	0.43[a]	0.1[a]	3.08[a]	0.27.9
P-value	0.005	0.001	0.003	0.0003
SEM	0.03	0.01	0.25	0.02

The means within the same column with at least one common letter, do not have significant difference (P>0.05).
SEM: standard error of the means.

Table 4 Thyroid hormones (ng/mL), blood gases (mmHg) levels and blood mineral (mg/dL) of broiler chicks reared under normal (NT) and cold (CT) environmental temperatures

Treatments	Thyroid hormone		Blood gases				Blood mineral	
	T3	T4	pCo_2	pO_2	pH	O_2 saturation (%)	Calcium	Phosphorous
NT	1.64[b]	16.48[a]	53.6[b]	65.5[a]	7.39[a]	63.5[a]	8.7	6.2
CT	1.92[a]	10.94[b]	58.7[a]	55.4[b]	7.15[b]	51.6[b]	9.9	6.3
P-value	0.0001	0.0001	0.003	0.0008	0.04	0.0001	0.02	0.8
SEM	0.05	0.96	1.01	1.91	0.06	2.03	0.29	0.3

The means within the same column with at least one common letter, do not have significant difference (P>0.05).
SEM: standard error of the means.

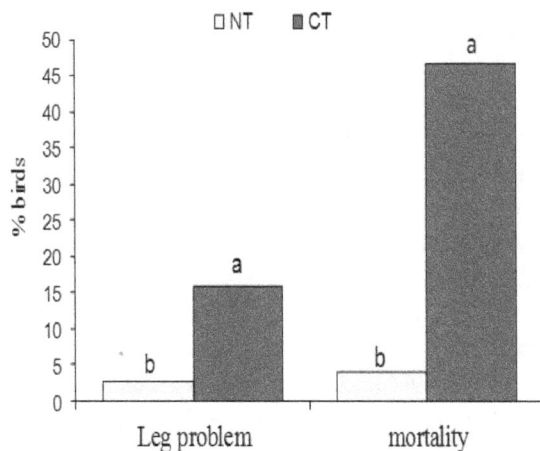

Figure 1 Leg problem and mortality of broiler chickens reared under normal (NT) and cold (CT) environmental temperatures

Plasma T3 level was higher and plasma T4 level was lower in CT exposed birds when compared to NT exposed birds. Blood phosphorous did not changed as effect of treatments but blood Ca level was significantly higher in CT exposed birds than NT exposed ones (P<0.05). According to the results of antioxidant status in Table 5, there were no significant differences for plasma MDA, total antioxidant capacity (TAC), SOD, GPX and uric acid levels between the treatments at day 42 of age (P<0.05).

The results of blood enzymes in both treatments showed similar activity for ALP, LDH, AST and NO enzymes (Table 6). The bone contents are shown in Table 7. There were no significant differences between the treatments for ash, Ca, P and moister in both the tibia and femur (P<0.05).

The birds of both treatments had the same tibia length but femur length was shorter in CT exposed birds (Figure 2; P<0.05).

Table 5 Plasma antioxidant activity of female broiler chicks reared under normal (NT) and cold (CT) environmental temperature

Treatments	SOD (U/g Hb)	GPX (U/g Hb)	MDA (nmol/mL)	TAC (nmol/L)	Uric acid (mg/dL)
NT	1078.1	33.14	1.82	1.86	3.16
CT	617.1	23.57	2.46	2.25	4.98
P-value	0.1	0.1	0.1	0.18	0.15
SEM	142.9	2.96	1.17	0.14	0.62

SOD: super oxide dismutase; GPX: glutathione peroxidase; TAC: total antioxidant capacity and MDA: malondialdehyde.
SEM: standard error of the means.

Table 6 Blood enzyme activity (U/L) of broiler chicks reared under normal (NT) and cold (CT) environmental temperatures

Treatments	ALP (U/L)	LDH (U/L)	AST (U/L)	NO (μm)
NT	1840	3770	297.5	19.34
CT	1375	1918	256.6	20.97
P-value	0.14	0.09	0.36	0.49
SEM	154.2	557.7	31.9	38.7

ALP: alkaline phosphatase; LDH: lactate dehydrogenase; AST: aspartate amino transferas and NO: nitric oxidase.
SEM: standard error of the means.

Table 7 Tibia and femur contents of broiler chicks reared under normal (NT) and cold (CT) environmental temperature

Treatments	Ash (%)		Calcium (%)		Phosphorous (%)		Moister (%)	
	Femur	Tibia	Femur	Tibia	Femur	Tibia	Femur	Tibia
NT	45.6	47.2	18.2	31.9	16.7	18.9	16.8	31.9
CT	41.2	53.2	21.8	30.6	17.8	15.5	22.2	30.6
P-value	0.34	0.51	0.13	0.59	0.69	0.10	0.12	0.13
SEM	2.2	186	1.1	13.6	1.3	9.1	1.7	71.3

SEM: standard error of the means.

The diameter of both tibia and femur was smaller in CT exposed birds as compared to NT exposed birds (Figure 3; P<0.05). Furthermore, CT birds had a higher incidence of leg problems than NT ones (15.99% *vs.* 2.67%) during the whole period (Figure 1; P<0.05).

The high ascites mortality in present experiment was in consistent with that of Daneshyar *et al.* (2012) and Fathi *et al.* (2011). Cold temperature together with high altitude (1670 m) in present experiment is the reason of high ascites mortality.

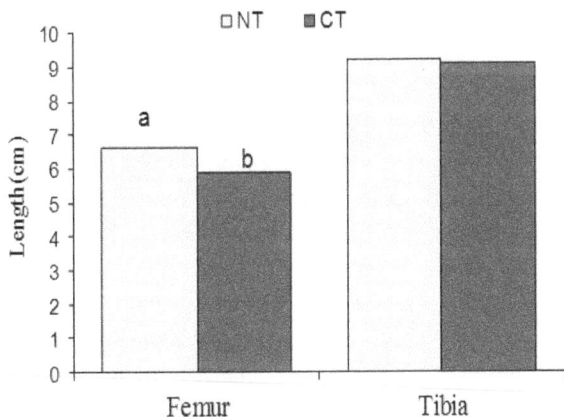

Figure 2 Femur and tibia length of broiler chickens reared under normal (NT) and cold (CT) environmental temperatures

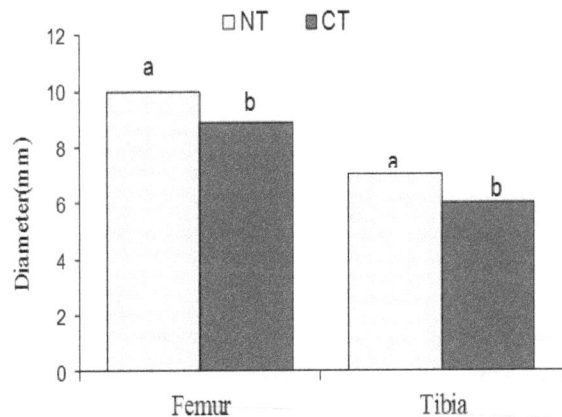

Figure 3 Femur and tibia diameter of broiler chickens reared under normal (NT) and cold (CT) environmental temperatures

In the experiment, ascites symptoms were noticeable from day 20 onward. During wk 4, ascites signs were observed as the abdominal cavity fluid, heart enlargement and mortality in cold exposed group, which was approved by high mortality rate (46.6%) and low grow rate (1752 g). The above mentioned negative effects of ascites supported by the results of many previous researches (Ipek and Sahan, 2006; Guo *et al.* 2007; Daneshyar *et al.* 2009).

Oxygen pressure and atmospheric pressure is lower at high altitude and this phenomenon along with cold environment increases the oxygen needs in order to maintain their body temperature (Wideman and Tackett, 2000). Lower feed efficiency and body weight gain in CT birds closely associated with cold temperature and utilization of energy to heat production (Wideman, 1988). This means that energy partitioning is different between the CT and NT birds and the amount of energy that contribute to grow as the protein and fat mas in NT birds was used for heat production in CT birds.

The greater RV/TV ratio and ventricular septum thickness of CT birds indicates the ascites development in recent experiment. Our results are supported by the results of previous studies which were showed significant difference between healthy and ascitic birds in heart parameter and hematology indexes (Daneshyar et al. 2009; Daneshyar et al. 2007; Guo et al. 2007; Ipek and Sahan, 2006). The right ventricle: total ventricle ratio above 0.27 also strongly indicates the start of ascites development, as proposed by Balog (2003). In agreement with our finding, failure function, hypertrophy and enlargement of right ventricle was shown in ascetic birds and therefore RV/TV ratio enhanced consequently (Daneshyar et al. 2007; Daneshyar et al. 2012; Huchzermeyer, 2012; Ocak, 2006). Our finding supported by previous researchers who reported that cold induced ascitic birds had significantly higher blood pCO2 and lower pO2 (Moayyedian et al. 2011; Van As et al. 2010), lower blood pH (Hafshejani et al. 2012; Closter et al. 2009) lower O2 saturation (Daneshyar et al. 2007), higher plasma T3 and lower plasma T4 level (Scheele et al. 1992; Moayyedian et al. 2011) than control birds. Blood gases and thyroid hormones can be used for ascetic detection whereas pCO2 and pO2 can predict the ascites development. Oxygen demands for cold exposed broilers increased in order to maintain their body temperature (Wideman and Tackett, 2000) or because of high metabolic rate, which leads to hypoxemia and subsequently development of pulmonary hypertension.

In ascetic birds, body tissue forced with decreased blood pO2 and increased blood pCO2 levels are the signs of hypoxemia condition that appeared in response to high growth rate, hypoxemia and subsequently development of pulmonary hypertension (Gupta, 2011). Furthermore, the low blood pH of ascetic birds declines the affinity of hemoglobin for oxygen in the lung and release of oxygen to the tissues as known Bohr effect (David et al. 1966). Lower blood pH in CT birds might be related to high metabolic rate and consequently decreased oxygen affinity to hemoglobin and hemoglobin saturation in the lung (Issacks et al. 1986).

Thyroid hormones have important role for in controlling metabolic rate, (Decuypere et al. 2000; Lin et al. 2008) and are linked with ascites susceptibility in broiler chickens under adverse environmental conditions (Hassanzadeh et al. 2004; De Smit et al. 2005). The T3 is the main hormone that is related to body temperature regulation and stimulates the body metabolism and increases the chickens growth (Yahav, 2000; Hafshejani et al. 2012). Our results are supported by Scheele et al. (1992) and Moayyedian et al. (2011) that indicated the birds which were exposed to cold

temperature (CT) showed lower plasma T4 levels and higher plasma T3 levels compared with the birds that reared at normal temperature (NT). In different results, Luger et al. (2001) detected that plasma T3 and T4 levels decreased while broilers exposed to low ambient temperature. The plasma T3 concentration is positively correlated with oxygen consumption in broilers (Bobek et al. 1977; Gabarrou et al. 1997). However, T3 concentration in the ascetic birds, which were exposed to cold temperature was significantly higher because of high oxygen demands and increase of metabolic rate (Stojevi et al. 2000). The decreased plasma T4 content can be explained by a negative feedback of T3 on the hypothalamus resulting in a decreased thyroid releasing hormone secretion, therefore lower T4 release (Moayyedian et al. 2011) and increase in shifting of T4 to T3 conversion, in cold exposing chickens.

MDA concentration is an important index for lipid peroxidation and oxidative damage caused by ROS in the cell (Iqbal et al. 2002) and plays an important role in ascites incidence. Some researchers suggested that heart failure in broilers with hypoxia and subsequent ascites syndrome can be associated with ROS production during oxidative stress (Fathi et al. 2011). Cold stress is one of the effective factors for inducing hypoxemic conditions in tissue of cold exposed birds that can generate ROS (Park and Kehre, 1991; Dawson et al. 1993; Han et al. 2004b) which may cause lipid peroxidation in the membrane of the cells resulting in injury of organs such as lung, heart and liver (Arab et al. 2006). MDA level is the main index of lipid peroxidation and therefore numerically higher plasma MDA content in CT-exposed birds can indicates the mild ROS production and lipid peroxidation in ascetic birds.

In the health body, protective enzymatic systems such as SOD and GPX are the natural body antioxidants that play important role to neutralize free radicals function (Duthie et al. 1989). SOD catalyzes the conversion of superoxide to hydrogen peroxide that is in turn reduced by GPX to water (Iqbal et al. 2002). Although plasma SOD and GPX activities have not been affected by ascites in present study, some researchers have reported the changes of the mentioned enzymes in ascetic birds. For example, Han et al. (2005) and Wang et al. (2012) detected the lower plasma SOD and GPX activity in cold exposed birds. Fathi et al. (2011) reported the higher GPX and SOD activity in plasma and liver tissue of cold induced ascites. Iqbal et al. (2002), indicated the lower reduced glutathione (GSH) level and higher GSSG/GSH ratio, as consequence of increased in GPX activity in lung mitochondria of ascetic birds. The possible inconsistency could be related to mild peroxidation (numerically lower plasma MDA content) of CT birds in present study, which had not affected the antioxidant enzyme activities.

None of the plasma enzymes was affected by ascites which was consistent with the results of previous literature that detected no significant changes for LDH (Daneshyar *et al.* 2009; Khajali and Qujeq, 2005) or AST (Daneshyar *et al.* 2009; Tankson *et al.* 2002) activity under cold temperature.

Although Han *et al.* (2005) indicated decreased nitric oxide synthase activity in cold temperature. We observed no changes of calcium and phosphorous contents in the tibia and femur in our study but blood calcium levels and leg problems were higher in ascetic birds. Moreover, the diameter and length of both tibia and femur bones were shorter in cold reared birds.

Although there is no report for the bone metabolism of broiler chickens under cold induced ascites and according to our knowledge this study was the first in this field, these changes could have two reasons. One possible reason is the high blood calcium content in CT exposed birds which might be related to low calcium needs for lower growth rate and shorter bones.

Second possible reason is the low calcium deposition to bone structures in CT exposed birds which might had led to smaller length and diameter of bones and consequently leg disorder incidence. Camacho-Escobar *et al.* (2011) who indicated high rate of leg disorders in ascetic birds confirm high percentage of leg problem. Low size of tibia and femur in ascetic birds could be related to lower calcium absorption.

The results of Wolfenson *et al.* (1987) approve namely the reduction of calcium absorption in cold-stressed birds. Lower size of bones in ascetic birds can be results of mineral imbalance that caused the high leg disorder in ascetic birds. Other reason can be discussed by high metabolism rate in cold temperature, which may leads to lower plasma pH and acidosis in CT-exposed birds. Acidosis may affect negatively the availability of 1,25-dihydroxycholecalciferol (Julian, 1998) and causes slow growth plate by chondrocytes proliferation, tibia and femur disorder and in final leg disorder in ascetic birds.

CONCLUSION

According to the results of present study, cold-induced ascites changes the thyroid hormone metabolism, blood gases and heart parts indices and reduces the performance and increase the mortality in broiler chickens. Furthermore, it leads to lower plasma pH, acidosis and possibly low availability of 1,25-dihydroxycholecalciferol which causes lower calcium absorption and deposition to bones and consequently smaller length and diameter of bones and leg problems in female broiler chickens.

ACKNOWLEDGEMENT

The authors would like to appreciate Mr Amir Mansor Vatankhah for his helps during the laboratory determination of blood indices.

REFERENCES

Arab H.A., Jamshidi R., Rassouli A., Shams G. and Hassanzadeh M.H. (2006). Generation of hydroxyl radicals during ascites experimentally induced in broiler. *Br. Poult. Sci.* **47(2)**, 216-222.

Aviagen. (2014). Ross 308: Broiler Nutrition Specification. Aviagen Inc., Huntsville, Alabama.

Balog J.M. (2003). Ascites syndrome (pulmonary hypertension syndrome) in broiler chickens: are we seeing the light at the end of the tunnel? *Avian. Poult. Biol. Rev.* **14**, 99-126.

Bobek S., Jastrzebski M. and Pietras M. (1977). Age related changes in oxygen consumption and plasma thyroid hormone concentration in the young chicken. *Gen. Comp. Endocrinol.* **31**, 169-174.

Bradshaw R.H., Kirkden R.D. and Broom D.M. (2002). A review of the aetiology and pathology of leg weakness in broilers in relation to welfare. *Avian. Poult. Biol. Rev.* **13**, 45-103.

Camacho-Escobar M.A., García-López J.C., Suárez-Oporta M.E., Pinos-Rodríguez J.M., Arroyo-Ledezma J.A. and Sánchez-Bernal E.I. (2011). Effects of feed intake restriction and micronutrients supplementation on ascites mortality and leg characteristics of broilers. *J. Appl. Anim. Res.* **39(2)**, 97-100.

Closter A.M., Van As P., Groenen M.A., Vereijken A.L., Van Arendonk J.A. and Bovenhuis H. (2009). Genetic and phenotypic relationships between blood gas parameters and ascites related traits in broilers. *Poult. Sci.* **88**, 483-490.

Daneshyar M., Kermanshahi H. and Golian A. (2009). Changes of biochemical parameters and enzyme activities in broiler chickens with cold-induced ascites. *Poult. Sci.* **88**, 106-110.

Daneshyar M., Kermanshahi H. and Golian A. (2007). Changes of blood gases, internal organs weight and performance of broiler chickens with cold induced ascites. *Res. J. Boil. Sci.* **2(7)**, 729-735.

Daneshyar M., Kermanshahi H. and Golian A. (2012). The effects of turmeric supplementation on antioxidant status, blood gas indices and mortality in broiler chickens with T3-induced ascites . *Br. Poult. Sci.* **53(3)**, 379-385.

David B.S., Maurizio E.A. and Jeffries W. (1966). The oxyzen Bohr effects in mouse hemoglobin. *Arch. Biochem. Biophys.* **113**, 725-729.

Dawson Y.L., Gores G.L. and Nieminen A.L. (1993). Mitochondria as a source of reactive oxygen species during reductive stress in rats hepatocytes. *Am. J .Physiol.* **264**, 961-967.

De Smit L.,Tona K., Bruggeman V., Onagbesan O., Hassanzadeh M., Arckens L. and Decuypere E. (2005). Comparison of three lines of broilers differing in ascites susceptibility or growth rate, 2. Egg weight loss, gas pressures, embryonic heat production, and physiological hormone levels. *Poult. Sci.* **84**, 1446-1452.

Decuypere E., Buyse J. and Buys N. (2000). Ascites in broiler chickens: exogenous and endogenous structural and functional causal factors. *Worlds. Poult. Sci. J.* **56,** 367-376.

Duthie G.G., Wahle K.W.J. and James W.P.T. (1989). Oxidants, Antioxidant's and Cardiovascular Disease. *Nutr. Res. Rev.* **2,** 52-62.

Fathi K.N., Ebrahim Nezhad Y., Aghdam Shahryar H., Daneshyar M. and Tanha T. (2011). Antioxidant enzyme activities characterization in pulmonary hypertension syndrome (PHS) in broilers. *Res. J. Biol. Sci.* **6(3),** 118-123.

Gabarrou J.F., Duchump C., Williams J. and Geraert P.A. (1997). A role of thyroid hormones in the regulation of dietinduced thermogenesis in birds. *Br. J. Nutr.* **78,** 963-973.

Guo J.L., Zheng Q.H., YIN Q.Q., Cheng W. and Jiang Y.B. (2007). Study on mechanism of ascites syndrome of broilers. *Am. J .Anim. Vet .Sci.* **2(3),** 62-65.

Gupta A.R. (2011). Ascetic syndrome in poultry: a review. *Worlds. Poult. Sci. J.* **67(3),** 457-468.

Hafshejani E.F., Gholami Ahangaran M. and Hosseni E. (2012). Study of blood cells, blood gases and thyroid hormones in broiler chickens suspected of ascites syndrome. *Glob. Vet.* **8(1),** 18-21.

Han B., Yoon S.S., Su J.L, Han H.R., Qu W. and Nigussie F. (2005). Effect of low ambient temperature on the concentration of free radicals related to ascites in broiler chickens. *Asian-Australas. J. Anim. Sci.* **18(8),** 1182-1187.

Han B., Yoon S.S., Su J.L., Han H.R., Wang M., Qu W.J. and Zhong D.B. (2004b). Effects of selenium, copper and magnesium on antioxidant enzymes and lipid peroxidation in bovine fluorosis. *Asian-Australas J. Anim. Sci.* **17(12),** 1695-1699.

Hassanzadeh H., Bozargmehri F., Buyse J., Bruggeman V. and Decuyper E. (2004). Effect of chronic hypoxia during embryonic development on physiological functioning and on hatching and post hatching parameters related to ascites syndrome in broiler chickens. *Avian. Pathol.* **33,** 558-564.

Huchzermeyer F.W. (2012). Broilers ascites: a review of the ascites work done at the poultry section of the Onderstepoort Veterinary Institute 1981-1990. *Worlds. Poult. Sci. J.* **68,** 41-50.

Ipek A. and Sahan U. (2006). Effects of cold stress on broiler performance and ascites susceptibility. *Asian-Australas J. Anim. Sci.* **19(5),** 734-738.

Iqbal M., Cawthon D., Beers K., Wideman R.F. and Bottje W.G. (2002). Antioxidant enzyme activities and mitochondrial fatty acids in pulmonary hypertension syndrome (PHS) in broilers1. *Poult. Sci.* **81,** 252-260.

Issacks R., Goldman P. and Kim C. (1986). Studies on avian erythrocyte metabolism XIV. Effect of CO_2 and pH on P50 in the chicken. *Am. J. Physiol.* **250,** 260-266.

Julian R.J. (1993). Ascites in poultry: a review article. *Avian. Pathol.* **22,** 419-454.

Julian R.J. (1998). Rapid growth problems: ascites and skeletal deformities in broilers. *Poult. Sci.* **77,** 1773-1780.

Katrina M.M., Michael G.E. and David A.W.A. (2001). Rapid, simple spectrophotometric method for simultaneous detection of nitrate and nitrite. *Nitric Oxide: Biol. Chem.* **5(1),** 62-71.

Khajali F. and Qujeq D. (2005). Relationship between growth and serum lactate dehydrogenase activity and the development of ascites in broilers subjected to skip-a-day feed restriction. *Int. J. Poult. Sci.* **4,** 317-319.

Lin H., Decuypere E. and Buyse J. (2008). Effect of thyroid hormones on the redox balance of broiler chickens. *Asian-Australas J. Anim. Sci.* **21,** 794-800.

Luger D., Shinder D., Rzepakovsky V., Rusal M. and Yahav S. (2001). Association between weight gain, blood parameters, and thyroid hormones and the development of ascites syndrome in broiler chickens. *Poult. Sci.* **80,** 965-971.

Maxwell M.H. and Robertson G.W. (1997). World broiler ascites survey 1996. *Poult. Int.* **36,** 16-19.

Maxwell M.H., Alexander I.A., Robertson J.W., Mitchell M.A. and McCorquodale C.C. (1995). Identification of tissue hypoxia in the livers of ascitic and hypoxia induced broilers using trypan blue. *Br. Poult. Sci.* **36,** 791-798.

Moaayyedian H., Asasi K., Nazifi S., Hassanzadeh M. and Ansari-Lari M. (2011). Relationship between venous blood gas parameters, thyroid hormone levels and ascites syndrome in broiler chickens exposed to cold temperature . *Iran J. Vet. Res.* **12(1),** 31-38.

Ocak F. (2006). Ascites in broilers. *J. Health. Sci.* **15(1),** 46-50.

Olkowski A.A., Korver D., Rathgeber B. and Classen H.L. (1999). Cardiac in index, oxyzen delivery and tissue oxyzen extraction in slow and fast growing chickens and in chickens with heart failure and ascites: a comparative study. *Avian. Pathol.* **28,** 137-146.

Oviedo-Rondón E.O., Ferket P.R. and Havestein G.B. (2006a). Understanding long bone development in broilers and turkeys. *Avian. Poult .Biol .Rev.* **17(3),** 77-88.

Oviedo-Rondón E.O., Ferket P.R. and Havestein G.B. (2006b). Nutritional factors that affect leg problems in broilers and turkeys. *Avian. Poult. Biol. Rev.* **17(3),** 89-103.

Park Y. and Kehre J.P. (1991). Oxidative changes in hypoxicre oxygenated rabbit heart: a consequence of hypoxia rather than reoxygenation. *Free Radic. Res. Commun.* **14,** 179-185.

SAS Institute. (2002). SAS®/STAT Software, Release 9.1. SAS Institute, Inc., Cary, NC. USA.

Scheele C.W., Decuypere E., Vereijken P.F.G. and Schreurs F.J. (1992). Ascites in broilers. 2. Disturbances in the hormonal regulation of metabolic rate and fat metabolism. *Poult. Sci.* **71,** 1971-1984.

Stojevi Z., Milinkovic-Tur S. and C. Katarina. (2000). Changes in thyroid hormones concentrations in chicken blood plasma during fattening. *Veterinarski Arh.* **70(1),** 31-37.

Tankson J.D., Thaxton J.P. and Vizzier-Thaxton Y. (2002). Biochemical and immunological changes in chickens experiencing pulmonary hypertension syndrome caused by *Enterococcus faecalis. Poult. Sci.* **81,** 1826-1831.

Tekeli A. (2014). Effects of ascites (pulmonary hypertension stndrome) on blood gas, blood oximetry parameters and heart section of broilers grown at high altitude. *J. Anim. Plant. Sci.* **24(4),** 998-1002.

Van As P., Elferink M.G., Closter A.M., Vereijken A., Bovenhuis H., Crooijmans R.P.M.A., Decuypere E. and Groenen M.A.M. (2010). The use of blood gas parameters to predict ascites susceptibility in juvenile broilers. *Poult. Sci.* **89,** 1684-1691.

Wang Y., Guo Y., Ning D., Peng Y., Cai H., Tan J., Yang Y. and Liu D. (2012). Changes of hepatic biochemical parameters and proteomics in broiler with cold-induced ascites. *J. Anim. Sci. Biotechnol.* **3(41),** 1-9.

Widema R.F. and Tackett C. (2000). Cardio-pulmonary function in preascitic (hypoxemic) or normal broilers reared at warm or cold temperatures: effect of acute inhalation of 100% oxygen. *Poult. Sci.* **79,** 257-264.

Wideman R.F.J.R. and Bottje W.G. (1993). Nutrition and Technical Symposium Proceedings. Novous International Inc., St. Louis, Missouri.

Wideman R.F.J.R. (1988). Ascites in poultry. *Monsanto Nutr. Update.* **6,** 1-7.

Williams B., Solomon S., Waddington D., Thorp B. and Farquharson C. (2000). Skeletal development in the meat-type chicken. *Br. Poult. Sci.* **41,** 141-149.

Wolfenson D., Sklan D., Graber Y., Kedar O., Bengal I. and Hurwitz S. (1987). Absorption of protein, fatty acids and minerals in young turkeys under heat and cold stress. *Br. Poult. Sci.* **28(4),** 739-742.

Yahav S. (2000). Relative humidity at moderate ambient temperatures: its effect on male broiler chickens and turkeys. *Br. Poult. Sci.* **41,** 94-100.

8

Insoluble Fibers Affected the Performance, Carcass Characteristics and Serum Lipid of Broiler Chickens Fed Wheat-Based Diet

K. Shirzadegan[1*] and H.R. Taheri[1]

[1] Department of Animal Science, Faculty of Agriculture, University of Zanjan, Zanjan, Iran

*Correspondence E-mail: k.shirzadegan@znu.ac.ir

ABSTRACT

The current study was conducted to survey the influences of addition of alfalfa meal (AM), rice bran (RB) and wood shaving (WS) in wheat-based diets [contain soluble non starch polysaccharides (NSP)] on performance carcass characteristic and serum lipids of broilers from 11 to 42 d of age. Seven hundreds 10-d-old male Ross 308 chicks were placed into 35 pens and allocated to seven wheat-soybean meal-based dietary treatments which were a control (CT) diet (without any fiber source) and six fiber-included diets consisting of three sources of fiber (AM, RB and WS) and two levels of fiber inclusion (3 and 6%) in a 3 × 2 factorial arrangement. According to the results, the average daily gain (ADG), average daily feed intake (ADFI), feed conversion ratio (FCR), corrected-FCR (C-FCR) and crop, proventriculus, gizzard (relative full weights) and heart, abdominal fat, liver, breast and thigh (relative weights) and serum low density lipoprotein (LDL) were affected by different types of fibers (P<0.05). The inclusion of insoluble fibers in wheat-based diet improved ADG and FCR in broilers, so that, the highest amount of ADG and the lowest amount of C-FCR was related to 3% WS contained diet (P<0.05). The highest amount of gizzard weight was depended to 6% WS contained diet (P<0.05) and the lowest breast and thigh weight were also related to RB diets (P<0.05), but the aforementioned treatments had no effect on the different intestine sections weight (P>0.05). In general, the inclusion of 3 to 6% insoluble fibers, except for 6% WS, in wheat-based diet improved growth performance in broiler chickens.

KEY WORDS   broilers, carcass traits, insoluble fiber, wheat.

INTRODUCTION

Wheat based diet is mainly used in some parts of the world such as Europe and Canada (Mahammadi Ghasem Abadi *et al.* 2014). Variations in nutritive values and nature of NSP of wheat result in inefficiencies in diet formulations, particularly in terms of energy and amino acid (Mahammadi Ghasem Abadi *et al.* 2014). NSPs as components of several poultry feed ingredients have been the interest of researches in recent years. These polysaccharides include a large variety of molecules that divided to main groups of soluble NSP (S-NSP) and insoluble NSP (I-NSP) (Shakouri *et al.*

2006). Some previous studies (Jiménez-Moreno *et al.* 2013; Taheri *et al.* 2016) showed that anti-nutritive effects of S-NSP on intestinal microflora, gut histomorphology, excreta moisture and performance. In fact, S-NSPs produce high digesta viscosity in intestine and inhibit digestion and absorption of nutrients such as lipids, protein and starch and decreased the performance (Hetland *et al.* 2004; Kalmendal *et al.* 2011). In contrast, most research conducted on poultry feeding, I-NSP has been considered a diluent of the diet (Rezaei *et al.* 2011). It has been demonstrated, based on research conducted in recent years, that the inclusion of moderate amounts of different fiber sources in the diet improves digestive organ development (Mateos *et al.* 2012) and increases HCl, bile acids and enzyme secretion (Molist *et al.* 2009; Svihus *et al.* 2004). These changes might result in improvements in nutrient digestibility (Rogel *et al.* 1987), growth performance (González-Alvarado *et al.* 2007; Taheri *et al.* 2016), increase in gizzard weight and gizzard contents and increase in apparent metabolizable energy corrected for nitrogen (AMEn) (Amerah *et al.* 2009; Jiménez-Moreno *et al.* 2009), gastrointestinal tract (GIT) health and eventually, animal welfare (Mateos *et al.* 2012). Consequently, the effects of dietary fiber (DF) on aforementioned factors will differ depending on the source and inclusion level of the fiber (Molist *et al.* 2009) as well as on the nature of the basal diet (Jiménez-Moreno *et al.* 2013). The alfalfa, wood shaving and rice bran are as good source of insoluble fiber, which can be used in poultry nutrition.

Alfalfa is widely used in animal feeding. It is moderately rich in protein, vitamins K, A and minerals but has low levels of energy (Mansoub and Pooryousef Myandoab, 2012). Dehydrated alfalfa meal generally is used at very low levels in poultry feeding, due primarily to its high fiber and low energy content (Jiang *et al.* 2012). The use of alfalfa in diets for monogastric animals is limited (Tkacova *et al.* 2011) by its high fiber content (above 7%). Alfalfa is well balanced in amino acids and rich in vitamins, carotenoids and saponins (Jiang *et al.* 2012). Carotenoids are polyenoicterpenoids having conjugated trans double bonds. They include carotenes (β-carotene and lycopene), which are polyene hydrocarbons, and xanthophylls (lutein, zeaxanthin, capsanthin, canthaxanthin, astaxanthin, and violaxanthin) having oxygen in different form (Tkacova *et al.* 2011). Mansoub and Pooryousef Myandoab (2012) reported that inclusion of alfalfa meal to the layer diets at the levels of 50, 100, 150 and 200 g/kg had no effect on feed intake (FI), but an addition of more than 5% decreased egg production (EP). Guclu *et al.* (2004) indicated that addition of 90 g/kg alfalfa meal into the laying quail diets had no adverse effect on performance and increased some of egg quality parameters.

Wood is essentially composed of cellulose, hemicelluloses, lignin, and some extractives compounds. The close to significant effect of wood shavings on total bile salt content of gizzard chyme in all layers, as well as the positive effect of wood shavings on jejuna amylase activity (U/g feed) observed in birds fed on whole wheat diets, indicate that stimulation of digestive processes may occur also in older birds (Hetland *et al.* 2003).

The rice bran is a powdery fine and fluffy material that consists seeds or kernels, in addition to particles of pericarp, seed coat, aleurone, germ and fine starchy endosperm. Rice bran is rich in B-vitamins and tocopherols and its nutrient density and profiles of amino acids and fatty acids,

including 74% of unsaturated fatty acids, are superior to cereal grains (Adrizal and Ohtani, 2002). Both rice bran protein and fat are of relatively high biological value (Ersin Samli *et al.* 2006). The results of a study carried out with poultry using rice bran at 0, 5, 10, 15, 20 and 25% levels indicated that use of 25% of mixed rice bran in layer diets had no adverse effect on the productive performance (Ersin Samli *et al.* 2006). With regard to effects of dietary inclusion of insoluble fibers on growth performance and digestive traits of birds, the aim of this research was to study the influence of AM, RB and WS in the wheat-based diet on growth performance, carcass traits and serum lipids of broilers in the grower phase.

MATERIALS AND METHODS

Experimental design and handling

The present study was conducted from June to July of 2015, when the climate condition was relative hot and humid. A total of 700 commercial broiler chicks (10-d-old male, Ross 308 strain) were divided into 35 groups, which consisted of 5 replications of 20 birds each and fed to seven wheat-based diets treatments which were a control diet (without any fiber source) and six fiber-included diets consisting of three sources (AM, RB and WS) and two levels of fiber inclusion (3 and 6%) in a 3 × 2 factorial arrangement. They were arranged using a completely randomized design (CRD). The photoperiod was 23 h light: 1 h dark, throughout the experiment. The rearing room (1.5×1.5 m) was provided with fans and coolers to adjust the environmental temperature to 24 ± 1 °C and this temperature were almost stable for total of experiment. Prophylactic measures against the most common infectious diseases were conducted.

The experimental starter diets contained 3098 to 3102 kcal ME and 2147 to 2152 % CP (not showed) and the grower diets (Table 1) were formulated to meet the requirements of broiler chickens according to recommendations of Aviagen for Ross 308 broilers (Aviagen, 2014).

Chemical evaluation of diets

The diets were analyzed for nutrient concentrations according to the standard procedures of AOAC (2000). Fiber sources and diets were analyzed for S-NSP and I-NSP according to the procedure of Englyst *et al.* (1994).

Growth performance

After an adaptation period, the ADG, ADFI and FCR were determined by pen from 25 to 42 d of age. Data were corrected by subtracting the amount of added WS from the total feed consumption (Hetland *et al.* 2003) to obtain C-FCR.

Table 1 Ingredient composition, nutrient content and particle size distribution of the experimental diets (% as-fed basis, unless otherwise indicated) from 25 to 42 d of age

Ingredient	Control	Alfalfa meal		Rice bran		Wood shaving	
		3%	6%	3%	6%	3%	6%
Wheat grain	40.0	40.0	40.0	40.0	40.0	40.0	40.0
Corn grain	20.57	17.40	14.60	18.00	15.10	11.45	8.25
Soybean meal	28.79	28.06	27.10	28.27	27.99	29.90	31.13
Alfalfa meal	-	3.0	6.0	-	-	-	-
Rice bran	-	-	-	3.0	6.0	-	-
Wood shaving	-	-	-	-	-	3.0	6.0
Soybean oil	6.60	7.60	8.44	6.70	6.90	8.60	10.60
Dicalcium phosphate	1.41	1.40	1.39	1.39	1.37	1.42	1.43
Calcium carbonate	0.96	0.88	0.79	0.99	1.00	0.97	0.96
Common Salt	0.16	0.18	0.16	0.20	0.20	0.20	0.20
Sodium bicarbonate	0.29	0.27	0.28	0.25	0.24	0.27	0.27
L-threonine	0.10	0.10	0.10	0.10	0.10	0.10	0.09
DL-methionine	0.33	0.34	0.35	0.33	0.33	0.34	0.34
L-lysine HCl	0.27	0.27	0.29	0.27	0.27	0.25	0.23
Mineral premix[1]	0.25	0.25	0.25	0.25	0.25	0.25	0.25
Vitamin premix[2]	0.25	0.25	0.25	0.25	0.25	0.25	0.25
Calculated analysis[3]							
AME_n (kcal/kg)	3198	3200	3200	3200	3200	3200	3200
CP	19.49	19.51	19.47	19.46	19.51	19.47	19.48
Methionine + cystine	0.91	0.91	0.91	0.91	0.91	0.91	0.91
Lysine	1.16	1.16	1.16	1.16	1.16	1.16	1.16
Determined analysis							
Dry matter (DM)	91.4	90.3	90.1	91.5	90.8	90.7	90.0
Gross energy (kcal/kg)	4026	4057	4086	4012	4007	4102	4191
Crude protein (CP) (N×6.25)	19.04	19.08	19.13	19.10	19.22	19.06	19.10
Ether extract	8.14	9.10	9.92	8.52	8.99	9.82	11.53
Calcium	0.80	0.81	0.81	0.80	0.80	0.80	0.80
Total phosphorus	0.66	0.65	0.64	0.63	0.63	0.65	0.64
Crude fiber	2.87	3.42	3.91	3.12	3.36	4.91	7.80
Neutral detergent fiber	11.93	13.37	14.05	12.76	13.60	14.08	16.26
Acid detergent fiber	4.80	5.44	5.97	5.23	5.68	6.90	8.99
S-NSP	2.24	2.28	2.35	2.21	2.20	2.20	2.18
I-NSP	13.99	14.94	15.65	14.82	15.56	16.63	19.07

[1] Provided the following per kilogram of diets: Manganese ($MnSO_4H_2O$): 75 mg; Zinc (ZnO): 85 mg; Iron ($FeSO_4 7H_2O$): 50 mg; Copper ($CuSO_4 5H_2O$): 10 mg; Selenium (Na_2SeO_3): 0.2 mg; Iodine (Iodized NaCl): 0.8 mg and Choline: 250 mg.
[2] Provided the following per kilogram of diets: vitamin A: 9000 IU; Cholecalciferol: 2000 IU; vitamin E: 36 IU; vitamin B_{12}: 0.015 mg; Menadione: 2 mg; Riboflavin: 6.6 mg; Thiamine: 1.8 mg; Pantothenic acid: 7.3 mg; Niacin: 30 mg; Folic acid: 1 mg; Biotin: 0.1 mg and Pyridoxine: 3 mg.
[3] According to NRC (1994).
S-NSP: soluble non-starch polysaccharide and I-NSP: insoluble non-starch polysaccharide.

Before weighing, the chicks were fasted for six h to ensure a consistent gut fill among all birds (Shirzadegan *et al.* 2014).

Carcass characteristics

Subsequent to the weighing of broiler chickens at the end of the experiment (42 days), three birds were selected from each of the replicate groups per treatments. These birds were slaughtered by severing the bronchial vein to determine some measurements of carcass yield involve internal organs (crop, heart, proventriculus, liver, abdominal fat, gizzard, small intestine, caecum and colon) as full digestive tract and external organs (carcass, breast and thigh). The weights of these organs were expressed as percentages of live body weight (Shirzadegan *et al.* 2014).

Serum metabolites

At the 42nd days of the experiment, blood samples (three bird each replicate) from the brachial veins of all chickens were collected for analysis. After coagulation, the blood samples were centrifuged at 3000 rpm for 10 min and the obtained serum was stored in a freezer at -20 °C for further analysis. In the blood serum, the following parameters were determined: total cholesterol (TC), triglyceride (TG) and low-density lipoproteins (LDL) by spectrophotometric method using commercially available kits (Parsazmun, Iran) (Shirzadegan *et al.* 2014).

Statistical analysis

Data were analyzed as a completely randomized design with seven treatments (diets) using the GLM procedure of

SAS (SAS, 2003). Differences among treatments were considered significant at (P<0.05). Significant differences between means (diets) were separated by Fisher's Least Significant Difference test (protected t-test). In addition, the effects of fiber source (AM, RB and WS), inclusion level (3 and 6%) of fiber and the interaction between source and level of fiber were studied as a 2 × 3 factorial arrangement analysis.

RESULTS AND DISCUSSION

Growth performance

The effect of fiber source and level of fiber on ADFI, ADG, FCR and C-FCR are shown in Table 2. Birds fed the RB or the WS diets showed the higher ADFI (P<0.001) than those fed the CT diet, however, the 3% AM diet decreased the ADFI compared to the CT diet and the ADFI was higher for birds fed the WS diet than for birds fed the RB diet. Birds fed the fiber-included diets showed the higher ADG (P<0.001) compared to the CT diet. The ADG was higher (P<0.01) for birds fed the WS or the RB diets than for birds fed the AM diet.

Feeding the fiber-included diets, except the 6% WS, improved (P<0.001) the FCR compared to the CT diet. The birds fed the AM diet showed the lowest FCR (P<0.001) compared to other fiber sources, and FCR was higher for birds fed the WS diet than for birds fed the RB diet. Birds fed the fiber-included diets showed the lower C-FCR (P<0.001) than those fed the CT diet. The C-FCR was higher (P<0.001) for birds fed the WS or the RB diets than for birds fed the AM diet.

Totally, an increase in the inclusion level of fiber in the diet increased the FCR and the C-FCR, but the effects were more pronounced for the WS diets than for the AM or the RB diets. Likewise, the interaction effect (P<0.001) was observed on growth performance traits (ADFI, ADG, FCR and C-FCR) between the fiber source and the inclusion level of fiber.

Organs weight

Influences of dietary fibers on full digestive tract weight (Table 3) and other carcass characteristic (Table 4) of broiler chicks are presented in below. Results indicated that the effects of treatments on carcass traits were statistically significant at 42 days old (P<0.05).

However, the crop, proventriculus, gizzard (full relative weights) and heart, abdominal fat, liver, breast and thigh (relative weights) were affected by different sources of fibers, so that, the highest amount of gizzard weight was depended to 6% WS contained diet (P<0.05) and the lowest abdominal fat weight was related to 6% AM group (P<0.05).

The lowest breast and thigh weight were also related to RB diets (P<0.05), but the treatments had no significant effect on the carcass yield, small intestine, ceca and colon weights of birds (P>0.05).

Serum lipids

Results of effects of dietary fibers on serum metabolites of broilers are presented in Table 5. These findings indicate that different inclusion of dietary fibers did not significantly (P>0.05) effect on the blood TG and TC concentrations, but plasma LDL percentages were affected (P<0.05) by treatments. According to results, the highest serum LDL at 40 days old was observed in the chicks fed on ration 3% WS compared to control. Moreover, the LDL level of birds fed 6% WS was lower than 6% RB (P<0.05), whereas, the other fiber-included diets not had different to control group (P>0.05).

According to the results, the decreased feed intake of the birds on the diets containing wheat compared to the fiber included diets probably is because of increasing the intestinal digesta viscosity that causes increase of feed retention time in the gastrointestinal tract (Rogel et al. 1987; Mateos et al. 2012). Since, there is a relationship between rate of feed passage through the gut and feed consumption in young chickens (Taheri et al. 2016), inclusion of this S-NSP in the diet leads to less feed intake. In the present study, it was found that the inclusion of wood shavings increased ADFI and decreased the FCR and C-FCR in chickens compared to other fiber sources. Similarly, Shakouri et al. (2006) found that diluting the diet with cellulose increased feed intake, whereas, Hetland et al. (2003) found that diluting a wheat-based diet with oat hulls had no influence on the weight gain and feed intake, but, in turkeys, diluting a maize-based diet with wood shavings was reported to improve the gain/feed during the critical first 7 d posthatch period (Jiang et al. 2012).

Taheri et al. (2016) also showed the broiler chickens receiving 4 and 8 % wheat bran in a barley-based diet had lower FCR than those fed the barley-based diet. It has been suggested that the birds are able to maintain normal weight gain when fed diets diluted with insoluble fibre by the increased capacity of the digestive system and/or faster passage through the digestive tract (Hetland et al. 2004). Consequently, the higher feed intake in birds fed on diets diluted with fibers in the present study can probably be explained by the larger gizzard sizes observed and faster gut emptying (Kalmendal et al. 2011). A large, well-developed gizzard improves gut motility and may increase cholecystokinin release (Svihus et al. 2004), which in turn stimulates the secretion of pancreatic enzymes and gastro-duodenal refluxes, as a result, the growth performance could be better in birds.

Table 2 Effect of fiber source and level of fiber on growth performance of broilers from 25 to 42 d

Factors	ADFI (g)	ADG (g)	FCR	C-FCR*
Diets				
Control	161.5[d]	77.1[d]	2.08[b]	2.08[a]
Alfalfa meal-3%	152.5[e]	80.0[cd]	1.91[e]	1.91[d]
Alfalfa meal-6%	158.4[d]	81.4[c]	1.95[d]	1.95[c]
Rice bran-3%	171.3[bc]	85.0[b]	2.01[c]	2.01[b]
Rice bran-6%	167.5[c]	82.3[bc]	2.03[c]	2.03[b]
Wood shaving-3%	173.4[b]	90.0[a]	1.92[de]	1.87[e]
Wood shaving-6%	179.9[a]	81.5[c]	2.20[a]	2.08[a]
SEM (n=5)	1.96	1.10	0.012	0.012
Fiber source (FS)				
Alfalfa meal	155.5	80.7	1.93	1.93
Rice bran	169.4	83.7	2.02	2.02
Wood shaving	176.7	85.8	2.06	1.97
SEM (n=10)	1.38	0.78	0.008	0.008
Inclusion level of FS (g/kg)				
30	165.8	85.0	1.95	1.93
60	168.6	81.7	2.06	2.02
SEM (n=15)	1.16	0.65	0.007	0.007
Probability				
Diets	0.01	< 0.0001	< 0.0001	< 0.0001
Fiber source	< 0.0001	0.0005	< 0.0001	< 0.0001
Inclusion level of FS	0.09	0.001	< 0.0001	< 0.0001
Fiber source × inclusion level of FS	0.03	0.0006	< 0.0001	< 0.0001

The means within the same column with at least one common letter, do not have significant difference (P>0.05).
ADFI: average daily feed intake; ADG: average daily gain; FCR: feed conversion ratio.
* C-FCR: corrected-FCR, data were corrected by subtracting the amount of added wood shavings from the total feed consumption (Hetland et al. 2003).
SEM: standard error of the mean.

Table 3 Effect of fiber source and level of fiber on the relative weight (% live BW) of full digestive tract of broilers at 42 d of age

Factors	Crop	Proventriculus	Gizzard	Duodenum	Jejunum	Ileum	Total intestine	Ceca	Colon
Diets									
Control	0.83[cd]	0.62[c]	3.11[d]	1.13	2.57	1.67	7.26	1.25	0.64
Alfalfa meal-3%	0.82[d]	0.61[c]	3.74[b]	1.08	2.43	1.48	6.78	1.15	0.63
Alfalfa meal-6%	0.90[a]	0.69[ab]	3.77[b]	1.11	2.53	1.53	7.01	1.21	0.63
Rice bran-3%	0.86[abcd]	0.66[abc]	3.17[cd]	1.10	2.54	1.60	7.04	1.17	0.62
Rice bran-6%	0.88[abc]	0.66[abc]	3.48[bc]	1.12	2.47	1.63	7.13	1.27	0.64
Wood shaving-3%	0.84[bcd]	0.64[c]	3.64[b]	1.15	2.58	1.48	7.06	1.22	0.64
Wood shaving-6%	0.90[a]	0.70[a]	4.17[a]	1.14	2.39	1.65	7.13	1.30	0.66
SEM (n=5)	0.019	0.018	0.125	0.017	0.099	0.061	0.153	0.059	0.018
Fiber source (FS)									
Alfalfa meal	0.86	0.65	3.75	1.10	2.48	1.51	6.9	1.18	0.63
Rice bran	0.87	0.66	3.32	1.11	2.51	1.62	7.1	1.22	0.63
Wood shaving	0.87	0.67	3.91	1.14	2.48	1.56	7.1	1.26	0.65
SEM (n=10)	0.014	0.013	0.090	0.012	0.068	0.042	0.116	0.042	0.013
Inclusion level of FS									
3%	0.84	0.64	3.52	1.11	2.52	1.52	7.0	1.18	0.63
6%	0.90	0.68	3.81	1.12	2.47	1.60	7.1	1.26	0.64
SEM (n=15)	0.011	0.010	0.073	0.010	0.055	0.034	0.095	0.034	0.011
Probability									
Diets	0.02	0.009	< 0.0001	0.12	0.79	0.16	0.50	0.57	0.91
Fiber source	0.75	0.50	0.0004	0.03	0.96	0.21	0.42	0.43	0.64
Inclusion level of FS	0.001	0.005	0.01	0.42	0.52	0.10	0.35	0.12	0.58
Fiber source × inclusion level of FS	0.27	0.11	0.18	0.39	0.35	0.46	0.87	0.96	0.74

The means within the same column with at least one common letter, do not have significant difference (P>0.05).
SEM: standard error of the mean.

The lower weight gain and higher FCR of the broilers fed control diet are also predictable due to less feed intake and less nutrients utilization because of high viscosity of the chime in birds (Shakouri et al. 2006), but chickens fed the coarse insoluble fiber (such as WS or RB diets) showed the higher ADFI than those fed the control diet, which it may be due an increase in the flow rate of feed in GIT.

The results of this study was in agreement with results of Gonzalez-Alvarado et al. (2007) and Jiménez-Moreno et al. (2009) when oat hulls, rice hulls or sunflower hulls were included in low fiber diets for broilers and suggests that the beneficial effects of dietary fiber on feed/gain ratio and growth performance of broilers were due primarily to an increase in diet digestibility and well-developed gizzard which resulted in more nutrients available for growth. Furthermore, in conjunction to better FCR in birds fed RB and AM diets, Guclu *et al.* (2004) indicated that adding alfalfa meal into the laying quail diets (90 g/kg) had no adverse effect on performance and increased some of egg quality parameters. Tkacova *et al.* (2011) also reported that broilers fed diets with high doses of alfalfa (60 g/kg diet) had lower feed conversion. However, it was reported that inclusion of alfalfa in diets of laying hens reduced performance expressed in terms of body weight and egg mass (EM) (Jiang *et al.* 2012).

However, results of Ersin Samli *et al.* (2006) experiment showed that rice bran could be included up to 10% without any adverse effect on laying performance, egg quality and digestive organs. The higher ADG and better FCR showed in experimental birds in present study may be due to mainly vitamins and essential nutrients found in rice bran.

Results of this experiment also indicated that addition of I-NSP to diet affected some characteristics of gastrointestinal tract of broilers.

In this way, there was a significant effect of NSP solubility on cecum weight, with pigs fed the S-NSP diet having heavier cecum than those on the I-NSP diets on both days 6 and 14 (Wellock *et al.* 2008). As well as, asserted the treatments containing an equal or higher ratio of cellulose resulted in increased intestine weight of broilers at 21 days of age.

This finding is conflicting with other researchers who have shown that cellulose usage decreases intestine weight in younger broilers (Sklan, 2003). Notably, broilers adapt to high levels of insoluble fibers by enlarging the GIT (Gonzalez-Alvarado *et al.* 2010). As a result, their weights increase. The relative empty weights of jejunum, ileum and small intestine of broilers were also affected by dietary insoluble fiber (Amerah *et al.* 2009). This adaptation is a rapid attempt to increase the absorptive surface area of the GIT in response to the lower diffusion rates, and it occurs by increasing the digesta viscosity levels (Jiménez-Moreno *et al.* 2009; Mateos *et al.* 2012). However, the breast, thigh and intestinal sections in our study were not improved by administration of fiber in diet.

Moreover, reported, abdominal fat decreased at 20% rice bran inclusion in the diet. Oladunjoye and Ojebiyi (2010) also noted that body fat content of chicken reduced when 40% oat hull was included in their diet. Gizzard weight and the length of small intestine, large intestine, duodenum and ileum were however increased significantly at 20% rice bran inclusion level.

This can be attributed to high fiber in this diet. Similarly reported, access to coarse hulls (such as wood shaving) stimulated gizzard development (Mateos *et al.* 2012), which this is in agreement with previous studies to fiber source such as wood shaving (Amerah *et al.* 2009) and alfalfa (Jiang *et al.* 2012).

Table 4 Effect of fiber source and level of fiber on the relative weight (% live BW) of carcass, breast, thigh + drumstick, abdominal fat, heart and liver in broiler at 42 d of age

Factors	Carcass	Breast	Thigh + drumstick	Abdominal fat	Heart	Liver
Diets						
Control	72.7	27.6[ab]	25.2[a]	1.95[abc]	0.45[a]	2.58[a]
Alfalfa meal-3%	73.3	27.7[a]	25.1[a]	1.92[bc]	0.41[c]	2.43[ab]
Alfalfa meal-6%	73.5	27.2[ab]	24.6[ab]	1.86[c]	0.40[c]	2.38[b]
Rice bran-3%	72.5	26.7[bc]	23.9[b]	1.99[ab]	0.42[bc]	2.31[b]
Rice bran-6%	72.7	27.1[ab]	24.6[ab]	2.00[ab]	0.42[bc]	2.40[b]
Wood shaving 3-%	73.4	27.3[ab]	25.1[a]	2.02[ab]	0.44[ab]	2.31[b]
Wood shaving-6%	72.3	26.1[c]	24.9[a]	2.03[a]	0.41[c]	2.30[b]
SEM (n=5)	0.39	0.31	0.29	0.036	0.008	0.056
Fiber source (FS)						
Alfalfa meal	73.4	27.4	24.9	1.89	0.40	2.40
Rice bran	72.6	26.9	24.2	2.00	0.42	2.36
Wood shaving	72.8	26.7	25.0	2.03	0.42	2.31
SEM (n=10)	0.26	0.21	0.21	0.025	0.006	0.040
Inclusion level of FS						
3%	73.1	27.3	24.7	1.98	0.42	2.35
6%	72.8	26.8	24.7	1.97	0.41	2.36
SEM (n=15)	0.21	0.17	0.17	0.020	0.005	0.032
Probability						
Diets	0.14	0.02	0.048	0.02	0.002	0.02
Fiber source	0.045	0.046	0.047	0.002	0.049	0.23
Inclusion level of FS	0.39	0.049	0.99	0.70	0.04	0.88
Fiber source × inclusion level of FS	0.12	0.04	0.15	0.50	0.18	0.50

The means within the same column with at least one common letter, do not have significant difference (P>0.05).
SEM: standard error of the mean.

Table 5 Effect of fiber source and level of fiber on serum lipid parameters (mg/dL) in broiler at 40 days old

Diets	TG	TC	LDL
Control	103.8	135.8	45.2[bc]
Alfalfa meal-3%	124.6	156.8	58.6[ab]
Alfalfa meal-6%	112.0	149.2	54.4[abc]
Rice bran-3%	134.6	146.0	52.0[bc]
Rice bran-6%	115.4	151.0	57.6[ab]
Wood shaving-3%	116.0	140.0	67.8[a]
Wood shaving-6%	112.6	138.0	42.6[c]
SEM (n=5)	7.71	6.22	4.87
Fiber source			
Alfalfa meal	118.3	153.0	56.5
Rice bran	125.0	148.5	54.8
Wood shaving	114.3	139.0	55.2
SEM (n=10)	5.66	4.55	3.54
Inclusion level			
3%	125.1	147.6	59.5
6%	113.3	146.1	51.5
SEM (n=15)	4.63	3.71	2.89
Probability			
Diets	NS	NS	0.02
Fiber source	NS	NS	NS
Inclusion level	0.046	NS	0.044
Fiber source × level	NS	NS	0.02

TG: triglyceride; TC: total cholesterol and LDL: low density lipoprotein.
The means within the same column with at least one common letter, do not have significant difference (P>0.05).
SEM: standard error of the mean.

The coarse hull particles are retained in the gizzard until they are ground to certain critical size that allows them to pass through the pyloric sphincter (Hetland et al. 2003). This leads to an increase in the volume of the organ's contents and a muscular adaptation to meet the greater demand for grinding. Showed that the increase in grinding activity of the gizzard, increase in digestive secretions, HCL, bile acid and endogenous enzymes together with a better mixing of digestive juices with the digesta (González-Alvarado et al. 2007) affect the functioning of the GIT and may modify microbial growth at specific digestive organ sites, might explain the positive effects of I-NSP on the digestibility of dietary components (Rezaei et al. 2011; Mateos et al. 2012). This study, the highest proventriculus and gizzard weights and the better growth performance were observed in fiber included diets that it could be refer to aforementioned subject.

The result of this study showed that inclusion of different I-NSP in diet significantly altered the blood metabolites of broilers in grower phase. In depending to this result, Sarikhan et al. (2009) reported that at starter period, blood lipids levels were significantly affected by levels of insoluble fibers in grower phase. In their research, serum TG levels were significantly lower in broilers fed diets containing 0.75% insoluble fibers and higher levels of high density lipoprotein (HDL) and lower concentrations of very low density lipoprotein (VLDL) observed in the serum of 0.50 and 0.75% insoluble fibers dietary groups.

Taheri et al. (2016) was also showed the broilers receiving 12% wheat bran in a barley-based diet had lower TC than those fed the corn-based diet. In contrast to our study, it showed that the blood TC and TG concentration were significantly reduced in bird fed diets supplemented to alfalfa powder (Oladunjoye and Ojebiyi, 2010).

The main reason of cholesterol and triglyceride reduction in serum of chicks may be due to substances like carvacrol and thymol (Mansoub and Pooryousef Myandoab, 2012) and high levels (2 to 3% dry matter) of saponins present in alfalfa, which have showed hypocholesterolemic, anticarcinogenic, anti-inflammatory, hypocholesterolemic and antioxidant activity (Jiang et al. 2012). Furthermore, indicated that rice bran insoluble fiber function as lower agent of TC and coronary artery disease. These effects are due to tocols (0.2%) of crude rice bran, which about 70 % are tocotrienols.

Tocol possess potent antioxidant activity and decreases serum cholesterol and hepatic cholesterol synthesis by the suppression of hydroxyl methyl glytaryl coenzyme A reductase (Ju and Vali, 2005). However, addition of fiber to diets in our study increased plasma LDL (3% WS diet) and abdominal fat, which it may be because of increase in nutrient passage rate and little degradation of bile salts by gut microorganisms and higher cholesterol absorption in broilers intestine.

CONCLUSION

Generally, based on the results of growth performance, the inclusion of 3 or 6% insoluble fiber, except for 6% WS, in wheat-based diet improved the ADG, FCR and gizzard function in broilers simultaneously from 25 to 42 d of age. The beneficial effect of insoluble fiber inclusion in the diet may be related to some extent to the longer feed passage in the upper part of the gastrointestinal tract, faster feed passage rate in the distal part of the gastrointestinal tract and the increase of nutrient retention. However, more research is needed to determine all the roles of I-NSPs on broilers performance.

ACKNOWLEDGEMENT

The authors sincerely thank Department of Animal Science, Faculty of Agriculture, University of Zanjan for this project.

REFERENCES

Adrizal O. and Ohtani S. (2002). Defatted rice bran non starch polysaccharides in broiler diets: effects of supplements on nutrient digestibilities. *J. Poult. Sci.* **39,** 67-76.

Amerah A.M., Ravindran V. and Lentle R.G. (2009). Influence of insoluble fibre and whole wheat inclusion on the performance, digestive tract development and ileal microbiota profile of broiler chickens. *Br. Poult. Sci.* **50,** 366-375.

AOAC. (2000). Official Methods of Analysis. Vol. I. 17th Ed. Association of Official Analytical Chemists, Arlington, VA, USA.

Aviagen. (2014). Ross 308: Broiler Nutrition Specification. Aviagen Ltd., Newbridge, UK.

Englyst H.N., Quigley M.E. and Geoffrey J.H. (1994). Determination of dietary fibre as non starch polysaccharides with gas-liquid chromatographic, high performance liquid chromatographic or spectrophotometric measurement of constituent sugars. *Analyst.* **119,** 1497-1509.

Ersin Samli H., Senkoylu N., Akyurek H. and Agma A. (2006). Using rice bran in laying hen diets. *J. Center. Europ. Agric.* **7,** 135-140.

Gonzalez-Alvarado J.M., Jimenez-Moreno E., Lazaro R. and Mateos G.G. (2007). Effects of type of cereal, heat processing of the cereal, and inclusion of fiber in the diet on productive performance and digestive traits of broilers. *Poult. Sci.* **86,** 1705-1715.

Gonzalez-Alvarado J.M., Jimenez-Moreno E., Gonzalez-Sanchez D., Lazaro R. and Mateos G.G. (2010). Effect of inclusion of oat hulls and sugar beet pulp in the diet on productive performance and digestive traits of broilers from 1 to 42 d of age. *Anim. Feed Sci. Technol.* **162,** 37-46.

Güçlü B., İşcan K.M., Uyanık F., Eren M. and Ağca A.C. (2004). Effect of alfalfa meal in diets of laying quails on performance, egg quality and some serum parameters. *Arch. Anim. Nutr.* **58,** 255-263.

Hetland H., Svihus B. and Krögdahl A. (2003). Effects of oat hulls and wood shavings on digestion in broilers and layers fed diets based on whole or ground wheat. *Br. Poult. Sci.* **44,** 275-282.

Hetland H., Choct M. and Svihus B. (2004). Role of insoluble non-starch polysaccharides in poultry nutrition. *World. Poult. Sci. J.* **60,** 415-422.

Jiang J.F., Song X.M., Huang X., Zhou W.D., Wu J.L., Zhu Z.G., Zheng H.C. and Jiang Y.Q. (2012). Effects of alfalfa meal on growth performance gastrointestinal tract development of growing duck. *Asia-Australas J. Anim. Sci.* **25,** 1445-1450.

Jiménez-Moreno E., Gonzalez-Alvarado J.M., de Coca Sinova A., Lazaro R. and Mateos G.G. (2009). Effects of source of fibre on the development and pH of the gastrointestinal tract of broilers. *Anim. Feed Sci. Technol.* **154,** 93-101.

Jiménez-Moreno E., Gonzalez-Alvarado J.M., de Coca Sinova A., Lazaro R. and Mateos G.G. (2013). Oat hulls and sugar beet pulp in diets for broilers. Effects on the development of the gastrointestinal tract and on the structure of the jejunal mucosa. *Anim. Feed Sci. Technol.* **182,** 44-52.

Ju Y.S. and Vali S.R. (2005). Rice bran oil as a potential resource for biodiesel: a review. *J. Sci. Ind. Res.* **64,** 866-882.

Kalmendal R., Elwinger K., Holm L. and Tauson R. (2011). High fibre sunflower cake affects small intestinal digestion and health in broiler chickens. *Br. Poult. Sci.* **52,** 86-96.

Mahammadi Ghasem Abadi M.H., Riahi M., Shivazad M., Zali A. and Adibmoradi M. (2014). Efficacy of wheat based *vs.* corn based diets on growth performance, carcass traits, blood parameters, jejunum morphological development, immunity, cecal microflora and excreta moisture in broiler chicks. *Iranian J. Appl. Anim. Sci.* **4(1),** 105-110.

Mansoub N.H. and Pooryousef Myandoab M. (2012). Effect of dietary inclusion of alfalfa (*Medicago sativa*) and black cumin (*Nigella sativa*) on performance and some blood metabolites of Japanese quail. *Res. Open. Anim. Vet. Sci.* **2(1),** 7-9.

Mateos G.G., Jiménez-Moreno E., Serrano M.P. and Lázaro R.P. (2012). Poultry response to high levels of dietary fiber sources varying in physical and chemical characteristics. *J. Appl. Poult. Res.* **21,** 156-174.

Molist F., Gomezde Segura A., Gasa J., Hermes R.G., Manzanill E.G., Anguit M. and Perez J.F. (2009). Effects of the insoluble and soluble dietary fibre on the physicochemical properties of digesta and the microbial activity in early weaned piglets. *Anim. Feed Sci. Technol.* **149,** 346-353.

NRC. (1994). Nutrient Requirements of Poultry, 9th Rev. Ed. National Academy Press, Washington, DC., USA.

Oladunjoye I.O. and Ojebiyi O.O. (2010). Performance characteristics of broiler chicken (*Gallus gallus*) fed rice (*Oriza sativa*) bran with or without roxazyme. *Int. J. Anim. Vet. Adv.* **2(4),** 135-140.

Rezaei M., Karimi Torshizi M.A. and Rouzbehan Y. (2011). The influence of different levels of micronized insoluble fiber on broiler performance and litter moisture. *Poult. Sci.* **90,** 2008-2012.

Rogel A.M., Balnave D., Bryden W.L. and Annison E.F. (1987). Improvement of raw potato starch digestion in chicks by feeding oat hulls and other fibrous feedstuffs. *Australian J. Agric. Res.* **38,** 629-637.

Sarikhan M., Shahryari H.A., Nazeradl K., Gholizadeh B. and Behesht B. (2009). Effects of insoluble fiber on serum biochemical characteristics in broiler. *Int. J. Agric. Biol.* **11**, 73-76.

SAS Institute. (2003). SAS®/STAT Software, Release 9.2. SAS Institute, Inc., Cary, NC. USA.

Shakouri M.D., Kermanshahi H. and Mohsenzadeh M. (2006). Effect of different non starch polysaccharides in semi purified diets on performance and intestinal microflora of young broiler chickens. *Int. J. Poult. Sci.* **5(6)**, 557-561.

Shirzadegan K., Fallahpour P., Nickkhah I. and Taheri H.R. (2014). Black cumin (*Nigella sativa*) supplementation in the diets of broilers influences liver weight and its enzyms. *Irania. J. Appl. Anim. Sci.* **5(1)**, 173-178.

Sklan D., Smirnov A. and Plavnik I. (2003). The effect of dietary fibre on the small intestine and apparent digestion in the turkey. *Br. Poult. Sci.* **44**, 735-740.

Svihus B., Juvik E., Hetland H. and Krogdahl A. (2004). Causes for improvement in nutritive value of broiler chicken diets with whole wheat instead of ground wheat. *Br. Poult. Sci.* **45**, 55-60.

Taheri H.R., Tanha N. and Shahir M.H. (2016). Effects of wheat bran inclusion in barley-based diet on villus morphology of jejunum, serum cholesterol, abdominal fat and growth performance of broiler chickens. *J. Livest. Sci. Technol.* **4(1)**, 9-16.

Tkacova J., Angelovicova M., Mrazova L., Kliment M. and Kral M. (2011). Effect of different proportion of lucerne meal in broiler chickens. *Anim. Sci. Biotechnol.* **44**, 141-144.

Wellock I.J., Fortomaris P.D., Houdijk J.G.M., Wiseman J. and Kyriazakis I. (2008). The consequences of non-starch polysaccharide solubility and inclusion level on the health and performance of weaned pigs challenged with enterotoxigenic *Escherichia coli*. *Br. J. Nutr.* **99**, 520-530.

9

Effects of Peppermint (*Mentha piperita*) and Aloe vera (*Aloe barbadensis*) on Ileum Microflora Population and Growth Performance of Broiler Chickens in Comparison with Growth Promoter Antibiotic

B. Darabighane[1*], F. Mirzaei Aghjeh Gheshlagh[1], B. Navidshad[1],
A. Mahdavi[2], A. Zarei[3] and S. Nahashon[4]

[1] Department of Animal Science, Faculty of Agricultural Science, University of Mohaghegh Ardabili, Ardabil, Iran
[2] Department of Animal Husbandry, Faculty of Veterinary Medicine, Semnan University, Semnan, Iran
[3] Department of Animal Science, Karaj Branch, Islamic Azad University, Karaj, Iran
[4] Department of Agricultural Science, Tennessee State University, Nashville, USA

*Correspondence E-mail: b.darabighane@uma.ac.ir

ABSTRACT

This research was conducted to compare the effects of two medicinal plants (peppermint and aloe vera) and antibiotic growth promoter on ileum microflora population and growth performance of broiler chickens. In this experiment, 375 one-day old male broiler chickens (Ross 308) were used on a completely randomized design with 5 dietary treatments which were replicated 5 times with 15 birds per replicate. The experimental treatments were: 1) the control diet (basal diet with no additive); 2) basal diet + 10 g/kg dry peppermint leaves (DPL); 3) basal diet + 10 g/kg aloe vera gel (AVG); 4) basal diet + 5 g/kg DPL + 5 g/kg AVG and 5) basal diet + 10 ppm virginiamycin. Growth performance parameters were evaluated during the starter, grower and finisher periods and the populations of *Lactobacillus* and *Escherichia coli* bacteria was determined on the 42nd day of age. The maximum number of *Lactobacillus* bacteria was observed in the ileum of broilers fed diets containing DPL + AVG (P<0.05). The birds receiving the medicinal plant supplements had lower *Escherichia coli* population in comparison with control birds (P<0.05). The antibiotic fed broilers had higher feed intake, body weight gain, and dressing percentage than the control treatment and those fed medicinal herbs. The broilers fed medicinal herbs had higher feed intake, body weight gain and dressing percentage than the control (P<0.05). The birds fed mixture of DPL + AVG has better growth performance and dressing percentage. This study suggests that using a mixture of peppermint and aloe vera as a feed supplement of broiler chickens could be a potential alternative for antibiotic growth promoter.

KEY WORDS aloe vera, broiler chicken, growth performance, microflora, peppermint.

INTRODUCTION

The antibiotic growth promoters are a group of feed additives which inhibit intestinal pathogens and improve the growth performance of poultry. However, in recent years the excessive use of antibiotic growth promoters has been a major concern due to increased antibiotic resistance in human and inclination of users to eat healthy foods (Castanon,

2007). At the moment, extensive efforts on finding alternatives to antibiotic growth promoters are under way. Probiotics, prebiotics, organic acids or medicinal plants have been suggested as alternatives for antibiotic growth promoters (Dibner and Richards, 2005). Among the feed additives used for poultry and food animals, medicinal plants have drawn higher attention; however, this attention is not solely limited to recent years. A review of the past events suggests that even ancient communities used medicinal plants to treat diseases of poultry and other animal. On the other hand, because medicinal plants are readily available than other feed additives, they are more highlighted. At the moment, studies concerning the influence of medicinal plants in broiler chickens suggests that these substances could be a viable substitute for antibiotic growth promoters (Diaz-Sanchez et al. 2015).

In this regard, Demir et al. (2003) reported that chickens fed thyme and garlic, had better growth performance than those receiving antibiotic growth promoters. It seems that because of their antibiotic properties, some medicinal plants could improve growth performance and health status of birds. On the other hand, these plants might exert multiple effects due to containing numerous effective compounds (Yang et al. 2009). In broader sense, medicinal plants have been proposed as alternatives to antibiotic growth promoters.

In the present research, two medicinal plants (i.e. peppermint and aloe vera) were used. Previous studies have suggested that these two plants have antibiotic properties. Peppermint (Mentha piperita) is planted in Europe, North America, Canada and other regions. It is highly significant in nutritional and pharmaceutical industries (McKay and Blumberg, 2006). Peppermint has been reported to possess certain anti-oxidant, anti-tumor, anti-allergic, anti-viral and anti-bacterial properties (McKay and Blumberg, 2006). Aloe vera (Aloe barbadensis) is one of the oldest medicinal plants known ever. It grows in both tropical and subtropical regions. Previous reports have shown that aloe vera presumably contains anti-bacterial, anti-viral, anti-fungal, anti-inflammatory and immunomodulatory properties (Christaki and Florou-Paneri, 2010). However, there are few studies concerning the effect of peppermint and aloe vera on performance of broiler chickens. Recently, Mehri et al. (2015) added peppermint powder (Mentha piperita) to feeds of Japanese quail.

In earlier studies, Ghazaghi et al. (2014) added Mentha spicata powder to the feed of the Japanese quail as well. In both studies, the number of Lactobacillus bacteria in the intestine increased significantly, whereas the number of E. coli bacteria decreased. However, the growth performance of Japanese quails did not change in comparison with the control treatment. Studies have also shown that feeding aloe vera could change broiler's intestinal microflora population such that the populations of Lactobacillus and Bifidobacteria bacteria increase while the population of E. coli bacteria decrease (Darabighane et al. 2012; Lin et al. 2005). Jafarzadeh et al. (2015) reported that in Japanese quails, aloe vera gel powder increased the number of Lactobacillus bacteria and reduced the number of E. coli bacteria; how-

ever, the aloe vera gel powder did not significantly improve growth performance of Japanese quails.

Other researchers reported the positive influence of peppermint and aloe vera on growth performance of chickens, however, they used peppermint essential oils in the feed (Emami et al. 2012), ethanolic extract of peppermint in drinking water (Nanekarani et al. 2012) or added aloe vera gel powder (Alemi et al. 2012) or aloe vera gel to the feed (Darabighane et al. 2011) of broiler chickens. The differences in the influence of medicinal plants on growth performance broiler chickens might be due to the type of Peppermint processing product (powder, extract or essential oil), products of aloe vera processing (powder or gel) and their incorporation into the feed or drinking water of broiler chickens. Considering the fact that diet is one of the most significant factors influencing the population of intestinal microflora and because microflora in the digestive system might significantly affect the health and growth performance of birds (Yang et al. 2009; Pan and Yu, 2014) The use of feed additives (e.g. medicinal plants) has the potential to change the microflora population of digestive system and realize maximum growth rate of birds. Therefore, the objective of present study was to compare the effect of dietary peppermint powder, aloe vera gel or a mixture of these two medicinal plants and antibiotic growth promoter on intestinal microflora population and growth performance of broiler chickens.

MATERIALS AND METHODS

Experimental treatments, diets and management

In the present study, 375 one-day old male broiler chickens (Ross 308) were used on a completely randomized design with 5 treatments, 5 replicates, each consisting of 15 broilers. The experimental treatments included the control diet (basal diet with no additive), a treatment that received basal diet + 10 g/kg DPL, basal diet + 10 g/kg AVG, basal diet + 5 g/kg DPL + 5 g/kg AVG and basal diet + 10 ppm virginiamycin. The broiler rations were formulated based on nutrient requirements of the Ross 308, using UFFDA software for starter (0-10 days), grower (11-24 days) and finisher (25-42 days) phases. Table 1 shows the ingredient and chemical composition of the starter, grower and finisher experimental diets.

The vaccination schedule was based on veterinarian's recommendation. The broiler chickens were fed ad libitum and they received a lighting regimen of 23 h light: 1 h darkness. The initial temperature was 32 °C which gradually reduced according to breeding standards. Control parameters, such as temperature, humidity, light and ventilation, were the same for all treatments.

Table 1 Ingredients and chemical composition of the experimental basal diets

Ingredients	Starter (1-10 d)	Grower (11-24 d)	Finisher (25-42 d)
Corn (g/kg)	560.8	596.8	654.1
Soybean meal (g/kg)	325.2	312.4	259.3
Corn gluten meal (g/kg)	35.9	0	0
Soybean oil (g/kg)	27.2	46.7	43.9
Dicalcium phosphate (g/kg)	21.9	19.4	18
Calcium carbonate (g/kg)	13.3	10.8	10.6
Sodium bicarbonate (g/kg)	1.4	0.8	2.3
Salt (g/kg)	2.2	2.7	1.9
Vitamin premix[1] (g/kg)	2.5	2.5	2.5
Mineral premix[2] (g/kg)	2.5	2.5	2.5
L-lysine HCl (g/kg)	3.3	1.9	1.8
DL-methionine (g/kg)	3	2.9	2.5
L-threonine (g/kg)	0.8	0.6	0.6
Calculated composition			
Metabolizable energy (kcal/kg)	2850	3000	3100
Crude protein (%)	25.15	22.10	19.60
Ca (%)	0.16	0.98	0.94
Available phosphorous (%)	0.49	0.47	0.43
Methionine + cysteine (%)	1.08	0.9	0.8
Lysine (%)	1.40	1.21	1.07

[1] Vitamin premix provided per kilogram of diet: vitamin A: 10000 IU; vitamin D_3: 5000 IU; vitamin E: 50 IU; vitamin K_3: 3 mg; Thiamine: 3 mg; Riboflavin: 9 mg; Nicotinic acid: 50 mg; Pantothenic acid (D-calcium pantothenate): 15 mg; vitamin B_6: 4 mg; D-biotin: 0.1 mg; Folic acid: 2 mg; vitamin B_{12}: 0.02 mg and Choline (choline chloride): 1000 mg.
[2] Mineral premix provided per kilogram of diet: iron ($FeSO_4 \cdot 7H_2O$): 55 mg; Iodine (Ca $(IO_3)_2$): 1.3 mg; Manganese ($MnSO_4 \cdot H_2O$): 120 mg; Zinc (ZnO): 100 mg; Copper ($CuSO_4 \cdot 5H_2O$): 16 mg and Selenium (Na_2SeO_3): 0.3 mg.

Preparation of medicinal plants

An essential amount of aloe vera leaves was supplied from a local farm and after washing the leaves with distilled water, they were cut by scalpel blade to extract the inner gel. The gel was turned into a homogeneous liquid by mixing device (Mwale et al. 2006).

The liquid was then mixed with the experimental feed. peppermint was supplied from a peppermint farm and after washing with distilled water, the peppermint leaves were air-dried in a dark room with ambient temperature (for 5 days at 28 °C and 40% of relative humidity) (Ghazaghi et al. 2014).

After drying, the leaves were powdered by a milling device and added to the broiler feed.

Measurement of growth performance parameters

The growth performance of broiler chickens comprising body weight gain (BWG), feed intake (FI) and feed conversion ratio (FCR) for starter, grower and finisher periods and for the entire experimentation period were measured. During these periods, the number and weights of mortalities in broiler chickens were recorded to adjust the growth parameters (Darabighane et al. 2011).

In order to measure the dressing percentage, 2 birds were randomly selected out of each replicate during the 42nd day and sacrificed.

Measurement of ileum microflora

In 42nd day, two birds from each replicate were randomly selected and sacrificed by cervical dislocation. In order to determine the numbers of Lactobacillus and E. coli bacteria, one gram of ileal content of broiler chickens was sampled. The numbers of Lactobacillus and E. coli bacteria were measured according to the method of Salim et al. (2013).

Statistical analysis

The experimental results were analyzed using ANOVA and the SAS (2002). The comparison of means was done through Duncan's multiple range test at the level of 0.05.

RESULTS AND DISCUSSION

Ileum microflora

The numbers of Lactobacillus and E. coli bacteria colonies are shown in Table 2.

Table 2 Effect of ppeppermint, aloe vera and antibiotic on ileum microflora counts ($\log_{10}$ CFU/g) of broiler chicks

Type of bacteria	Dietary treatments					SEM
	CON	DPL	AVG	DPL + AVG	VM	
Lactobacillus	5.30^{bc}	5.70^{b}	5.52^{b}	6.35^{a}	4.92^{c}	0.23
E. coli	6.45^{a}	5.32^{b}	5.27^{b}	5.60^{b}	5.35^{b}	0.21

CON: control; DPL: dry peppermint leaves; AVG: aloe vera gel; DPL+AVG: mixture of DPL + AVG and VM: virginiamycin.
The means within the same row with at least one common letter, do not have significant difference (P>0.05).
SEM: standard error of the means.

The results showed that the *Lactobacilli* population was increased in birds fed a combination of peppermint-aloe vera (DPL+AVG). The *Lactobacilli* population was not differing between broilers receiving AVG and DPL than the control treatment. The lowest *Lactobacilli* population was observed in the antibiotic fed birds which was not significantly different from the control treatment. The control treatment had the highest *E. coli* bacteria population (P<0.05). There was no significant difference in *E. coli* bacteria population between birds fed antibiotic and those receiving the medicinal plants (i.e. DPL, AVG and DPL+AVG). However, the least number of *E. coli* was associated with the birds fed diets containing AVG.

Growth performance
The results of FI during the starter, grower and finisher periods as well as for the entire experimentation period are shown in Table 3. In the starter period, the antibiotic receiving birds had higher FI (P<0.05) than other treatments. Among those treatments which had received medicinal plants (i.e. DPL, AVG and DPL+AVG) the treatment receiving the mixture of DPL + AVG recorded higher FI. During the grower period, the highest and lowest FI were recorded in antibiotic receiving birds and the control birds, respectively (P<0.05). However, no significant difference was observed in FI of birds fed medicinal plants. In the finisher period, as well as the entire experiment period, the highest and lowest FI were observed in antibiotic fed and control birds, respectively (P<0.05). Among experimental treatments receiving medicinal plants, the FI of AVG treatment was insignificantly different from the other two treatments (DPL and DPL+AVG).

The results for BWG during starter, grower and finisher as well as whole the experiment period are shown in Table 4. In the starter period, higher BWG was recorded in antibiotic receiving birds (P<0.05), which was significantly superior to other treatments with exception of the birds which received a mixture of DPL + AVG. In the starter period, the control treatment had the lowest BWG (P<0.05). During the grower period, antibiotic receiving birds had higher BWG in comparison with other treatments (P<0.05). The broilers fed medicinal plants (i.e. DPL, AVG and DPL+AVG) showed a higher BWG than the control treatment (P<0.05). In the finisher period, no significant difference was observed in BWG of the experimental treatments

(P>0.05). The control treatment had lower BWG in comparison with other treatments (P<0.05).

During the total experiment period, the highest BWG was observed in antibiotic fed birds (P<0.05), but the difference was not significant from the AVG receiving birds. No significant differences were observed in BWG of birds fed the DPL + AVG or AVG. The least BWG for the entire rearing period was observed in the control treatment (P<0.05).

The results of FCR of starter and grower, finisher as well as the entire experimentation period are shown in Table 5. In the starter period, the lowest FCR was observed in birds receiving medicinal plants compared to the antibiotic fed or control treatment (P<0.05). During the grower period, differences in FCR between the control treatment and the treatments fed medicinal plants were not significant. However, birds in the antibiotic treatment had higher FCR (P<0.05). No significant difference was observed for FCR of experimental treatment during the finisher phase. In the entire experimental period, the highest FCR was observed in antibiotic-fed birds (P<0.05). Although lower FCR was calculated for AVG fed birds, but the differences in the bird fed the DPL + AVG and the control birds were not significant (P>0.05).

The dressing percentage of experimental birds at 42 day of age are shown in Table 6. Higher dressing percentage was recorded in antibiotic receiving birds than from the birds fed DPL and those in the control treatment (P<0.05). The birds which received medicinal plants (DPL, AVG and DPL+AVG) were not different in dressing percentage, but were significantly higher than the control treatment (P<0.05).

As a part of digestive ecosystem, microflora of digestive system exerts significant influence on health and growth performance of birds. Therefore, creating a healthy intestinal environment and increase of useful bacteria through antibiotic alternatives could significantly improve growth performance and increase immune competence (Huyghebaert et al. 2011).

The results of present study suggest that consuming peppermint, aloe vera and a mixture of peppermint-aloe vera could increase the number of *Lactobacillus* bacteria and decrease *E. coli* bacteria. In other words, increase in the number of *Lactobacillus* bacteria leads to reduced pH and limited growth of *E. coli* bacteria. The results of present experiment match those of previous works.

Table 3 Effect of peppermint, aloe vera and antibiotic on feed intake (g/bird) of broilers

Growth phase	Dietary treatments					SEM
	CON	DPL	AVG	DPL + AVG	VM	
Starter (1-10 d)	245.30[c]	247.20[c]	249.12[c]	257.50[b]	277.85[a]	5.98
Grower (11-24 d)	1146.42[c]	1180.92[b]	1192.37[b]	1189.75[b]	1232.45[a]	13.74
Finisher (25-42 d)	2672.50[d]	2776.87[c]	2783.87[bc]	2790[b]	2815.90[a]	24.73
The entire period	4064.22[d]	4205[c]	4225.37[bc]	4237.25[b]	4326.2[a]	42.27

CON: control; DPL: dry peppermint leaves; AVG: aloe vera gel; DPL+AVG: mixture of DPL + AVG and VM: virginiamycin.
The means within the same row with at least one common letter, do not have significant difference (P>0.05).
SEM: standard error of the means.

Table 4 Effect of peppermint, aloe vera and antibiotic on body weight gain (g/bird) of broilers

Growth phase	Dietary treatments					SEM
	CON	DPL	AVG	DPL + AVG	VM	
Starter (1-10 d)	200.25[d]	222.07[c]	226.07[bc]	228.50[ab]	232.77[a]	5.69
Grower (11-24 d)	727.57[d]	738.85[c]	748.87[b]	747.12[b]	755.87[a]	8.84
Finisher (25-42 d)	1401.25[b]	1452.37[a]	1460.30[a]	1454.25[a]	1462.50[a]	11.37
The entire period	2329.07[d]	2413.30[c]	2435.25[ab]	2429.87[bc]	2451.15[a]	21.52

CON: control; DPL: dry peppermint leaves; AVG: aloe vera gel; DPL+AVG: mixture of DPL + AVG and VM: virginiamycin.
The means within the same row with at least one common letter, do not have significant difference (P>0.05).
SEM: standard error of the means.

Table 5 Effect of peppermint, aloe vera and antibiotic on feed conversion ratio (FCR) of broilers

Growth phase	Dietary treatments					SEM
	CON	DPL	AVG	DPL + AVG	VM	
Starter (1-10 d)	1.225[a]	1.113[c]	1.102[c]	1.126[c]	1.194[b]	0.065
Grower (11-24 d)	1.575[b]	1.598[b]	1.592[b]	1.592[b]	1.630[a]	0.090
Finisher (25-42 d)	1.907	1.912	1.906	1.918	1.925	0.073
Total period	1.745[b]	1.742[b]	1.735[b]	1.743[b]	1.765[a]	0.071

CON: control; DPL: dry peppermint leaves; AVG: aloe vera gel; DPL+AVG: mixture of DPL + AVG and VM: virginiamycin.
The means within the same row with at least one common letter, do not have significant difference (P>0.05).
SEM: standard error of the means.

Table 6 Effect of peppermint, aloe vera and antibiotic on dressing percentage of broilers

Dietary treatments	CON	DPL	AVG	DPL + AVG	VM	SEM
Dressing percentage	68.42[c]	69.95[b]	71.25[ab]	71.27[ab]	72.10[a]	0.644

CON: control; DPL: dry peppermint leaves; AVG: aloe vera gel; DPL+AVG: mixture of DPL + AVG and VM: virginiamycin.
The means within the same row with at least one common letter, do not have significant difference (P>0.05).
SEM: standard error of the means.

Mehri *et al.* (2015) reported that adding peppermint powder (10, 20, 30 and 40 g/kg) to feed of Japanese quails could decrease the number of *E. coli* bacteria and increase the number of lactic acid bacteria.

In a similar pattern, Ghazaghi *et al.* (2014) added 3 percent of *Mentha spicata* powder to diet of Japanese quails. The antibacterial effects of peppermint essential oil are associated with menthol (Iscan *et al.* 2002) and antibacterial effects of *Mentha spicata* essential oil is related to carvone. The underlying mechanism of antibacterial property of main elements of those essences which contain Menthol and / or carvone is associated with their hydrophobic property and plasma membranous walls of microbes. Increased amount of some specific ions on or within plasma membrane exerts extensive influence on the driving force of protons, level of intercellular ATP and general activity of microbial cells. The phenol compounds not only invade the cytoplasmic membrane but also remove permeability of the membranes and releases the main intracellular elements.

Consequently, they could eliminate electron transfer, absorption of nutritional materials and synthesis of nucleic acid (Mehri *et al.* 2015).

In order to examine the effect of aloe vera on intestinal microflora, Darabighane *et al.* (2012) added aloe vera gel (1.5, 2 and 2.5 percent) to the feed of broiler chickens. Similarly, Lin *et al.* (2005) added aloe vera gel (0.1 percent) to the feed of broiler chickens. Both of these works found out that adding aloe vera increased the number of *Lactobacillus* bacteria and reduced the number of *E. coli* bacteria. Jafarzadeh *et al.* (2015) found out that adding the aloe vera gel powder (0.1, 0.2 and 0.3 percent) to the feed of Japanese quails could increase the number of *Lactobacillus* bacteria and reduce the number of *E. coli* bacteria so that the treatment which received 0.3 percent of aloe vera gel powder had significant increase in the number of *Lactobacillus* bacteria. This improved intestinal microflora population in aloe vera receiving broiler chickens could be due to presence of polysaccharides in the aloe vera.

One of these polysaccharides is Acemannan which is believed to have anti-bacterial characteristics (Christaki and Florou-Paneri, 2010).

In this experiment, the treatment receiving the DPL + AVG significantly increased the number of *Lactobacillus* bacteria compared with other treatments. This could be related to the synergistic effect of peppermint and aloe vera on its antibacterial property. The medicinal plants have different complicated compounds which assign these plants multiple characteristics. When some medicinal plants are consumed together, the synergic effect could reveal more beneficial effects.

In regard to significance of intestinal microflora in health and growth performance of the birds, one should pay attention to the role of intestinal bacterial population on increased potential of immune system of the body. The bacteria which produce lactic acid could contribute to production of antibodies and phagocytic activity against pathogens in intestine and other tissues. Therefore, increased resistance against pathogens could contribute to higher growth of broiler chickens. Another noteworthy point in regard to intestinal microflora population is that intestinal microflora could change intestinal morphology (Pan and Yu, 2014).

Due to anti-microbial properties, medicinal plants can positively influence intestinal morphology and improve absorption of nutritional materials as well as improved growth performance of broiler chickens.

In alignment with the results on FI, in starter, grower and finisher periods as well as for the entire experiment period, the antibiotic receiving treatment had higher FI and BWG than the other experimental treatments. In this regard, Ramiah *et al.* (2014) reported that broiler chickens that had received virginiamycin antibiotic had significantly higher FI and BWG than control treatment. Miles *et al.* (1984) and Miles *et al.* (2006) also reported that adding virginiamycin antibiotic to the feed of broiler chickens increases their body weight. Emami *et al.* (2012) conducted a test on the effects of peppermint essential oil on broiler chickens when compared with administering virginiamycin antibiotic and prebiotics. They suggested that those broiler chickens which had received virginiamycin in their feed had higher FI and body weight than other treatments at 42nd day. The body weight increase of broiler chickens could be associated with higher intake of feed and better absorption of nutritional materials (Belay and Teeter, 1994). Of course, morphological change of intestine (ratio of height of villus to depth of crypt) could be another reason of positive influence of virginiamycin on growth performance of these broiler chickens (Emami *et al.* 2012).

Among the experimental treatments received medicinal plants (DPL, AVG and a DPL+AVG), the highest FI was observed in birds fed AVG or the mixture of DPL + AVG.

The results on BWG also suggest that the BWG of treatments receiving AVG and DPL + AVG was higher than those received DPL and the control treatment. Although DPL increased FI and BWG than the control treatment, but the values of these variables were lower than those treatments which had received AVG, DPL + AVG and the antibiotic. Ocak *et al.* (2008) added 0.2 percent of peppermint powder to feed, Gurbuz and Ismael (2016) added 1.5 percent of peppermint into feed, Dosti *et al.* (2014) put 15 g/kg of peppermint into the feed, Nanekarani *et al.* (2012) added 0.3 percent of ethanolic extract of peppermint in drinking water and Emami *et al.* (2012) poured 200 mg per kilogram of peppermint to the feed; all of them reported that peppermint contributes to growth performance of broiler chickens. In contrast, some studies offer reports on lack of influence of peppermint on growth performance. Mehri *et al.* (2015) added peppermint powder (10, 20, 30 and 40 g/kg) to the rations of Japanese quails. In the same vein, Khaligh Gharetappe *et al.* (2015) adding 0.4 percent of peppermint leaves powder to the feed of broiler chicken had no significant influence on growth performance.

In this experiment, broiler chickens fed the AVG had higher growth performance than the control treatment. Similar finding was reported in previous studies where aloe vera powder or gel was added into feed or aloe vera gel was added into drinking water. Alemi *et al.* (2012) added 0.75 and 1 percent of aloe vera gel powder to the feed of broiler chickens and observed a better growth performance for treatments containing the supplements than the control. In addition, Mmereole (2011) added 1 percent of aloe vera leaves powder to the feed of chickens and reported higher body weight than the control treatment. The addition of 1.5, 2 and 2.5 percent of aloe vera gel to the feed of broiler chickens could increase FI, BWG and FCR in comparison with control treatment (Darabighane *et al.* 2011). In addition, adding aloe vera gel (1.8 percent) to the drinking water of broiler chickens could enhance body weight and FCR in comparison with control treatment (Hassanbeigy-Lakeh *et al.* 2012).

Despite increased FI, BWG and dressing percentage of those treatments which received medicinal plants, their FCR was not different from the control treatment. This finding was also observed for grower, finisher and total experiment period. It seems that some medicinal plants might change the bacterial population of intestine and health status of digestive system because of their antibacterial properties. As a result, the growth performance of these broiler chickens improves. In regard to tests of influence of medicinal plants on poultry and animal, one should note that the environmental factors such as weather, soil composition, and growth stage of a plant could affect the amount of effective materials of plants. In addition, method of preparing the powder, extract or essential oil or herbal parts could influence the effective materials. In conducted tests of influence of medicinal plants on poultry, it is observed that researchers usually use medicinal plants as complement for the mixture or in combination with drinking water. This factor could also influence the growth performance, immune system or intestinal microflora.

CONCLUSION

Based on the findings of the present experiment, one could conclude that adding the powdered leaves of peppermint, aloe vera gel or a mixture of peppermint-aloe vera to the feed of broiler chickens could improve ileum microflora (increase the number of *Lactobacillus* bacteria, reduce the number of *E. coli* bacteria) and improve growth performance of broiler chickens. Despite of the fact that antibiotic receiving treatment had higher growth performance than the birds fed medicinal plants, the application of a mixture of peppermint-aloe vera (5 g/kg dry peppermint leaves and 5 g/kg aloe vera gel) as feed additive could be a suitable alternative for antibiotic growth promoters.

ACKNOWLEDGEMENT

The authors would like to thank an anonymous reviewer for his/her thorough review and highly appreciate the comments, which significantly contributed to improving the quality of the article.

REFERENCES

Alemi F., Mahdavi A., Ghazvinian K., Ghaderi M. and Darabighane B. (2012). The effects of different levels of aloe vera gel powder on antibody titer against Newcastle disease virus and performance in broilers. Pp. 47 in Proc. Int. Poult. Sci. Forum. Georgia World Congress Center, Atlanta, Georgia.

Belay T. and Teeter R.G. (1994). Virginiamycin effects on performance and saleable carcass of broilers. *Appl. Poult. Res.* **3**, 111-116.

Castanon J. (2007). History of the use of antibiotic as growth promoters in European poultry feeds. *Poult. Sci.* **86**, 2466-2471.

Christaki E.V. and Florou-Paneri P.C. (2010). Aloe vera: a plant for many uses. *J. Food Agric. Environ.* **8**, 245-249.

Darabighane B., Zarei A., Shahneh A.Z. and Mahdavi A. (2011). Effects of different levels of aloe vera gel as an alternative to antibiotic on performance and ileum morphology in broilers. *Italian J. Anim. Sci.* **10**, 36.

Darabighane B., Zarei A. and Shahneh A.Z. (2012). The effects of different levels of aloe vera gel on ileum microflora population and immune response in broilers: a comparison to antibiotic effects. *J. Appl. Anim. Res.* **40**, 31-36.

Demir E., Sarica Ş., Özcan M. and Sui Mez M. (2003). The use of natural feed additives as alternatives for an antibiotic growth promoter in broiler diets. *Br. Poult. Sci.* **44**, 44-45.

Diaz-Sanchez S., D'Souza D., Biswas D. and Hanning I. (2015). Botanical alternatives to antibiotics for use in organic poultry production. *Poult. Sci.* **94(6)**, 1419-1430.

Dibner J. and Richards J. (2005). Antibiotic growth promoters in agriculture: history and mode of action. *Poult. Sci.* **84**, 634-643.

Dosti A., Taherpour K., Nasr J. and Ghasemi H. (2014). The comparative effects of dietary peppermint (*Mentha pipperita*), probiotic and prebiotic on growth performance and serum biochemical parameters of broilers performance. *Anim. Sci. J. (Pajouhesh and Sazandegi).* **101**, 91-100.

Emami N.K., Samie A., Rahmani H. and Ruiz-Feria C. (2012). The effect of peppermint essential oil and fructooligosaccharides, as alternatives to virginiamycin, on growth performance, digestibility, gut morphology and immune response of male broilers. *Anim. Feed Sci. Technol.* **175**, 57-64.

Ghazaghi M., Mehri M. and Bagherzadeh-Kasmani F. (2014). Effects of dietary *Mentha spicata* on performance, blood metabolites, meat quality and microbial ecosystem of small intestine in growing Japanese quail. *Anim. Feed Sci. Technol.* **194**, 89-98.

Gurbuz Y. and Ismael I. (2016). Effect of peppermint and basil as feed additive on broiler performance and carcass characteristics. *Iranian J. Appl. Anim. Sci.* **6**, 149-156.

Hassanbeigy-Lakeh Z., Roustaee Ali-Mehr M. and Haghighian-Roudsari M. (2012). Effect of aloe gel on broiler performance. Pp. 973-977 in Proc. 5th Iranian Congr. Anim. Sci. Isfahan, Iran.

Huyghebaert G., Ducatelle R. and Van Immerseel F. (2011). An update on alternatives to antimicrobial growth promoters for broilers. *Vet. J.* **187**, 182-188.

Iscan G., KIrimer N., Kürkcüoglu M.N., Baser H.C. and DEMIrci F. (2002). Antimicrobial screening of *Mentha piperita* essential oils. *J. Agric. Food Chem.* **50**, 3943-3946.

Jafarzadeh A., Darmani-kuhi H., Ghavihossein-zadeh N. and Roostaei Ali-mehr M. (2015). Effect of different levels of aloe vera gel powder on performance, intestinal microflora and gastrointestinal organs in Japanese quills (*Coturnix japonica*). *Anim. Sci. J. (Pajouhesh and Sazandegi).* **106**, 231-242.

Khaligh Gharetappe F., Hassanabadi A., Semnaninezhad H. and Nassiry M. (2015). The Effect of dietary tarragon (*Artemisia dracunculus*) and peppermint (*Mentha piperita*) leaveson growth performance and antibody response of broiler chickens. *Iranian J. Appl. Anim. Sci.* **5**, 403-409.

Lin J., Zhang F.Y., Xu Y., Ting Z.X. and Po Y. (2005). Effects of gel, polysaccharide and acemannan from aloe vera on broiler gut flora, microvilli density, immune function and growth performance. *Chinese J. Vet. Sci.* **25**, 668-671.

McKay D.L. and Blumberg J.B. (2006). A review of the bioactivity and potential health benefits of peppermint tea (*Mentha piperita*). *Phytother. Res.* **20**, 619-633.

Mehri M., Sabaghi V. and Bagherzadeh-Kasmani F. (2015). Peppermint (*Mentha piperita*) in growing Japanese quails diet: performance, carcass attributes, morphology and microbial populations of intestine. *Anim. Feed Sci. Technol.* **207**, 104-111.

Miles R., Butcher G., Henry P. and Littell R. (2006). Effect of antibiotic growth promoters on broiler performance, intestinal growth parameters, and quantitative morphology. *Poult. Sci.* **85**, 476-485.

Miles R., Janky D. and Harms R. (1984). Virginiamycin and broiler performance. *Poult. Sci.* **63**, 1218-1221.

Mmereole F. (2011). Evaluation of the dietary inclusion of aloe vera as an alternative to antibiotic growth promoter in broiler production. *Pakistan J. Nutr.* **10**, 1-5.

Mwale M., Bhebhe E., Chimonyo M. and Halimani T. (2006). The *in vitro* studies on the effect of aloe vera and aloe spicata on the control of coccidiosis in chickens. Int. J. Appl. Res. Vet. Med. **4**, 128-133.

Nanekarani S., Goodarzi M., Heidari M. and Landy N. (2012). Efficiency of ethanolic extract of peppermint (*Mentha piperita*) as an antibiotic growth promoter substitution on performance, and carcass characteristics in broiler chickens. *Asian Pac. J. Trop. Biomed.* **2**, 1611-1614.

Ocak N., Erener G., Burak Ak F., Sungu M., Altop A. and Ozmen A. (2008). Performance of broilers fed diets supplemented with dry peppermint (*Mentha piperita*) or thyme (*Thymus vulgaris)* leaves as growth promoter source. *Czech J. Anim. Sci.* **53**, 169.

Pan D. and Yu Z. (2014). Intestinal microbiome of poultry and its interaction with host and diet. *Gut Microbes.* **5**, 108-119.

Ramiah S.K., Zulkifli I., Rahim N.A.A., Ebrahimi M. and Meng G.Y. (2014). Effects of two herbal extracts and virginiamycin supplementation on growth performance, intestinal microflora population and fatty acid composition in broiler chickens. *Asian-Australas J. Anim. Sci.* **27**, 375-382.

Salim H., Kang H., Akter N., Kim D., Kim J., Kim M., Na J., Jong H., Choi H. and Suh O. (2013). Supplementation of direct-fed microbials as an alternative to antibiotic on growth performance, immune response, cecal microbial population and ileal morphology of broiler chickens. *Poult. Sci.* **92**, 2084-2090.

SAS Institute. (2002). SAS®/STAT Software, Release 9.1. SAS Institute, Inc., Cary, NC. USA.

Yang Y., Iji P. and Choct M. (2009). Dietary modulation of gut microflora in broiler chickens: a review of the role of six kinds of alternatives to in-feed antibiotics. *Worlds Poult. Sci. J.* **65**, 97-114.

10

Effects of Mega Doses of Phytase on Growth Performance, Bone Status and Nutrient Excretion of Broilers Fed Diets Containing High Levels of Rice Bran

K.G.S.C. Katukurunda[1*], N.S.B.M. Atapattu[2] and P.W.A. Perera[2]

[1] Department of Food Science and Technology, Faculty of Applied Science, University of Sri Jayewardenepura, Gangodawila, Nugegoda, Sri Lanka
[2] Department of Animal Science, Faculty of Agriculture, University of Ruhuna, Mapalana, Kamburupitiya, Sri Lanka

*Correspondence E-mail: kgsckatukurunda@sci.sjp.ac.lk

ABSTRACT

Phytate in poultry rations containing rice bran (RB) critically reduce poultry performance while increasing N and P excretion. The objective of this study was to determine whether the anti-nutritive problems associated with higher inclusions of rice bran in poultry rations (up to 40%) could be mitigated with mega doses of phytase. Twenty days old male broiler chicks (n=180) in 60 floor pens were fed on 10 dietary combinations of a completely randomized factorial design (2×5). Two dietary rice bran (RB) levels (20 or 40%) and five levels (0, 1000, 2000, 3000 and 4000 FTU/kg diet) of phytase (Natuphos 500) were main factors. Cage-wise daily feed/water intakes and body weights on day 28, 35 and 42 were determined. Weekly and total weight gain and feed conversion ratio were determined. Cr_2O_3 mixed diets were fed from day 35 to determine illeal digestibility of crude protein, phosphorus (P) and dry matter. On day 42, following a 12-hour fast, two birds from each pen were humanely slaughtered to determine visceral organ weights. Fat free tibia ash contents and latency-to-lie test done on day 28, 35 and 42 were used as bone parameters. Growth performance parameters were not enhanced significantly due to mega doses of phytase. Negative effects like body weight reductions affected latency to lie time increments and further confirmed by insignificancy of tibia ash increments. Phytase significantly improved the crude protein digestibility. The optimum levels of phytase for the best crude protein digestibility with 20% and 40% dietary rice bran were 3000 and 4000 FTU/kg, respectively. Digestibility values of P and dry matter also affected. Supplementation of mega doses of phytase improved illeal crude protein digestibilities but not growth performances and bone status. Mega doses of phytase did not mitigate the adverse effects of 40% rice bran included broiler diets.

KEY WORDS broilers, crude protein, mega doses, phytase, rice bran.

INTRODUCTION

Between 60% and 70% of the total production cost of poultry is associated with the cost of feedstuffs and thus cost of poultry feeds is of great economic importance (Attia *et al.* 2012). Cereals and their by products are the main energy source in commercial poultry diets. During recent years price of cereals, particularly maize increased sharply since its use for biofuel production. Consequently, attention should be paid to increase the use of cereal by products such as rice bran in poultry feeding to reduce the cost of feed production. Rice bran is widely used in poultry rations mainly in Asian region. It has been widely accepted that the inclusion of rice bran in excess of 20% in the diets results in significant reductions in growth performance while increasing the excretion of N and P (Arabi, 2013; Rutherfurd *et al.* 2002; Cowieson *et al.* 2004; Wu *et al.* 2004). Phytic acid present in rice bran complexes with minerals such as calcium, magnesium, iron and zinc and reduces the availability of P and other minerals which complex with it (Reddy *et al.* 1989). Phytic acid complexes with proteins, amino acids and proteolytic enzymes making the aminoacid less available (Selle *et al.* 2000). Higher levels of rice bran in diets reported to reduce protein and dry matter digestibility as well (Piyaratne *et al.* 2008; Samli *et al.* 2006; Gallinger *et al.* 2004).

Since both phytates and fiber acts as critical anti nutrients, supplementation of broiler finisher diets containing higher levels of rice bran with exogenous phytase is hypothesized to remove adverse effects of phytates. Since diets formulated to have higher levels of rice bran contain relatively a higher amount of phytate, higher levels of exogenous phytase than what is normally used in industry may be needed to hydrolyze all or much of the phytates. A number of studies (Sooncharernying, 1991; Wu *et al.* 2004; Edwards *et al.* 1988; Attia *et al.* 2012) have evaluated to impacts of mega doses of phytase in poultry, but none has used it in rice bran based diets. The objective of this study was to determine whether the anti-nutritive problems which associated with higher inclusions of rice bran in poultry rations up to 40% could be mitigated with mega doses of exogenous phytase.

MATERIALS AND METHODS

Birds and cage arrangement
A total of 200 broiler chicks (strain Cobb) were purchased from a local hatchery (Kekanadura, Sri Lanka) and brooded and managed according to the standard management practices. Chicks were fed on a commercial broiler starter diet (Prima Sri Lanka) until day 20. On day 20, 180 birds were weighed and allocated into 60 floor pens (75 cm×75 cm×75 cm) so that between cage weight variation was minimum.

Ration composition
Pens were randomly allocated into 10 dietary combinations of a completely randomized design, in 2 × 5 factorial arrangement. The main factors of the experiment were two dietary rice bran levels (20 or 40%) and five levels of phytase (0, 1000, 2000, 3000 and 4000 FTU/kg diet). Natuphos 500 (Baden Aniline and Soda Factory Corporation, Germany) used as the phytase source. The ingredient compositions of the diets are shown in Table 1. The maize meal level reduced with increasing dietary RB inclusion level. Fresh rice bran was collected from a local mill and immediately transported to the laboratory. Rice bran was heat stabilized (100 °C for 30 minutes). Rations were formulated using trial and error method on EXCEL and verified by using the software CUFA TOTALFEED.

Table 1 Ingredient compositions and calculated CP, CF, P and ME contents of two rations containing either 20 or 40% rice bran

Ingredient %	Level of dietary rice bran %	
	20	40
Yellow maize meal	44.4	23.905
Rice bran	20	40
Soya oil meal	23.8	24.45
Coconut oil	4.42	5.8
Fish meal	4.96	3.225
Dicalcium phosphate	0.94	0.95
Shell grit	0.94	1.14
DL-methionine	0.04	0.03
L-lysine HCl	0	0
Salt	0.25	0.25
Vitamin and mineral premix (local market)	0.25	0.25
Total diet composition	100.000	100.00
Crude protein (%)	20	20
Metabolizable energy (kcal/kg)	3139	3140
Non phytate P (%)	0.35	0.35
Ca (%)	0.9	0.9
Crude fibre (%)	4.9	6.8

Management of birds
Cages were provided with paddy husk litter. Each cage had a feeder and a drinker. Experimental diets and water was given *ad libitum* from day 20-42. Cage-wise daily feed and water intakes were calculated. Water intake was corrected for evaporation losses.

Growth performance parameters
Birds were weighed on day 28, 35 and finally on day 42. Mortality percentage, live weight on day 28, 35, 42, weekly and total weight gain and feed conversion ratio were determined as growth performance parameters. Cage-wise daily feed intakes were calculated as feed offered feed left over.

Determination of bone status
The fat free tibia ash contents and latency-to-lie test were used as bone status determination criteria. Right after killing left tibia of one bird from each cage was removed and frozen for analysis of fat free tibia ash. The latency to lie test was done on 28, 35 and 42 days on one randomly selected bird from each pen, on each day as described by Weeks *et al.* (2002).

Determination of nutrient digestibility values
Indigestible marker assisted illeal crude protein and dry matter digestibility and, the availability, dry matter and the availability of P were determined. All diets contained 2 g of Cr_2O_3/kg as an inert marker.

Chromic oxide mixed diets were fed from day 35-42. One randomly selected bird from each pen was killed on day 41 and illeal contents (extending from Mackel's diverticulum to a point of 40 mm proximal to the ileo-caecocolic junction) were collected immediately after slaughtering. Illeal contents were pooled according to ten dietary treatment combinations. Six randomly drawn samples from each composite were analyzed for Cr, CP and P. Illeal level digestibility/availability values were determined. Chromium, CP and P contents of the feed and illeal samples were determined according to the procedures given by AOAC (1995).

Apparent ileal nutrient digestibility %= [{(Nt/Cr)$_d$ - (Nt/Cr)$_i$} / (Nt/Cr)$_d$] × 100

Where:
(Nt/Cr)$_d$: ratio of nutrient to Cr in the feed.
(Nt/Cr)$_i$: ratio of nutrient to Cr in illeal digesta.

On day 42, following a 12-hour fast, two birds from each cage were humanely slaughtered. The carcasses were then manually eviscerated. Abdominal fat pad (including fat surrounding gizzard, bursa of F×abricius, cloaca and adjacent muscles) was weighed individually and calculated as a percentage of live weight of the bird.

Statistical analysis
The experiment was analyzed as a completely randomized design in 2 × 5 factorial arrangements. Minitab 17 (2013) was used for the analyses. Main effects of rice bran and phytase and interactions were determined. Effects were considered significant when (P<0.05) and regression equations (linear, cubic and sigmoid) were determined for phytase levels. In growth performance data analysis, cage means were used as replicates. In digestibility data analysis, samples taken from each composite (composite illeal samples from each treatment combination) were served as the replicates. In bone ash data analysis each bird was served as replicates. Latency to lie data analysis values of each bird selected for the test (minutes to lie) were served as replicate values.

RESULTS AND DISCUSSION

Growth performance, latency to lie and bone status
Effects of two dietary rice bran levels and five levels of phytase on growth performance, latency to lie and bone status of broiler chicken are shown in Table 2. Performance parameters such as live weight on day 23 and 42, weight gain, feed conversion ratio (FCR) were significantly infe

rior for the birds fed 40% RB compared to those fed 20% RB. Feed intake not significantly influenced by dietary treatments.

In contrast, Deniz et al. (2007) concluded that feed intake was significantly reduced with the increasing level of rice bran in broiler diets. Feed conversion ratio was significantly increased with 40% dietary RB. None of the above mentioned growth parameters was influenced by phytase or RB phytase interaction.

Probably due to lower body weight, birds fed 40% RB showed significantly longer latency to lie time compared to those fed 20% RB on day 28 and 35. Phytase supplementation also resulted in longer latency to lie time. The latency to lie time on day 42 also increased significantly when phytase supplemented diets were fed.

However, RB level had no significant effect on the latency to lie on day 42. Tibia ash content was not significantly affected by RB. Gallinger et al. (2004) showed that the feed conversion ratio and tibia ash were more sensitive than weight gain to detect anti-nutritive factors in rice bran. They have further reported that high concentrations of rice bran (in excess of 20%) resulted in a significant reduction in body weight.

Gallinger et al. (2004) suggest that rice bran should be included in broiler diets at a level between 10 and 20% if strategies are not used to decrease the anti-nutritive activity. The short period of the trial possibly reduced the chances of significant reduction in tibia ash content in the present experiment.

Nutrient digestibility values
Effects of increasing levels of phytase supplementation on CP, P and dry matter digestibility of broiler chicken fed either 20 or 40% dietary rice bran are shown in Table 3. 20% RB fed chicks showed significantly (P<0.05) higher CP digestibility than those fed 40% RB. Meanwhile, phytase supplementation also increased the CP digestibility. Positive effects of supplemental phytase on protein digestibility and illeal % N digestibility have been reported elsewhere (Cowieson et al. 2004; Rutherfurd et al. 2002; Managi et al. 2009). There was a significant cubic relationship between incremental phytase and CP digestibility (Figure 1).

At both 20 and 40% RB levels incremental phytase increased the CP digestibility. However, always, CP digestibility of 40% RB diet was lower than that of 20% diet. At both RB levels, the maximum CP digestibility was reported with 4000 FTU/kg phytase. The relationship between phytase and CP digestibility at 20% RB could best be expressed as a quadratic relationship (Figurre 2), whereas the same for 40% RB was a cubic relationship (Figure 3).

Table 2 Effects of increasing levels of phytase supplementation on growth performance, latency to lie and bone status of broiler chicken fed either 20 or 40% dietary RB

Interaction effects		Live weight on day (g)		Weight gain (g)	Feed intake (g)	FCR	Latency to lie (min) on day			Tibia ash (%)
RB	Phytase (FTU/kg)	23	42				28	35	42	
	0	907	2424	1517	2218	1.46	158	74.8	17.1	40.6
	1000	953	2398	1445	1915	1.35	161	79.0	19.0	44.1
20	2000	947	2307	1360	2459	1.84	169	79.8	21.3	40.0
	3000	937	2157	1219	2152	1.82	168	78.8	17.1	41.8
	4000	885	2411	1525	2183	1.45	175	82.3	18.0	42.2
	0	852	2147	1295	2281	1.77	120	53.8	12.1	45.6
	1000	936	2138	1201	2137	1.83	167	75.0	25.6	43.8
40	2000	914	2008	1093	2242	2.06	150	65.0	18.6	43.5
	3000	877	2155	1278	2297	1.81	144	63.0	20.0	45.9
	4000	911	2205	1294	1944	1.57	144	59.5	25.3	44.3
Main effects										
RB	20	926	2339[a]	1413[a]	1801	1.6[b]	167[a]	79[a]	19	42
	40	898	2131[b]	1232[b]	1711	1.81[a]	144[b]	63[b]	20	45
	0	879	2286	1406	2250[b]	1.62[bc]	142[a]	65	15[c]	43
	1000	945	2268	1323	2026[c]	1.58[bc]	153[a]	76	23[ab]	44
Enzyme	2000	931	2158	1227	2351[a]	1.98[a]	159[a]	73	20[bc]	42
	3000	907	2156	1249	2225[b]	1.82[ab]	153[a]	69	18[bc]	44
	4000	895	2308	1409	2064[c]	1.51[c]	115[b]	71	31[a]	43
Significance level										
SEM		44.0	85	89	155	0.16	8.2	3.8	2.3	3.9
RB		NS	**	**	NS	*	***	***	NS	NS
Phytase		NS	NS	NS	NS	NS	*	*	*	NS
RB × phytase		NS	NS	NS	NS	NS	NS	NS	*	NS

RB: rice bran and FCR: feed conversion ratio.
The means within the same column with at least one common letter, do not have significant difference (P>0.05).
* (P<0.05); ** (P<0.01) and *** (P<0.001).
NS: non significant.
SEM: standard error of the means.

Accordingly, as far as CP digestibility is concerned, 3000 FTU/kg of diet can be recommended as the best phytase level for diets having 20% RB. Use of 4000 FTU is recommended for the diets having 40% RB.

Compared to control, 1000, 2000, 3000 and 4000 FTU phytase/kg increased the CP digestibility 16, 19, 21 and 24% (Figure 3). In contrast, Kornegay (1996), Kornegay (1999) standard response curves for phytase and protein digestibility, the linear increment of N digestibility at 750 and 450 FTU/kg phytase diets were 0.4 and 2.9%. Moreover Namkung and Leeson (1999) reported that phytase significantly increased total N digestibility by 2.5% when fed 1149 FTU/kg phytase included maize soybean diet. Ravindran et al. (2000a); Ravindran et al. (1999) and Ravindran et al. (2000b) reported a 2.7% of N digestibility increase in low phosphorus (P) diet and 1.8% of total N digestibility in adequate P diet with wheat and sorghum. Moreover Ravindran et al. (1999) reported 2.7, 2.8 and 4.7% of total N digestibilities due to of phytase (600 FTU/kg), xylanase (6600 EXU/kg) and both in combination (600 FTU/kg+6600 EXU/kg), respectively. However Zhang et al. (1999) reported that phytase did not increased the digestibilities of amino acids.

In this experiment the mega doses of phytase increased the CP digestibility in chicks by several magnitudes than above studies. The use of relatively poor quality diets having at least 20% RB may be a one reason for better response for phytase. Numerous studies (Arabi, 2013; Rutherfurd et al. 2002; Cowieson et al. 2004; Wu et al. 2004) have shown that phytase increased the amino acids availability and in-turn the performance. And reduce the excretion of N with faeces (Aggrey et al. 2002; Cowieson et al. 2004; Rutherfurd et al. 2002).

However, it needs to be noted that though phytase increased the digestibility values of CP, the growth performance values were not altered significantly. It is hypothesized that phytase mitigated the effects of high RB on protein digestibility but not the other adverse effects.

Rutherfurd et al. (2002) reported that inclusion of phytase in poultry diets with rice bran did not significantly alter the amino acid digestibility, except glycine. Furthermore they concluded that it is possible that the phytate in rice bran plays minor role in the amino acid-matrix interactions that occur within the feedstuff or that the phytate molecules associated with proteins were inaccessible to phytase because of static interference.

Table 3 Effects of increasing levels of phytase supplementation on CP, P and dry matter digestibility of broiler chicken fed either 20 or 40% dietary RB

Treatment factors		Digestibility / availability (%) (%)		
RB	Phytase (FTU/kg)	Crude protein	P	Dry matter
20	0	64.7	78.0	77.3
	1000	76.1	74.2	73.5
	2000	78.2	69.7	69.3
	3000	80.6	81.3	81.2
	4000	82.2	79.2	79.1
40	0	59.2	79.3	79.1
	1000	72.1	76.0	75.8
	2000	74.6	75.9	75.1
	3000	75.4	76.1	75.6
	4000	81.6	79.7	79.8
Mean effects				
RB	20	76	77	76
	40	73	77	77
Enzyme	0	62[d]	79	78
	1000	74[c]	75	75
	2000	76[bc]	73	72
	3000	78[b]	79	78
	4000	82[a]	79	79
Significance level				
SEM		1.3	4.0	4.2
RB		***	NS	NS
Phytase		***	NS	NS
RB × phytase		NS	NS	NS

RB: rice bran.
The means within the same column with at least one common letter, do not have significant difference (P>0.05).
* (P<0.05); ** (P<0.01) and *** (P<0.001).
NS: non significant.
SEM: standard error of the means.

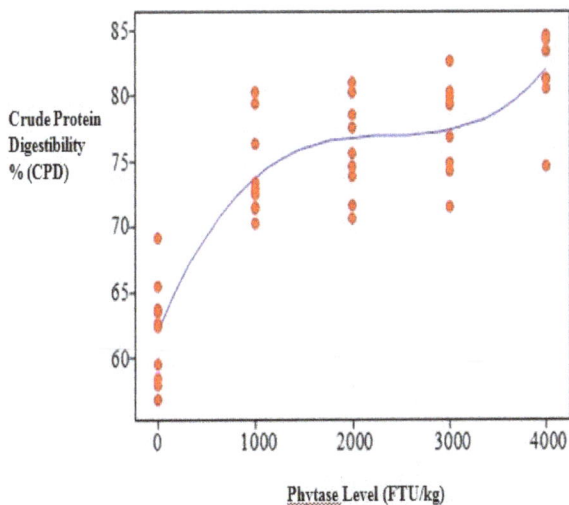

Figure 1 Relationship between incremental phyase level and CP digestibility (CPD, %), R-Sq: 80.4%, R-Sq (adj): 79.1%
$CPD = 62.04 + 0.01834\,Phy - (0.000008\,Phy)^2 + (0.000000\,Phy)^3$

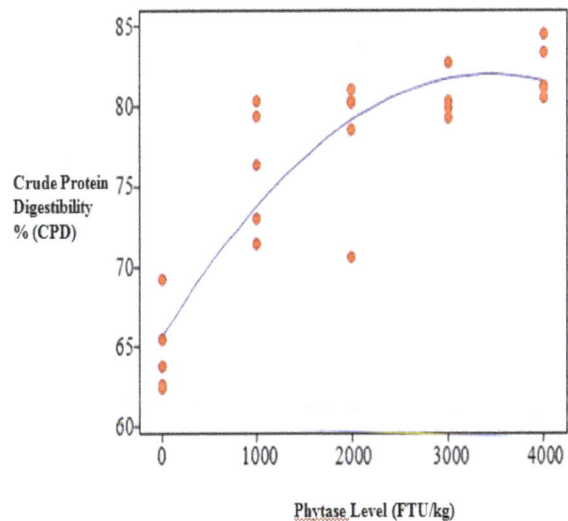

Figure 2 Crude protein digestibility (CPD, %) of 20% rice bran diets fed to broilers for phytase treatments, R-Sq: 79.7%, R-Sq (adj): 77.8%
$CPD = 65.66 + 0.009546\,Phy - (0.000001\,Phy)^2$

Contrary to above conclusions, the results of present experiment suggest that improvements in CP digestibility could be achieved when RB based diets are supplemented with phytase. Furthermore, for maximum CP digestibility higher level of phytase than what industry used is needed.

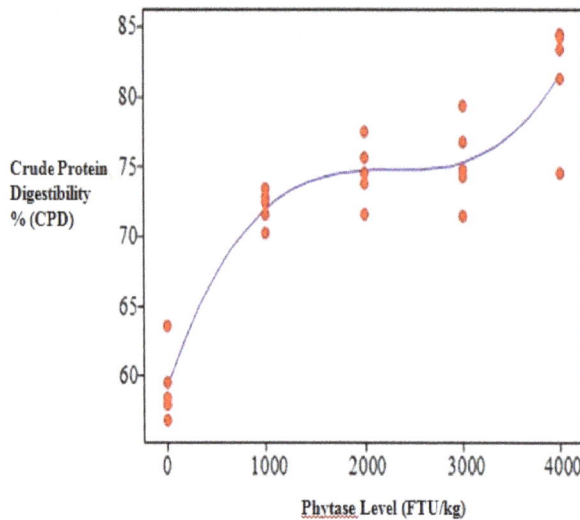

Figure 3 CP digestibility of 40% rice bran diets fed to broilers for phytase treatments, R-Sq: 89.8%, R-Sq (adj): 88.3%
$CPD = 59.28 + 0.02043\,Phy - (0.000009\,Phy)^2 + (0.000000\,Phy)^3$

Diets having higher level of RB needs even higher levels of phytase. Numerous studies (Arabi, 2013; Rutherfurd *et al.* 2002; Cowieson *et al.* 2004; Wu *et al.* 2004) have shown that phytase increased the amino acids availability and in turn the performance and reduce the excretion of N with faeces (Aggrey *et al.* 2002; Cowieson *et al.* 2004; Rutherfurd *et al.* 2002). It is not possible to conclude whether excretion of N was reduced due to phytase since faecal level digestibility was not determined.

However, no response in growth performance suggests that utilization of amino acids might have been low (Cowieson *et al.* 2004). However, Attia *et al.* (2012) have concluded that high fiber diets increased the exogenous losses of nutrients. Therefore, suitability of illeal digestibility measurements is questionable under the present experimental conditions. Phosphorous and dry matter digestibility values were not significantly influenced by RB or phytase levels.

Table 4 Effects of increasing levels of phytase supplementation on relative weights of gizzard, liver, Pancreas and small intestines and relative lengths of small intestines of broiler chicken fed either 20 or 40% dietary RB

Treatment factors		Relative weights (g)				Relative length of small intestine (cm)
RB	Phytase (FTU/kg)	Gizzard	Liver	Pancreas	Small intestine	
20	0	53	55.2	5.8	164	131
	1000	55.2	57.8	5.2	165	122
	2000	55.2	57.6	5.8	167	120
	3000	51	54.4	6.4	160	128
	4000	54.8	51.8	5.6	163	116
40	0	50.6	58.2	6	164	122
	1000	51.4	57.8	5.4	151	122
	2000	54	55.8	6.2	142	111
	3000	57	56.2	5.8	151	118
	4000	54.8	54.4	5.4	169	126
Mean effects						
RB	20	54	55	6	164[a]	123
	40	54	56	6	155[b]	119
Enzyme	0	52	57	6[ab]	164	126
	1000	53	58	5[b]	158	120
	2000	55	57	6[ab]	154	120
	3000	54	55	6[a]	156	121
	4000	55	53	6[ab]	166	119
Significance level						
SEM		1.05	0.8	0.12	1.88	1.22
RB		NS	NS	NS	NS	NS
Phytase		NS	NS	NS	NS	NS
RB × phytase		NS	NS	NS	NS	NS

RB: rice bran.
The means within the same column with at least one common letter, do not have significant difference (P>0.05).
* (P<0.05); ** (P<0.01) and *** (P<0.001).
NS: non significant.
SEM: standard error of the means.

Our findings are contradictory to many studies (Arabi, 2013; Olukosi et al. 2010; Piragozliev et al. 2008; Munir and Maqsood, 2013) who reported increased P availability and bone ash contents in broilers, due to supplemental phytase. Extreme dietary P levels, inorganic undigested P, endogenous losses of P; inbuilt phytase action in the gut (Munir and Maqsood, 2013) may be reasons for the absence of response to phytase within the experimental limits. RB × phytase interaction also not significant in this experiment.

Determination of relative weights of digestive organs

Effects of increasing levels of phytase supplementation relative weights of gizzard, liver, pancreas and small intestines and relative lengths of small intestines of broiler chicken fed either 20 or 40% dietary rice bran are shown in Table 4. Digestive organs parameters did not show any significant influences due to RB or Phytase. However Deniz et al. (2007) concluded that there was an increase in relative size of the digestive organs when rice bran was included in the diet.

Gallinger et al. (2004) have also reported that the broilers fed a diet with 40% rice bran had higher pancreas and intestine weights than control and suggested the presence of other anti-nutritive factors other than phytate. In this experiment, the weights of pancreases and small intestines showed numerically increased by phytase levels and RB, respectively and the statistical significance (P>0.05) were absent. The other parameters such as weights of liver, gizzard and relative length of small intestine were not significant with RB inclusion, phytase and interaction between RB and phytase.

According to Gallinger et al. (2004) the concentrations of rice bran in excess of 20% in the diet produce significant reductions of body weight. Furthermore, feed conversion ratio and tibia ash contents were inferior with diets containing more than 10% of rice bran.

In this experiment, phytase did not show the expected performance of animals. Many studies have failed to achieve positive responses due to phytase and have reported a range of explanations including the differences in breeds and genetics (Edwards et al. 1988; Aggrey et al. 2002) slaughter age (Edwards et al. 1988), feeding operations such as alternative time feeding (Edwards, 2004), particle size of feed (Kasim and Edwards, 2000; Kilburn and Edwards, 2001; Kilburn and Edwards, 2004) and inappropriate mixing (Shirley and Edwards, 2003) unconsidered nutrient levels such as Ca, P and vitamin D (Edwards and Veltman, 1983; Edwards, 2004; Edwards, 1993), fiber (Ballam et al. 1984) and dietary Ca and P (Qian et al. 1996; Mitchell and Edwards, 1996).

CONCLUSION

Dietary phytate in poultry rations can seriously reduce the performance of birds because of the anti nutritive problems which associated with the utilization of nutrients. In this experiment the growth performance parameters were not enhanced significantly though the mega doses of phytase have increased the crude protein digestibility. The optimum levels of phytase for highest crude protein digestibility were different between two rice bran levels. The 20% rice bran included diet gave its highest crude protein digestibility at 3000 FTU/kg while 40% rice bran included diet gave its maximum crude protein digestibility at 4000 FTU/kg. Negative effects of 40% RB on growth performance, bone status and digestibility indices of P were not mitigated by phytase supplementation. The supplementation of mega doses of phytase found to increase illeal crude protein digestibilities but not the growth performances and bone parameters.

ACKNOWLEDGEMENT

The authors wish to award their sincere gratitude to TURIS project of University of Ruhuna, Sri Lanka for their financial assistance and Mr. P.K. Lal and the academic supportive staff of the Department of Animal Science, Faculty of Agriculture, University of Ruhuna, Sri Lanka for their outstanding contribution.

REFERENCES

Aggrey S.E., Zhang W., Bakalli R.I., Pesti G.M. and Edwards Jr. H.M. (2002). Genetics of phytate phosphorus bio-availability in poultry. Pp. 277-279 in 7th World Congr. Genet. Appl. Livest. Prod. Montpellier, France.

AOAC. (1995). Official Methods of Analysis. 16th Ed. Association of Official Analytical Chemists, Arlington, VA, USA.

Arabi S.A.M. (2013). Effect of phytase on protein and phosphorous utilization for broiler chicks. Bakht Alruda Sci. J. 7, 145-159.

Attia Y.A., El-Tahawy W.S., Abd El-Hamid A.E.H.E., Hassan S.S., Nizza A. and Kelaway M.I.E. (2012). Effect of phytase with or without multi enzyme supplementation on performance and nutrient digestibility of young broiler chicks fed mash or crumble diets. Italian J. Anim. Sci. 11, 56-65.

Ballam G.C., Nelson T.S. and Kirby L.K. (1984). Effect of fiber and phytate source and of calcium and phosphorus level on phytate hydrolysis in the chick. Poult. Sci. 63, 333-338.

Cowieson A.J., Acamovic T. and Bedford M.R. (2004). The effects of phytase and phytic acid on the loss of endogenous amino acids and minerals from broiler chickens. Br. Poult. Sci. 45(1), 101-108.

Deniz G., Orhan F., Grencoglu H., Eren M., Gezen S.S. and Turkmen I.I. (2007). Effects of different levels of rice bran with and without enzyme on performance and size of the digestive organs of broiler chickens. *Rev. Méd. Vét.* **158(7)**, 336-343.

Edwards Jr H.M. (2004). Effect of phytase on the utilization of phytate phosphorus energy and protein. Pp. 1-5 Proc. 2nd Mid Atlantic Nutr. Conf. Timonium, Maryland.

Edwards Jr. H.M. and Veltmann Jr. J.R. (1983). The role of calcium and phosphorus in the etiology of tibial dyschondroplasia in young chicks. *J. Nutr.* **113**, 1568-1575.

Edwards Jr H.M., Palo P., Sooncharernying S. and Elliot M.A. (1988). Factors influencing the bioavailability of phytate phosphorus to chickens. *Bioavailability.* **88**, 2.

Edwards Jr H.M. (1993). Dietary 1,25-dihydroxycholecalciferol supplementation increases natural phytate phosphorus utilization in chickens. *J. Nutr.* **123**, 567-577.

Gallinger C.I., Suarez D.M. and Irazusta A. (2004). Effects of rice bran inclusion on performance and bone mineralization in broiler chicks. *J. Appl. Poult. Res.* **13**, 183-190.

Kasim A.B. and Edwards Jr H.M. (2000). Effect of sources of maize and maize particle sizes on the utilization of phytate phosphorus in broiler chicks. *Anim. Feed Sci. Technol.* **86**, 15-26.

Kilburn J. and Edwards Jr H.M. (2004). The effect of particle size of commercial soybean meal on performance and nutrient utilization of broiler chicks. *Poult. Sci.* **48**, 584-392.

Kilburn J. and Edwards Jr H.M. (2001). The response of broilers to the feeding of mash or pelleted diets containing maize of varying particle sizes. *British Poult. Sci.* **42**, 484-492.

Kornegay E.T. (1996). Effect of Natuphos phytase on protein and amino acid digestibility and nitrogen retention of poultry. Pp. 493-514 in Phytase in Animal Nutrition and Waste Management. M.B. Coelho and E.T. Kornegay, Eds. Mount Olive, New Jersey.

Kornegay E.T., Denbow D.M. and Zhang Z. (1999). Infuence of microbial phytase supplementation of a low protein-amino acid diet on performance, ileal digestibility of protein and amino acids and carcass measurements of finishing broilers. Pp. 557-572 in Phytase in Animal Nutrition and Waste Management. M.B. Coelho and E.T. Kornegay, Eds. Mount Olive, New Jersey.

Managi M.K., Sands J.S. and Coon C.N. (2009). Effect of phytase on illeal amino acid digestibility, nitrogen retention and AMEn for broilers fed diets containing low and high phytate phosphorus. *Int. J. Poult. Sci.* **8(10)**, 929-938.

Mitchell R.D. and Edwards Jr H.M. (1996). Additive effects of 1,25-dihydroxycholecalciferol and phytase on phytate phosphorus utilization and related parameters in broiler chickens. *Poult. Sci.* **75**, 111-119.

Munir K. and Maqsood S. (2013). A review on role of exogenous enzyme supplementation in poultry production. *Emir. J. Food Agric.* **25(1)**, 66-80.

Namkung H. and Leeson S. (1999). Effect of phytase enzyme on nitrogen-corrected apparent metabolizable energy and the ileal digestibility of nitrogen and amino acids. *Poult. Sci.* **78**, 1317-1319.

Olukosi O.A., Cowieson A.J. and Adeola O. (2010). Broiler responses to supplementation of phytase and admixture of carbohydrases and protease in maize soya bean meal diets with or without maize distillers' dried grain with solubles. *British Poult. Sci.* **51(3)**, 434-443.

Piragozliev V., Oduguwa O., Acamovic T. and Bedford M.R. (2008). Effects of dietary phytase on performance and nutrient metabolism in chickens. *British Poult. Sci.* **49(2)**, 144-154.

Piyaratne M.K.D.K., Atapattu N.S.B.M., Mendis A.P.S. and Amarasinghe A.G.C. (2008). Effects of balancing rice bran based diets for up to four amino acids on growth performance of broilers. *Trop. Agric. Res. Ext.* **12(2)**, 57-61.

Qian H., Kornegay E.T. and Denbow D.M. (1996). Phosphorus equivalence of microbial phytase in turkey diets as influenced by calcium to phosphorus ratios and phosphorus levels. *Poult. Sci.* **75**, 69-81.

Ravindran V., Cabahug S., Selle P.H. and Bryden W.L. (2000a). Response of broiler chickens to microbial phytase supplementation as infuenced by dietary phytic acid and non-phytate phosphorus levels, II. Effects on apparent metabolisable energy, nutrient digestibility and nutrient retention. *British Poult. Sci.* **41**, 193-200.

Ravindran V., Selle P.H. and Bryden W.L. (1999). Effects of phytase supplementation, individually and in combination, with glycanase on the nutritive value of wheat and barley. *Poult. Sci.* **78**, 1588-1595.

Ravindran V., Selle P.H., Ravindran G., Morel P.C.H., Kies A.K. and Bryden W.L. (2000b). Microbial phytase improves performance, metabolizable energy and ileal amino acid digestibility of broilers fed a lysine defcient diet. *Poult. Sci.* **80**, 338-344.

Reddy N.R., Pierson M.D., Sathe S.K. and Salunkhe D.K. (1989). Phytates in cereals and legumes, CRC Press, Boca Raton, Florida.

Rutherfurd S.M., Chung T.K. and Moughan P.J. (2002). The effect of microbial phytase on ileal phosphorus and amino acid digestibility in the broiler chicken. *British Poult. Sci.* **43**, 598-606.

Samli H.E., Senkoylu N., Akyurek H. and Agma A. (2006). Using rice bran in laying hen diets. *J. Cent. Eur. Agric.* **7(1)**, 135-140.

Selle P.H., Ravindran V., Caldwell R.A. and Bryden W.L. (2000). Phytate and phytase: consequences for protein utilization. *Nutr. Res. Rev.* **13**, 255-278.

Shirley R.B. and Edwards Jr H.M. (2003). Graded levels of phytase past industry standards improve broiler performance. *Poult. Sci.* **82**, 671-680.

Sooncharernying S. (1991). Studies on the utilization of phytate phosphorus from corn and soybean meal diets by broiler chickens. Ph D. Thesis. University of Georgia, Athens, Georgia.

Weeks C.A., Knowles T.G., Gordon R.G., Kerr A.E., Peyton S.T., and Tillbrook N.T. (2002). New method for objectively assessing lameness in broiler chickens. *Vet. Record.* **151**, 762-764.

Wu Y.B., Ravindran V., Thomas D.G., Birtles M.J. and Hendriks W.H. (2004). Influence of phytase and xylanase, individually or in combination, on performance, apparent metabolizable energy, digestive tract measurements and gut morphology in broilers fed wheat-based diets containing adequate level of phosphorus. *British Poult. Sci.* **45(1),** 76-84.

Zhang X., Roland D.A., McDaniel G.R. and Rao S.K. (1999). Effect of Natuphos phytase supplementation to feed on performance and ileal digestibility of protein and amino acids in broilers. *Poult. Sci.* **78,** 1567-1577.

11

The Effect of Energy Sources and Levels on Performance and Breast Amino Acids Profile in Cobb 500 Broiler Chicks

S.M. Akbari[1], A.A. Sadeghi[1*], M. Amin Afshar[1], P. Shawrang[2] and M. Chamani[1]

[1] Department of Animal Science, Science and Research Branch, Islamic Azad University, Tehran, Iran
[2] Nuclear Agriculture Research School, Nuclear Science and Technology Research Institute, Atomic Energy Organization of Iran, Karaj, Iran

*Correspondence E-mail: a.sadeghi@srbiau.ac.ir

ABSTRACT

The present study conducted to investigate the effect of energy sources and levels on performance and breast meat amino acids profile in Cobb 500 broiler chicks. A total of 600 1-day-old Cobb 500 broiler chicks with an average weight of 39 ± 0.50 g were randomly divided into five treatments. Each treatment was further divided into four replicates. Chicks were fed a basal diet based on corn and energy level was same as Cobb 500 manual as control group, the basal diet with 3% lesser energy than control (T1), the basal diet with 6% lesser energy than control (T2), the basal diet based on corn and fat level according to Cobb 500 instruction manual (T3), the basal diet based on corn and fat with 3% upper energy (T4) for 42 days. Results showed that the best and the worst performance were for T4 and T2, respectively. Also, feed intake of chicks increased significantly in T4. Body weight gain was also significantly higher in the treated group with the basal diet based on corn and fat with 3% upper energy (T4). As result relevant although the lowest feed conversion ratio was for control and T3 on 14 and 28 days but also it was at the lowest on T3 on 42 days. The breast muscle function and amino acid profiles showed that there were significant effects between arginine (Arg), glutamic acid (Glu), proline (Pro), alanine (Ala), aspartic acid (Asp), serine (Ser), glycine (Gly), isoleucine (Ilo), lysine (Lys), valine (Val) and phenylalanine (Phy) amino acids (P≤0.01). The highest level of amino acids (g/g DM basis) was for Glu, Asp, Arg, Lys, Ser, Phe and Pro. In conclusion, it seems that inclusion of higher energy level than broiler nutritional requirements recommendation for Cobb 500 chicks could give better performance and affect the quantity and quality of their breast meat and its amino acids profile. In order to achieve higher weights, more energy is needed than the recommendation of Cobb 500, but to have better feed conversion ratio the recommended energy level is the best.

KEY WORDS amino acids, breast meat, broiler, energy sources, performance.

INTRODUCTION

Poultry meat is a good source of protein having high biological value, high digestibility and relatively little fat content and it is a good source of iron and some of the B-complex vitamins and fat-soluble organic compounds (Marcu *et al.* 2013). It is well known that this huge demand for poultry meat has put pressure on breeders, nutritionists, and farmers to improve the growth rate of birds, feed efficiency and breast meat yield. One of the major challenges to researchers is the provision of alternative feeds for monogastric animals. Corn has remained the major energy source in compounded diets for poultry. The various uses to which corn is being committed, such staple food for man, brewing and confectionary, has placed additional cost constraints on its continued use in poultry diets. The solution is to explore the use of alternative feed ingredients, hitherto under exploited by poultry farmers (Durunne *et al.* 2015). Among the alternative feedstuff which could be used as energy sources for poultry diet even though it is lower in protein and other essential nutrients (Odukwe, 1994). Poultry meat and its products are important components of the diet of developed countries and their consumption is affected by various sensory properties such as color, tenderness and flavor (Resurreccion, 2002).

These changes have driven the poultry industry to put an emphasis on the improvement of breast meat yield and muscle mass development (Abdullah and Matarneh, 2010). Energy and protein are very important nutrients for broilers like other living creatures. Energy is required for body functioning and protein is an essential constituent of all tissues of animal body. Protein having major effect on growth performance of the bird is the most expensive nutrient in broiler diets (Kamran *et al.* 2004). The usefulness of a protein feedstuff for poultry depends upon its ability to supply a sufficient amount of the essential amino acids that the bird requires, as well as the protein digestibility and the level of toxic substances associated with it (Scanes *et al.* 2004). In general, vegetable protein sources are nutritionally unbalanced and poor in certain essential amino acids and this decreases their biological value as they may not furnish the required limiting amino acids needed by birds for egg and meat production. Poultry nutritionists have paid more attention to the use of animal protein sources to create a balanced diet (Akhter *et al.* 2008).

Fats or oils are energy rich feeds. Fats also provide varying quantities of the essential nutrient linoleic acid (Leeson and Summers, 2001). Another important role of fats in diet is its inhibition from de novo lipogenesis in broiler chickens (Wongsuthavas *et al.* 2011) that could increase energy efficiency in diets. Researchers (Haunshi *et al.* 2012) concluded that provision of 2600 kcal/kg ME and 16% crude protein would be ideal for optimum growth of Aseel birds during juvenile phase. However, to obtain better feed conversion ratio (FCR), feeding Aseel birds with diet having 2800 kcal/kg ME and 16% CP would be ideal. It was shown that feeding different levels of soybean oil in chicks fed with a NRC recommended protein level had no significant effect on weight gain in 7-21 days (Tabeidian *et al.* 2005). Another researcher (Maiorka *et al.* 2004) showed that feed intake was not significantly influenced by the way amino acid requirements were expressed. They also noted that feed intake was not influenced by dietary energy level. However, metabolizable energy intake was higher in birds fed diets containing 3200 kcal ME/kg. Other researcher (Nobre *et al.* 1994) evaluated four energy levels and reported no significant differences in carcass yield between treatments.

Batal and Parsons (2004) indicated that metabolizable energy corrected for nitrogen (Men), varies among carbohydrate sources and increases with age for most carbohydrate soybean meal diets. Murugesan *et al.* (2013) suggested that energy is used following the pattern of production and maintenance before storage requirements and that fat pad (energy storage) may be the most sensitive indicator of dietary energy status over short term in Hybrid Line W36 laying hens. El-Yamany *et al.* (2008) noted that replacement of different levels of linseed oil, sun flower oil, or olive oil from poultry fat in control diet of Japanese quail improved the digestibility of all nutrients crude protein and ethyl extract while crude fiber digestibility remained affected. Additionally, no significant effect was recorded on both edible giblets (gizzard, liver and heart) and offal (blood, feather, legs, head and viscera) percentage. Brake (1990) reported that 5% addition of fat to broiler breeder diets resulted in improved egg production and reduced feed intake. Due to the effects of different sources of fat and protein levels on performance and amino acid profile on poultry meat this study was performed to investigate the effect of energy sources and levels on performance and breast meat amino acids profile in Cobb 500 broiler chicks.

MATERIALS AND METHODS

Birds and diets

All procedures used in this experiment were approved by the Department Science and Research Branch, Islamic Azad University, Tehran, Iran. A total of 600 1-day-old Cobb 500 commercial broiler chickens with an average weight of 39 ± 0.50 g were randomly distributed into five dietary treatments with four replicates each. The birds were housed in groups of 5 in 20 pens under standard conditions of temperature, humidity and ventilation. The study was carried out at the poultry farm of Atomic Energy Organization, Karaj, Iran, for a period of 6 weeks. Chicks were fed the basal diet based on corn-soybean meal and energy level by Cobb 500 instruction manual as control group, basal diet with 3% lesser energy than control (T1), basal diet with 6% lesser energy than control (T2), basal diet based on corn and fat level according to Cobb 500 instruction manual (T3), basal diet based on corn and fat with 3% upper energy (T4). The experimental diets formulated using broiler performance and nutrition supplement Cobb 500 (2013). Diets and fresh water were provided *ad libitum* during the experiment (Tables 1, 2 and 3).

Data collection

The live body weight gains and feed consumption of chick were measured individually, while feed conversion efficiency was performed weekly.

Table 1 Composition (measured in %) of the experimental diets for broiler chicks (0-14 days old)

Ingredients	Control	T1	T2	T3	T4
Corn	63.38	63.33	62.13	58.27	54.30
Soybean meal	22.57	28.86	31.48	31.90	32.21
Corn gluten meal	9.17	3	0	2.7	4.7
Soybean oil	0	0	0	2.5	4.14
Di-calcium phosphate[1]	2.07	2.06	2.06	2.05	2.05
Oyster shell	1.06	1.03	1.01	1.01	1.01
NaCl	0.38	0.38	0.38	0.38	0.38
DL-methionine	0.24	0.33	0.35	0.30	0.26
L-lysine	0.53	0.40	0.36	0.31	0.36
L-threonine	0.10	0.11	0.13	0.08	0.09
Mineral and vitamin premix[2]	0.5	0.5	0.5	0.5	0.5
Filler[3]	0	0	1.60	0	0
Total	100	100	100	100	100
Calculated nutrient content					
Metabolizable energy (kcal/kg)	3035	2943	2853	3035	3120
Crude protein (CP) (%)	22	22	22	22	22
Ca (%)	0.90	0.90	0.90	0.90	0.90
P (%)	0.45	0.45	0.45	0.45	0.45
Met (%)	0.65	0.67	0.67	0.67	0.67
Lys (%)	1.18	1.18	1.18	1.18	1.18
Met + Ces (%)	0.98	0.98	0.98	0.98	0.98
Thr (%)	0.86	0.86	0.84	0.86	0.86
Trp (%)	0.18	0.20	0.20	0.21	0.21
Arg (%)	1.24	1.24	1.24	1.24	1.24
Val (%)	0.91	0.91	0.91	0.91	0.91
Na (%)	0.17	0.17	0.17	0.17	0.17
K (%)	0.70	0.80	0.82	0.81	0.81
Cl (%)	0.23	0.23	0.23	0.23	0.23

[1] Per kg contains: Ca: 23% and P: 18.5%.
[2] Provided per kilogram: vitamin A: 360000 IU; vitamin D_3: 800000 IU; vitamin E: 7200 IU; vitamin K_3: 800 mg; vitamin B_1: 720 mg; vitamin B_9: 400 mg; vitamin H_2: 40 mg; vitamin B_2: 2640 mg; vitamin B_3: 4000 mg; vitamin B_5: 12000 mg; vitamin B_6: 1200 mg; vitamin B_{12}: 6 mg; Choline choloride: 200000 mg; Manganese: 40000 mg; Iron: 20000 mg; Zinc: 40000 mg; Copper: 4000 mg and Iodine: 400 mg.
[3] Inert filler used to complete diet formulations to 100%.

To evaluate amino acid compositions of breast tissue on day 28, all birds were processed after a 4 h feed withdrawal period; two birds per treatment were randomly selected and killed in a commercial slaughter house according to the international animal rights.

Evaluation amino acid composition of breast tissue
The amino acid profile of the muscle samples was determined by the method of Antoine et al. (2001) cited by Okarini et al. (2013) and Okruszek et al. (2013) with slight modification. The sample obtained were 3 mg protein or 5 g chicken breast meat dry weight and then 1 mL 6 N HCL was added.

The mixture was ponged with N_2 and then heated in the oven at 110 °C for 24 hours. A sample was prepared by 6 N HCl hydrolysis then dissolved in 5 mL 0.01 N HCl and filtered using Millipore 0.45 μm filter. Mobile phase A was made up of 0.025 M sodium acetate buffer (pH 6, 5) 0, 5 g Na-EDTA, 90 mL methanol and 10 mL tetra hydro folate (THF) (80:10:9:1) prepared from analytical grade dissolving to 1 L water Hi Pure. The pH of the acetate buffer (A buffer) was adjusted to 6.5 using NaOH solution, and mobile phase B buffer contain 95% methanol on hi pure water. The mobile phases were ultra filtered through Millipore filter having a pore diameter of 0.45 μm (WHATMAN® diameter 25 mm) and degassed by spraying for 5 minutes with pure nitrogen. Gradient elution was generated using solvent delivery module Varian Pro Star Model Number 240 (Chromatography Systems, Walnut Creek, CA 94598 USA Made in USA), was used for controlling the gradient and flow rate (1.0 mL/min) of the mobile phases. OPA-thiol reagent was made up at least 24 hours before used by dissolving 50 mg o-phthaldialdehyde in 4 mL methanol and 0.025 mL mercaptoethanol was added. The mixture was thoroughly mixed, then 0.050 mL Brij-30 and 1 mL borax buffer solution were added. The OPA-thiol reagent stored in the dark bottle at temperature 4 °C in a tightly closed container.

The amino acid chicken breast meat were analyzed by cation exchange ICI Instrument high performance liquid chromatography (HPLC) with column Ultra Techsphere ODS 3 μ particle size, 4.6 mm × 7.5 cm PARKER 316 (SGE PTY. LTD Victoria Australia) and O-phthaldialdehyde (OPA) precolumn derivatisation.

Table 2 Composition (measured in %) of the experimental diets for broiler chicks (14-28 days old)

Ingredients	Control	T1	T2	T3	T4
Corn	69.22	68.8	67.7	65.31	59.31
Soybean meal	18	23.43	26.2	24	28.45
Corn gluten meal	8.2	3.24	0	4.2	2.9
Soybean oil	0	0	0	2	5
Di-calcium phosphate[1]	1.9	1.9	1.9	1.9	1.9
Oyster shell	1.05	1.05	1.05	1.05	1.05
NaCl	0.37	0.37	0.37	0.37	0.37
L-methionine	0.22	0.27	0. 3	0.25	0.25
L-lysine	0.44	0.34	0.28	0.32	0.22
L-threonine	0.10	0.10	0.12	0.10	0.05
Mineral and vitamin premix[2]	0.5	0.5	0.5	0.5	0.5
Filler[3]	0	0	1.58	0	0
Total	100	100	100	100	100
Calculated nutrient content					
Metabolizable energy (kcal/kg)	3108	3014	2921	3108	3201
Crude protein (CP) (%)	19	18.4	17.8	19	19.5
Ca (%)	0.84	0.84	0.84	0.84	0.84
P (%)	0.42	0.42	0.42	0.42	0.42
Met (%)	0.53	0.53	0.53	0.53	0.53
Lys (%)	1.05	1.05	1.05	1.05	1.05
Met + Ces (%)	0.89	0.89	0.89	0.89	0.89
Thr (%)	0.78	0.78	0.78	0.78	0.78
Trp (%)	0.21	0.22	0.22	0.22	0.22
Arg (%)	1.25	1.25	1.25	1.25	1.25
Val (%)	0.91	0.91	0.91	0.91	0.91
Na (%)	0.19	0.19	0.19	0.19	0.19
K (%)	0.72	0.81	0.85	0.81	0.86
Cl (%)	0.35	0.33	0.32	0.32	0.31

[1] Per kg contains: Ca: 23% and P: 18.5%.
[2] Provided per kilogram: vitamin A: 360000 IU; vitamin D_3: 800000 IU; vitamin E: 7200 IU; vitamin K_3: 800 mg; vitamin B_1: 720 mg; vitamin B_9: 400 mg; vitamin H_2: 40 mg; vitamin B_2: 2640 mg; vitamin B_3: 4000 mg; vitamin B_5:12000 mg; vitamin B_6: 1200 mg; vitamin B_{12}: 6 mg; Choline chloride: 200000 mg; Manganese: 40000 mg; Iron: 20000 mg; Zinc: 40000 mg; Copper: 4000 mg and Iodine: 400 mg.
[3] Inert filler used to complete diet formulations to 100%.

The amino acid standard (L-alanine, L-arginine, L-aspartic acid, L-glutamic acid, glycine, L-histidine, L-isoleucine, L-leucine, L-lysine, L-methionine, L-phenylalanine, L-serine, L-threonine, L-lysine, L-valine, cysteine and phenylalanine (PIERCE, Rockford Illinois 61105, USA) were used. Amino acids were analyzed in three replicates and each replicate of sample was obtained from 3 different whole chicken breast meats (Musculus Pectoralis Superficial is, left and right breast muscles of each bird).

The amino acid composition was expressed as g of amino acid per 100 g of raw breast meat. Membrane (WHATMAN® diameter 25 mm) followed by adding potassium-borate buffer (pH 10.4). A 5 µL quantity of hydrolyzes protein sample was added to 25 µL of OPA reagent and then injected after 1 minute of derivatization.

Statistical analysis

Data analysis was performed by using the general linear model procedure and the comparison of means was made through Duncan's, (1995) multiple range tests by using SAS 9.1 software (SAS, 2004).

RESULTS AND DISCUSSION

The effect of the energy source on performance of broiler chicks is shown in Table 4. Data from this study showed that there were significant differences between treatment about feed intake (FI), body weight (BW) and FCR at 14.28 and 42 days old (P≤0.01). Using of T1, T2 and T3 at 14 and 28 days old had decreased BW and FI compared to the control, but at 42 days old T1, T2, T3 and T4 increased FI and BW. According to the Table 4 data the feed conversion ratio of broiler chicks significantly (P<0.01) increased by feeding treatments. It was observed that the performance parameter showed significant (P≤0.01) differences throughout the experimental period.

As result revealed from Table 5 amino acid compositions of breast muscles were varied by using different levels of energy and protein (P≤0.01). Result showed that most of Arg obtained in the T4, T3 and T2, respectively. The results showed that the amount of energy and its kind could affect Arg content. Glu numerically at highest in group T4 with the highest energy levels and was based on corn and soybean oil.

Table 3 Composition (measured in %) of the experimental diets for broiler chicks (29-42 days old)

Ingredients	Control	T1	T2	T3	T4
Corn	70.18	71.47	71.5	65.07	62.4
Soybean meal	19.3	20	24.2	25.8	25
Corn gluten	6.23	4.2	0	1.5	3.5
Soybean oil	0	0	0	3.5	5
Di-calcium phosphate[1]	1.7	1.7	1.7	1.7	1.7
Oyster shell	0.92	0.92	0.92	0.90	0.90
NaCl	0.32	0.32	0.32	0.32	0.32
L-methionine	0.18	0.22	0.27	0.22	0.20
L-lysine	0.36	0.36	0.27	0.21	0.22
L-threonine	0.09	0.09	0.10	0.06	0.04
Mineral and vitamin premix[2]	0.5	0.5	0.5	0.5	0.5
Choline chloride	0.22	0.22	0.22	0.22	0.22
Filler[3]	0	0	0	0	0
Total	100	100	100	100	100
Calculated nutrient content					
Metabolizable energy (kcal/kg)	3185	3085	2990	3185	3275
Crude protein (CP) (%)	18	17.4	16.8	18	18.5
Ca (%)	0.76	0.76	0.76	0.76	0.76
P (%)	0.38	0.38	0.38	0.38	0.38
Met (%)	0.48	0.48	0.48	0.48	0.48
Lys (%)	0.95	0.95	0.95	0.95	0.95
Met + Ces (%)	0.74	0.74	0.74	0.74	0.74
Thr (%)	0.65	0.65	0.65	0.65	0.65
Trp (%)	0.16	0.16	0.17	0.17	0.17
Arg (%)	1.02	1.02	1.03	1.03	1.03
Val (%)	0.73	0.73	0.74	0.74	0.74
Na (%)	0.16	0.16	0.16	0.16	0.16
K (%)	0.6	0.6	0.7	0.71	0.7
Cl (%)	0.29	0.29	0.28	0.28	0.28

[1] Per kg contains: Ca: 23% and P: 18.5%.
[2] Provided per kilogram: vitamin A: 360000 IU; vitamin D_3: 800000 IU; vitamin E: 7200 IU; vitamin K_3: 800 mg; vitamin B_1: 720 mg; vitamin B_9: 400 mg; vitamin H_2: 40 mg; vitamin B_2: 2640 mg; vitamin B_3: 4000 mg; vitamin B_5: 12000 mg; vitamin B_6: 1200 mg; vitamin B_{12}: 6 mg; Choline choloride: 200000 mg; Manganese: 40000 mg; Iron: 20000 mg; Zinc: 40000 mg; Copper: 4000 mg and Iodine: 400 mg.
[3] Inert filler used to complete diet formulations to 100%.

All the treatments were significantly different in terms of Pro amino acid. Content of Ala and Ser had decreased significantly by using T1, T2, T3 and T4, respectively. Data showed that Asp increased significantly by treated groups. The differences between chicken groups studied were increased significantly for Gly ($P \leq 0.01$). The data obtained in this study showed that Leu remained unaffected and Ilu increased for the treatments T3 and T4, while, Lys and Val decreased compared to the control. The Phe content was increased by treatment except for T3 and T4. There were no significant differences for Tyr, Trp, His, Met, Cys, Thr, Gln, Asn amino acids profile between treatments.

Recent research showed that energy and protein manipulation in broiler's diet could have been able to change performance and carcass characteristics of broilers. In current study, using different levels of energy and protein could affect performance and amino acid profile in experimental Cobb 500 broilers. In our study, varying levels of energy and protein affected performance.

We observed significant depression in BW (for the first 28 days) for T1 and T2 while, FI presented the same results only for the last period (29-42 days) and was the highest for T4. Additionally, FCR was the highest for T2 in all ages. These results indicate that as dietary energy level increases BW and FI also increase, but FCR requires.

Recent studies by Eits *et al.* (2003) showed that increasing levels of a well balanced protein up to 27.0% considerably improved the performance of birds. Additionally, data provided by Temim *et al.* (2000) showed improved performance when raising the amino acid profile. While Ajuyah *et al.* (1993) noted that an increase in the saturation of the diet decreased the weight gain and, final weights, although controversial results have been reported elsewhere. Furthermore, Palmer (1993) showed that the improvement of performance and growth in poultry may be due to the inclusion of some fatty acids which may affect muscle protein synthesis and protein deposition through a prostaglandin-depend mechanism.

Table 4 The effect of the energy source on performance of broiler chicks

Treatments	14 days old			28 days old			42 days old		
	BW (g)	FI (g)	FCR	BW (g)	FI (g)	FCR	BW (g)	FI (g)	FCR
Control	341.80[a]	401.0[b]	1.17[e]	1279.6[b]	1757.5[bc]	1.36[e]	3617.5[b]	2160.3[b]	1.67[d]
T1	321.87[b]	407.5[b]	1.26[b]	1199.2[c]	1872.5[ab]	1.55[b]	3649[b]	2093.2[c]	1.74[b]
T2	303.85[c]	405.5[b]	1.32[a]	1086.1[d]	1857[ab]	1.62[a]	3722.5[ab]	2030.7[d]	1.82[a]
T3	329.97[ab]	404.2[b]	1.22[d]	1243.7[b]	1717.7[c]	1.37[d]	3670[b]	2163.1[b]	1.65[e]
T4	342.70[a]	428.5[a]	1.24[c]	1337.8[a]	1906.2[a]	1.42[c]	3826.5[a]	2260.5[a]	1.69[c]
SEM	7.182	4.902	0.025	42.388	35.99	0.054	38.46	36.58	0.031

BW: body weight; FI: feed intake and FCR: feed conversion ratio.
The means within the same row with at least one common letter, do not have significant difference (P>0.01).

Table 5 The effect of the energy source on breast muscles amino acid composition of broiler chicks at 28 days old (g/100 g dry weight)

Treatments \ Amino acids	Control	T1	T2	T3	T4	SEM
Arg	3.818[c*]	4.645[bc]	5.807[a]	5.108[ab]	5.960[a]	0.392
Glu	7.237[b]	6.150[c]	5.310[c]	7.930[ab]	8.820[a]	0.623
Pro	3.110[c]	2.645[d]	2.000[e]	3.860[b]	4.337[a]	0.417
Ala	4.520[a]	3.833[b]	3.310[c]	2.602[d]	2.140[e]	0.424
Asp	4.210[d]	4.878[c]	6.480[a]	5.763[b]	6.250[ab]	0.426
Ser	4.720[a]	4.128[b]	3.528[c]	2.812[d]	2.320[e]	0.432
Gly	2.140[c]	2.282[c]	3.752[a]	3.208[b]	3.600[a]	0.333
Ilu	2.523[b]	1.962[c]	1.620[c]	2.873[a]	3.145[a]	0.282
Leu	1.425[a]	1.407[a]	1.380[a]	1.563[a]	1.257[a]	0.049
Lys	4.940[a]	4.450[b]	3.720[c]	3.120[d]	2.642[e]	0.419
Val	2.580[a]	2.120[b]	1.153[d]	1.688[c]	1.263[d]	0.266
Phe	3.230[b]	3.923[a]	4.210[a]	2.608[c]	2.247[c]	0.373
Tyr	1.083	1.103	1.045	1.070	1.082	0.01
Trp	0.418	0.395	0.365	0.443	0.407	0.012
His	0.565	0.585	0.520	0.458	0.492	0.023
Met	0.860	1.240	2.120	1.610	1.750	0.216
Cys	0.560	0.620	0.710	0.500	0.410	0.012
Thr	0.135	0.185	0.213	0.160	0.172	0.012
Gln	0.115	0.130	0.137	0.165	0.180	0.012
Asn	0.175	0.262	0.335	0.332	0.280	0.03

The means within the same row with at least one common letter, do not have significant difference (P>0.01).

While Marcu et al. (2013) demonstrated that the economic efficiency of Cobb 500 broiler chickens growth was positively influenced by the growth performance and the recorded viability. Furthermore, Danicke et al. (1997) reported that the performance was higher for broilers that received a soybean oil in diet that was supplemented with an enzyme compared with unsupplemented tallow based diet, while significant interactions were reported between dietary fat type and carbohydrate addition (Danicke et al. 2000).

The results of present study revealed that FI was affected by energy level in all three periods. These results are not completely in accordance with the results obtained by Abudabos (2014) that showed during the starter period, FI was not affected by fat source, energy level or enzyme supplementation or their interactions (P>0.05) but FI was influenced by the energy level of the diet at finisher period (P<0.05).

These differences might be the result of differences in energy level and fat content of the diets, because Pesti (1982) mentioned that chicks fed diets with soybean or corn oil consumed more ME than chicks fed comparable diets that had a low fat content. Also, Murugesan et al. (2013) showed that there were no significant interactions in BW, FI and FE among dietary and challenge treatments. They showed that direct comparison of the excess energy contributed by the 3% diets provided an average of 69% increase over the energy value derived from the equations. They mentioned that the vegetable oils were fairly consistent, although the pure soybean oil resulted in slightly higher apparent metabolizable energy corrected for nitrogen (AMEn) than the crude corn oil. The poultry fat resulted in an AMEn value lower than the vegetable oils, in line with previous reports for fat of this nature. The soy methyl esters resulted in significantly higher dietary energy, resulting in an AMEn value slightly higher than the corn oil.

According to our results and the differences with the previously mentioned experiments, we can conclude that different energy sources and levels have the potential to affect the performance of broilers. Furthermore, our results showed that amino acid compositions of breast muscles were varied by using different levels of energy and protein (P≤0.01).

Furthermore, our results showed that amino acid compositions of breast muscles were varied by using different levels of energy and protein (P≤0.01). However, it did not correlate with trypsin inhibitor activity levels, indicating that other factors also affect amino acid digestibility of soybean meal. Furthermore Foltyn et al. (2013) reported that the replacement of soybean meal by raw full-fat soybean decreased amino acids coefficients of ileal apparent digestibility when the level of extruded full fat soybean was higher than 40 g/kg in feed, while Makkink et al. (1994) also reported that trypsin activity in the jejunum is affected by protein sources. Arg and Lys are amino acids that specifically cause the release of trypsinogen in a pancreatic tissue homogenate, because they contain sites of tryptic cleavage (Niederau et al. 1986). Clarke and Wiseman (2015) demonstrated that the coefficients of ileal apparent digestibility for individual amino acids varied widely among soybean samples.

Haunshi et al. (2012) determined that to obtain better FCR, feeding Aseel birds with diet having 2800 kcal/kg ME and 16% CP would be ideal. These studies further strength our hypothesis that energy requirements must be re-evaluated.

According to Nobakht et al. (2011) showed that digestibility of dietary fats is affected by its fatty acid profile. As unsaturated fats contain higher metabolizable energy, Crespo and Esteve Garcia (2002) have found better utilization of unsaturated compared to saturated fats. Some evaluations by Yu and Sim (1987) and Nash et al. (1996) did not find any differences in the performance of birds fed different oil sources. Nwoche et al. (2001) evidenced that 4% dietary inclusion of palm oil as the best inclusion level that will bring about an optimum growth in broilers, our results showed that FI is not affected by protein source for the first 28 days for the low energy level diets, something which is not true for the higher energy level diets. Similarly, Dari and Penz (1996) showed that feed intake was not significantly influenced by the way amino acid requirements were expressed, which is consistent with previously published results from experiments comparing diets based on alternative ingredients with low amino acid digestibility to diets based on ingredients with high amino acid digestibility (corn and soybean meal) and to diets with low digestibility ingredients supplemented with synthetic amino acids. Additionally, Maiorka et al. (2004) suggest that for-

mulation based on digestible amino acids is necessary if the diets contain protein sources that are not reliable in terms of amino acid digestibility. The response to formulation based on digestible amino acids was minimized when birds received the low energy level diet (2900 kcal ME/kg). These results are in line with our study and confirm our hypothesis that energy requirements must be re-evaluated.

Rohaeni (2015) suggested that feed treatment provides major effects on daily weight gain, feed intake and feed conversion and the best feed that generates the highest daily weight gain, carcass percentage and feed conversion is feed which contains 19% and 22% protein levels derived from an animal based protein source and feed intake using a plant based protein source, soybean meal which contains a 22% protein level is not recommended. This is consistent with studies reported by Kartikasari (2000) that the higher the protein level the feed contains, the more increasing the weight gain.

Saima et al. (2008) concluded that quality of feed ingredients impose direct effect on their available amino acids profile. Dozier et al. (2008) showed that feeding the high amino acid diets decreased feed consumption, improved feed conversion and increased total breast meat yield. Many researchers mentioned that dietary fat supplementation has been shown to improve feed conversion and decrease feed consumption of broiler chicks (Dale and Fuller, 1980; Saleh et al. 2004).

Our results showed that the amount of energy and its kind could affect the breasy muscle amino acids content for many amino acids such as Arg, Pro, Ala and Ser. Similarly, Leeson et al. (1996) showed that the increase in feed consumption associated with low energy diets can affect growth and meat yield with a concurrent increase in energy and amino acids intake and all other needed nutrients for tissue assimilation. Additionally, Vieira et al. (2004) evidenced that optimum dietary digestible Methionine + cysteine level depends on dietary protein level and should therefore be related to the protein content. Furthermore, in accordance with our results, Nahashon et al. (2005) showed that on 3100 and 3150 kcal of ME/kg of diet at 0 to 4 weeks exhibited greater (P<0.05) BW gain, greater carcass and breast weights (P<0.05) and lower (P<0.05) feed consumption and feed conversion ratios than those on a diet with 3050 kcal of ME/kg. Mean feed consumption of birds fed 25% crude protein diets was higher (P<0.05) than those on other dietary crude protein concentrations. In agreement with our results, Sturkie (1976) showed that birds on high-energy diets, often due to relatively high fat content, have on average lower feed consumption due to reduced rate of passage of digesta through the gastrointestinal tract. The higher feed intake for birds on higher crude protein (CP) found in our experiment is diets is consistent with the findings by Sengar (1987) but quite contrary to the report of Waldroup (1990) in which low CP diets significantly depressed appetite.

CONCLUSION

The results showed that the higher energy level than nutritional needs based on Cobb 500 broiler chickens requirements as specified in the manual, was effective on the expression of some amino acids and the quantity and quality of carcass meat, particularly breast muscle are affected. These effects indicate that the current nutritional recommendations are not sufficient for realizing the full genetic potential of current broiler strains for body weight gain but in terms of feed conversion ratio the most appropriate level of energy obtained from Cobb 500 broiler chickens requirements manual.

ACKNOWLEDGEMENT

The authors sincerely acknowledge the economical help provided by the Nuclear Science and Technology Research Institute, Atomic Energy Organization of Iran for assistance to run this test.

REFERENCES

Abdullah A.Y. and Matarneh S.K. (2010). Broiler performance and the effects of carcass weight, broiler sex, and post chill carcass aging duration on breast fillet quality characteristics, J. Appl. Poult. Res. **19**, 46-58.

Abudabos A.M. (2014). Effect of fat source, energy level and enzyme supplementation and their interactions on broiler performance. *South African J. Anim. Sci.* **44**, 280287.

Ajuyah A.O., Hardin R.T. and Sim J.S. (1993). Effect of dietary full-fat flax seed with and without antioxidant on the fatty acid composition of major lipid classes of chicken meats. *Poult. Sci.* **72**, 125-130.

Akhter S., Khan M., Anjum M., Ahmed S., Rizwan M. and Ijaz M. (2008). Investigation on the availability of amino acids from different animal protein sources in golden cockerels. *J. Anim. Plant. Sci.* **18**, 53-54.

Antoine F.R., Wie C.I., Littell R.C., Quinn B.P., Hogle A.D. and Marshall M.D. (2001). Free amino acid in dark- and white-muscle fish as determined by o-phthaldialdehyde precolumn derivatization. *J. Food Sci.* **66**, 72-77.

Batal A.B. and Parsons C.M. (2004). Utilization of various starch sources as affected by age in the chick. *Poult. Sci.* **83**, 1140-1147.

Brake J. (1990). Effect of four levels of added fat on broiler breeder performance. *Poult. Sci.* **69**, 1659-1663.

Clarke E. and Wiseman J. (2015). Effects of variability in trypsin inhibitor content of soya bean meals on true and apparent ileal digestibility of amino acids and pancreas size in broiler chicks. *Anim. Feed Sci. Technol.* **121**, 125-138.

Cobb 500. (2013). Cobb Broiler Performance and Nutrient Supplement Guide. Cobb-Vantress Inc., Siloam Springs, Arkansas.

Crespo N. and Esteve-Garcia E. (2002). Nutrient and fatty acid deposition in broilers fed different dietary fatty acid profiles. *Poult. Sci.* **81**, 1533-1542.

Dale N.M. and Fuller H.L. (1980). Effect of diet composition on feed intake and growth of chicks under heat stress. II. Constant *vs.* cycling temperatures. *Poult. Sci.* **59**, 1434-1441.

Danicke S., Simon O., Jeroch H. and Bedford M. (1997). Interactions between dietary fat type and xylanase supplementation when rye-based diets are fed to broiler chickens. 1. Physicochemical chyme features. *Br. Poult. Sci.* **38**, 537-545.

Danicke H., Jeroch H., Bottcher W. and Simon O. (2000). Interations between dietary fat type and enzyme supplementation in broiler diets with high pentosan contents: effects on precaecal and total trac digestibility on fatty acids, metabolizability of gross energy, digesta viscosity and weights of small intestine. *Anim. Feed Sci. Technol.* **84**, 279-294.

Dari R.L. and Penz J.R. (1996). The use of digestible amino acid and ideal protein concept in diet formulation for broilers. *Poult. Sci.* **75**, 67-73.

Dozier W.A., Kidd M.T. and Corzo A. (2008). Dietary amino acid responses of broiler chickens. *J. Appl. Poult. Res.* **17**, 157-167.

Duncan D.B. (1995). Multiple range and multiple f-test. *Biometrics.* **11**, 1-42.

Durunne C.S., Udedibie A.B.I. and Uchegbu M.C. (2015). Effect of dietary in clusion of Aanthoata macrophyla meal on the performance of starter chicks. *Nigerian J. Anim.* **32**, 268-273.

EI Yamany A.T., Hewida M.H., Abdel-Samee D. and EL-Ghamry A.A. (2008). Evaluation of using different levels and Sources or oil in growing Japanese quail diets. *American-Eurasian J. Agric. Environ.* **3**, 577-582.

Eits R.M., Kwakkel R.P., Verstegen M.W.A. and Emmans G.C. (2003). Responses of broiler chickens to dietary protein: effects of early life protein nutrition on later responses. *Br. Poult. Sci.* **44**, 398-409.

Foltyn M., Rada V., Lichovnikova M., Safarik I., Lohnisky A. and Hampel D. (2013). Effect of extruded full-fat soybeans on performance, amino acids digestibility, trypsin activity, and intestinal morphology in broilers. *Czech J. Anim. Sci.* **58(10)**, 470-478.

Haunshi S.A., Panda K., Rajkumar U., Padhi M.K., Niranjan M. and Chatterjee R.N. (2012). Effect of feeding different levels of energy and protein on performance of aseel breed of chicken during Juvenile Phase. *Trop. Anim. Health Prod.* **44**, 1653-1658.

Kamran Z., Mirza M.N., Haq A.U. and Mahmood S. (2004). Effect of decreasing dietary protein levels with of decreasing dietary protein levels with optimum amino acids profile on the performance of broilers. *Pakistan Vet. J.* **24**, 165-168.

Kartikasari L.R. (2000). The chemical compotion and the study of fatty acids of breast meat of broiler treated with methionine supplementation in the low protein containing diets. *Bullet. Anim. Sci.* **25(1)**, 33-39.

Leeson S., Caston L. and Summers J.D. (1996). Broiler response to diet energy. *Poult. Sci.* **75**, 529-535.

Leeson S. and Summers J.D. (2001). Nutrition of the Chicken. Published by University Books, Ontario, Canada.

Maiorka A., DahIke F., Santin E., Kessler A.M. and Penzjr A.M. (2004). Effect of energy levels of diets formulated on total or digestible amino acid basis on broiler performance. *Rev. Bras. Cienc. Avic.* **17(1),** 1-5.

Makkink C.A., Negulescu G.P., Guixin Q. and Verstegen M.W.A. (1994). Effect of dietary protein source on feed intake, growth, pancreatic enzyme activities and jejuna morphology in newly weaned piglets. *British J. Nutr.* **72,** 353-368.

Marcu A., Vacaru-Opris Dumitrescu I.G., Marcu A., Petculescu Ciochina L., Nicula M., Dronica D. and Kelciov B. (2013). Effect of diets with different energy and protein levels on breast muscle characteristics of broiler chickens. *Anim. Sci. Biotechnol.* **46(1),** 1-7.

Murugesan G.R., Kerr B.J. and Persia M.E. (2013). Evaluation of energy values oil sources when fed to broiler chicks. *Anim. Indust. Rep.* **659(1),** 55.

Nahashon S.N., Adefope N., Amenyenu A. and Wright D. (2005). Effects of dietary metabolizable energy and crude protein concentrations on growth performance and carcass characteristics of french guinea broilers. *Poult. Sci.* **84,** 337-344.

Nash D.M., Hamilton R.M.G., Sanford K.A. and Hulan H.W. (1996). The effect of dietary menhaden meal and storage on the omega-3 fatty acids and sensory attributes of egg yolk in laying hens. *Canadian J. Anim. Sci.* **76,** 377-383.

Niederau C., Grendell J.H. and Rothman S.S. (1986). Digestive end products release pancreatic enzymes from particulate cellular pools, particularly zymogen granules. *Biochim. Biophys. Acta.* **81,** 281-291.

Nobakht A., Tabatabaei S. and Khodaei S. (2011). Effects of different sources and levels of vegetable oils on performance,carcass traits and accumulation of vitamin E in breast meat bbroilers. *Curr. Res. J. Biol. Sci.* **3(6),** 601-605.

Nobre R.T.R., Silva D.J., Fonseca J.B., Silva M.A. and Lana G.R.Q. (1994). Efeito do nível de energia sobre a qualidade da carcaça de diferentes grupos genéticos de frangos de corte. *Revist. Brasileira Zootec.* **24,** 603-614.

Nwoche E.C., Ndubuisi E.C. and Iheukwumere F.C. (2001). Effect of dietary palm oil on the growth performance of broiler chicks. Pp. 23-27 in Proc. 6th Ann. Conf. Anim. Sci. Assoc. Maiduguri, Nigeria.

Odukwe C.A . (1994). The feeding value of composite cassava root meal for broiler chicks. Ph D. Thesis. University of Nigeria, Nsukka, Nigeria.

Okarini I.A., Purnomo H., Aulanni A.M. and Radiati L.E. (2013). Proximate total phenolic,antioxidant activity and amino acids profile of bali indigenous chicken, spent laying hen and broiler breast fillet. *Int. J. Poult. Sci.* **12(7),** 415-420.

Okruszek A., Woloszyn J., Haraf G., Orkusz A. and Werenska M. (2013). Chemical composition and amino acid profiles of goose muscles from native polish breeds. *Poult. Sci. Assoc.* **92,** 1127-1133.

Palmer R.J. (1993). Prostaglandins and the control of muscle protein synthesis and degradations leukotrienes essent, fatty

acids. *Poult. Sci.* **72,** 95-99.

Pesti G.M. (1982). Characterization of the response of male broiler chickens to diets of various proteins and energy contents. *Br. Poult. Sci.* **23,** 527-537.

Resurreccion A.V.A. (2002). Sensory aspects of consumer choices for meat and meat products. *Meat Sci.* **66,** 11-20.

Rohaeni E.S. (2015). The effects of the protein level from soybean meal and poultry meat meal on the growth of broiler chickens. *Livest. Res. Rural Dev.* Available at: http://www.lrrd.org/lrrd27/5/roha27084.html.

Saima M., Akhter M., Khan Z.U., Anjum M.I., Ahmed S., Rizwan M. and Ijaz M. (2008). Investigation on the availability amino acids from different animal protein in golden cockerels. *J. Anim. Plant. Sci.* **18,** 2-3.

Saleh E.A., Watkins S.E., Waldroup A.L. and Waldroup P.W. (2004). Consideration for dietary nutrient density and energy feeding programs for growing large male broiler chickens for further processing. *Int. J. Poult. Sci.* **3,** 11-16.

SAS Institute. (2001). SAS®/STAT Software, Release 9.1. SAS Institute, Inc., Cary, NC. USA.

Scanes C.G., Brant G. and Ensminger M.E. (2004). Turkeys and turkey *meat.* Publisher: Pearson Prentice Hall, Upper Saddle River, New Jersey.

Sengar S.S. (1987). Feed intake and growth rate pattern in White Leghorn chicks maintained on different planes of nutrition. *Poult. Advisor.* **20,** 23-27.

Sturkie P.D. (1976). Alimentary canal: anatomy, prehension, deglutition, feeding, drinking passage of ingesta and motility. Pp. 185-195 in Avian Physiology. P.D. Sturkie, Ed. Springer Verlag, New York.

Tabeidian A., Sadeghi G.H. and Pourreza J. (2005). Effect of dietary protein levels and soybean oil supplementation on broiler performance. *Int. J. Poult. Sci.* **4,** 799-803.

Temim S., Chagneau A.M., Guillaumin S., Michel J., Peresson R. and Tesseraud S. (2000). Does excess dietary protein improve growth performance and carcass characteristics in heat-exposed chickens? *Poult. Sci.* **79,** 312-317.

Vieira S.L., Lemme A., Goldenberg D.B. and Brugalli I. (2004). Responses of growing broilers to diets with increased sulfur amino acids to lysine ratios at two dietary protein levels. *Poult. Sci.* **83,** 1307-1313.

Waldroup P.W., Tidwell N.M. and Izat A.L. (1990). The effect of energy and amino acid levels on performance and carcass quality of male and female broilers grown separately. *Poult. Sci.* **69,** 1513-1521.

Wongsuthavas S., Yuangklang C., Vasupen K., Mitchaothai J., Alhaidary A., Mohamed H. and Beynen A.C. (2011). Fatty acid metabolism in broiler chickens fed diets either rich in linoleic or alpha-linolenic acid. *Asian J. Anim. Vet. Adv.* **6,** 282-289.

Yu M.M. and Sim J.S. (1987). Biological incorporation of N-3 polyunsaturated fatty acids into chicken eggs. *Poult. Sci.* **63,** 195-212.

12

Effects of Chicory Powder and Butyric Acid Combination on Performance, Carcass Traits and some Blood Parameters in Broiler Chickens

M. Faramarzzadeh[1], M. Behroozlak[2], F. Samadian[3] and V. Vahedi[4*]

[1] Department of Animal Science, Islamic Azad University, Shabestar Branch, Shabestar, Iran
[2] Department of Animal Science, Faculty of Agriculture, Payame Noor University, Tehran, Iran
[3] Department of Animal Science, College of Agriculture, Yasouj University, Yasouj, Iran
[4] Department of Animal Science, Moghan College of Agriculture and Natural Resources, University of Mohaghegh Ardabili, Ardabil, Iran

*Correspondence E-mail: vahediv@uma.ac.ir

ABSTRACT

The objective of this study was to determine the effect of different levels of chicory root and stem powder (CRSP) and butyric acid (BA) combination on the performance, carcass traits, relative weight of internal organs and some blood parameters in broiler chickens. Two hundred and forty one-day-old broilers (Ross 308) were used in a completely randomized design with four treatments with six replicates of 10 birds each. The treatments were: 1) basal diet without CRSP and BA (control group), 2) basal diet + 15 g/kg CRSP + 0.3% BA (CRSP15), 3) basal diet + 30 g/kg CRSP + 0.3% BA (CRSP30) and 4) basal diet + 45 g/kg CRSP + 0.3% BA (CRSP45). At day forty six, two birds per replicate (pen) were randomly selected and slaughtered for carcass traits and some blood parameter measurements. The results showed that body weight gain of birds fed CRSP45 diet was significantly (P<0.05) higher than those fed on the other treatment diets during starter (1-21 d), grower (22-46 d) and whole experimental period (1-46 d). The drumstick and breast yield as a percentage of live body weight were significantly (P<0.05) higher in birds fed CRSP45 diet compared with those fed the control diet. There were not significant effect (P>0.05) of dietary treatments on serum biochemical parameters and relative weights of abdominal organs excluding the liver.

KEY WORDS body weight gain, broiler chickens, butyric acid, carcass yield, chicory.

INTRODUCTION

The risk of antibiotic resistance by continuous administration of antibiotic growth promoters (AGPs) in animal feed led to a ban on sub-therapeutic use of AGPs in poultry diets in European Union since January 2006. One of the consequences of excluding AGPs from broiler diets will be inevitably change in the microbial ecology of intestinal tract in chicks (Knarreborg *et al.* 2002). Different alternatives for antibiotics are proposed such as organic acids, probiotics, prebiotics and phytogenic products (Dahiya *et al.* 2006).

Amongst the organic acids, short chain fatty acids (SCFA) such as butyric acid (BA) are considered as potential alternative to AGPs. Butyrate which is derived from the fermentation of non starch polysaccharides (NSP) is considered to be important for normal development of epithelial cells (Pryde *et al.* 2002). However, the levels of SCFA are quite low in the distal gastrointestinal (GI) tract and caeca of young broiler chickens (Van der Wielen, 2006) and so the young chicks may be the best candidates for dietary supplementation. It was shown that dietary inclusion of BA had no effect on body weight or weight gain, but birds consumed less feed when diets were supplemented with butyrate compared to the control birds (Lesson *et al.* 2005).

Prebiotics are certain substrates, such as dietary fiber that escape digestion in the foregut and reach the distal parts of animal intestine and are now a central issue in nutrition application. Recent studies have indicated that chicks require fiber as a natural prebiotic in the diet to stimulate the development of the upper GI tract (Gonzalez-Alvarado *et al.* 2008). However, it is also known that inclusion of fiber in poultry diet is associated with negative effects on performance and digestibility. The adverse effects of fibers are mainly related to the soluble NSP components in the dietary fiber fraction and increased digesta viscosity in the intestine (Bach knudsen, 2001).

A prebiotic effect of dietary fiber will help to reduce the antibiotic usage in livestock and this will reduce the risk of transferring the antibiotic resistance gene to human pathogens (Looft *et al.* 2012). Furthermore, dietary fiber has been associated with gut disorder management such as *Salmonella* infection in chickens and post-weaning diarrhea in pigs (Montagne *et al.* 2003).

Chicory (*Cichorium intybus*) is a perennial phytogenic herb that can produce nutritious and is a potential feed resource that could partly replace cereal grain fiber in animal feed and reduce feed cost (Ivarsson *et al.* 2011). The dietary fiber in chicory forage has high content of pectin (80-90 g/kg dry DM), a type of NSP with uronic acid which is highly soluble in comparison with other pectin sources (Ivarsson *et al.* 2011). Moreover, chicory root has a high content of oligofructose and inulin, which as a prebiotic can be used to manipulate the composition of microflora in the gut (Flickinger *et al.* 2003). Inulin content of whole chicory plant ranges from 150 to 200 g/kg (Flickinger *et al.* 2003). Inulin is a chain of fructans with non-soluble protein which has minimal side effects, and is a good source of energy in an animal's diet (Lunn and Buttriss, 2007). It also regulates appetite and lipid-to-glucose metabolism. The β $(2 \rightarrow 1)$ glycosidic bond is responsible for the inluin resistence against the digestive enzymes of the host. Previous studies have shown that inulin can promote the growth of beneficial microbes, such as lactic acid bacteria and *Bifidobacteria* (Patterson *et al.* 2010) and inhibit growth of pathogenic bacteria like *Escherichia coli* and *Salmonella* spp. (Xu *et al.* 2003) along the digestive tract. The addition of chicory root fructans either inulin or oligo fructose to broiler feed improved body weight gain, feed conversion, carcass yield and increased the small intestine length of female broilers (Yusrizal and Chen, 2003a). Furthermore, supplementation of inulin to broiler diets has been found to increase the concentrations of jejunal lactate and caecal butyrate (Rehman *et al.* 2006).

According to some studies, supplementation of only commercial prebiotic to broiler diet did not increase body weight gain (BWG) of broiler at 42 d, but in combination with BA the effect on BWG and serum chemical parameters get significant improvement compared to control group. In addition, SCFAs such as butyrate are also considered as potential alternatives to AGPs (Lesson *et al.* 2005). Chicory in addition to be as a fiber source and prebiotic, contains a number of medicinally important compounds, such as inulin, bitter sesquiterpene lactones, coumarins, flavonoids, and vitamins (Varotto *et al.* 2000). Therefore, the aim of this study was to investigate the impact of dietary inclusion of dried chicory root and stem powder and butyric acid on performance, carcass characteristics and blood parameters in broiler chickens.

MATERIALS AND METHODS

Birds and diets

The whole chicory plant were purchased from grocery, dried in the oven at 50 °C for 48 hr and then powdered. The chemical compositions of CRSP were determined in laboratory. The obtained proportion of DM, CP, CF, Ash, EE and NFE were 30.63, 6.60, 28.18, 11.85, 3.50 and 49.87 percent, respectively. Metabolizable energy (ME) of this powder was calculated by fallowing equation: $(8.62 \times CP) + (50.12 \times EE) + (37.67 \times NFE) = 2111$ kcal/kg. This value was used in diet formulation by UFFDA software. The encapsulated calcium butyrate (Greencab 70 Coated®) was provided by Nutri Concept Company, Fougeres, France.

Total of 240 one-day-old broiler chicks (Ross 308) with an initial BW of 45 ± 0.15 g were randomly divided into 24 experimental units of 10 birds each with six replicates per treatment for a total of four different treatments. Temperature and relative humidity was maintained within the optimum range. Lighting was 23 h light and 1 h darkness. The chicks were housed in floor pens (1.25×1.25 m). Starter diets were offered *ad libitum* from 1 to 21 days of age. Then finisher diets were offered *ad libitum* from 22 to 46 days of age.

These dietary treatments were formulated according to the Ross Broiler Nutrient Specifications manual to provide a similar nutrient profile with the exception of using two feed additives or a combined addition of these additives. The diets were: 1) basal diet as a control with no supplementation, 2) basal diet + 0.3% coated butyric acid + 15 g/kg chicory root and stem powder (CRSP15), basal diet + 0.3% coated butyric acid + 30 g/kg chicory powder (CRSP30) and basal diet + 0.3% coated butyric acid + 45 g/kg chicory powder (CRSP45). Ingredients and the composition of the experimental diets are shown in Table 1.

Table 1 Ingredients and composition of the dietary treatments

Diet ingredients	Starter period				Grower period			
	CON	CRSP15	CRSP30	CRSP45	CON	CRSP15	CRSP30	CRSP45
Corn grain (8% CP)	55.7	54.00	52.29	49.95	59.8	56.82	52.75	53.73
Soybean meal (43 % CP)	38.30	36.20	35.31	35.72	33.30	33.82	31.62	31.37
Corn gluten meal	0	2	3	3	0	0	2.00	2.50
Sunflower oil	1.56	1.52	1.62	2.07	3.08	3.76	3.50	3.77
Chicory powder	0	1.5	3.0	4.5	0	1.50	3.00	4.50
Dicalcium phosphate[1]	1.8	1.8	1.79	1.79	1.28	1.27	1.27	1.27
Calcium carbonate	1.31	1.31	1.31	1.30	1.38	1.37	1.37	1.36
Salt	0.37	0.36	0.36	0.36	0.32	0.32	0.32	0.32
Mineral and vitamin premix[2]	0.50	0.50	0.50	0.50	0.50	0.50	0.50	0.50
DL-methionine	0.32	0.31	0.30	0.31	0.18	0.19	0.17	0.17
L-lysine	0.05	0.10	0.11	0.10	0.07	0.05	0.10	0.11
L-threonine	0.10	0.10	0.10	0.10	0.10	0.10	0.10	0.10
Coated butyric acid	0	0.30	0.30	0.30	0	0.30	0.30	0.30
Calculated composition								
Metabolizable energy (kcal/kg)	2900	2900	2900	2900	3050	3050	3050	3050
Crude protein (%)	21.90	22.13	22.29	22.39	19.90	19.99	20.22	20.35
Ca (g/kg)	1.00	1.00	1.00	1.00	0.90	0.90	0.90	0.90
Available phosphorous (g/kg)	0.45	045	0.45	0.45	0.35	0.35	0.35	0.35
Methionine + cystine (g/kg)	0.99	0.92	0.99	0.99	0.80	0.80	0.80	0.80
Lysine (g/kg)	1.23	1.10	1.22	1.22	1.12	1.11	1.11	1.11

CON: basal diet without chicory root and stem powder (CRSP) and butyric acid (BA) (control group); CRSP15: basal diet + 15 g/kg CRSP + 0.3% BA; CRSP30: basal diet + 30 g/kg CRSP + 0.3% BA and CRSP45: basal diet + 45 g/kg CRSP + 0.3% BA.
[1] Dicalcium phosphate contained: 16% phosphorous and 23% calcium.
[2] Vitamin and mineral premix per kg of diet: vitamin A (retinol): 1500 IU; vitamin D_3 (cholecalciferol): 10 IU; vitamin E (tocopheryl acetate): 1 mg; vitamin K_3: 1.5 mg; Thiamine 4 mg; Riboflavin: 5 mg; Panthothenic acid: 10 mg; Folic acid: 0.015 mg; Cyanocobalamin: 20 mg; Niacin: 30 mg; Biotin: 80 mg; Choline chloride: 0.65 mg; Antioxidant: 100 mg; Co: 0.1 mg; Mn ($MnSO_4.H_2O$, 32.49% Mn): 4 mg; Cu ($CuSO_4.5H_2O$): 0.5 mg; I (KI, 58% I): 0.1 mg; Se ($NaSeO3$, 45.56% Se): 1520 mg and Ca: 100 mg.

Growth performance and carcass measurements

Body weight and feed intake (FI) were registered by pen at arrival and on d 21 and 46. Feed conversion ratio (FCR) was calculated and corrected for mortality. The performance is presented for the each rearing phase (1-21 d and 22 to 46 d) and for the overall experiment (1-46 d). After 12 h of fasting, forty-eight broilers (two birds from each group or pen) were selected randomly for carcass evaluations at 46 days of age. After slaughtering by a sharp knife for complete bleeding, their feathers were plucked mechanically and then eviscerated by hand. Whole carcase, abdominal fat pad (excluding the gizzard fat), empty gizzard, liver, heart and spleen were excised and weighed individually. Lengths of total gut (duodenum, jejunum, ileum and cecum) were recorded after ingesta were removed. The carcass yields and the weights of internal organs were calculated as a percentage of the preslaughter live body weights of broiler chickens.

Serum biochemical analysis

Blood samples were collected during slaughtering from birds in non heparinised tubes and then the blood was centrifuged at 3000×g for 10 min. to obtain serum (SIGMA 4-15 Lab Centrifuge, Germany). The Serum was collected and stored at -20 °C until analyzed for glucose, triglyceride, total cholesterol, high density lipoprotein (HDL) and low-density lipoprotein (LDL) cholesterol, using the Hitachi

911 autoanalyzer (Roche Diagnostics, Division of Hoffman-La Roche Limited, Quebec, Canada) according to the procedures recommended by the manufacturer of the kits (Pars-Azmoon Company, Tehran).

Statistical analysis

The experiment was carried out in a complete randomized design (CRD) and the data were subjected to one-way analysis of variance (ANOVA) according to the General liner Model (GLM) procedure of SAS version 9.1 (SAS, 2004). Differences between treatments were determined using the Duncan's multiple range test and reported as means ± SEM and (P<0.05) was considered the significant level.

RESULTS AND DISCUSSION

Growth performance

Body weight gain, FI and FCR of chicks during the experimental period are summarized in Table 2. Chicken BWG was affected markedly by supplementing 45 g/kg of chicory powder plus coated butyric acid during the first, second and overall period (d 1-21, d 22-46 and d 1-46, P<0.05). Other treatments did not have any significant effect on BWG in comparison with control group at second feeding phase (P>0.05), but in total period (1-46 days) lower level of herbal supplement plus butyric acid

(CRSP15) also have a significant difference from control group. No significant (P>0.05) differences due to dietary treatment effects were observed on feed intake in any of the feeding periods. According to Table 2, birds receiving diet supplemented with 45 g/kg chicory powder plus butyric acid (CRSP45) at first feeding phase (1-21 days) had a significantly lower FCR compared to the control group (P<0.05). But as broilers aged, treatments did not induce any significant impact on FCR in the second and whole period (1-46 days) although it was calculated to be higher in the control (P>0.05).

Carcass traits and internal organs weight
Relative weights of organs and body parts are given in Table 3. It shows that carcass percentage were higher in broilers fed supplemented diets than those given the control diet (P<0.05). The proportions (% of live body weight) of drumstick and breast were higher in the birds fed CRSP45 diet in comparison with the control and other groups. Apart from liver relative weight, none of dietary treatments produced significant differences in any of the measured abdominal parameters (P>0.05).

Blood parameters
Table 4 summarizes the impact of treatments on serum constituents at day 46 of age. Treatments did not induce any significant effect on the serum concentration of glucose, triglyceride, total cholesterol, LDL-c and HDL-c (P<0.05).

Recent researches focus on chicory powder that has major fiber components and acts as a potential prebiotic due to inulin-type fructans and oligofructose (Izadi *et al.* 2013). Addition of chicory root to diet improves the growth performance, egg production and the length of small intestine in poultry (Yusrizal and Chen, 2003b; Rehman *et al.* 2007). In the present study, addition of 1.5% and 4.5% chicory powder increased body weight gain of broilers compared to the control group. Results are in agreement with Yusrizal and Chen (2003a) who reported that birds received 1% oligofructose were heavier, especially female broilers (10%), compared to control. Another study demonstrated that adding fructooligosaccharide and inulin to the basal diet of broiler chickens had positive effects on performance. Although, a previous report showed that feeding moderate levels of inulin (5-10%) did not speed up growth rate (Ammerman *et al.* 1989).

Our results showed that dietary addition of 45 g/kg CRSP improved FCR only during the first feeding period in broilers (P<0.05). According to Liu *et al.* (2011), supplementing 6% chicory root powder to broiler diet improved BWG and FCR only in first phase (1-13 days) not during the overall period (d 1-27), that is in agreement with the present study. The profound effects of chicory powder plus BA supple-

mentation in the first feeding period (1-21 d) can be attributed to over sensitivity of broiler chickens to colibacillosis infection that are more frequently occur in the gut of newly hatched chickens (Calnek *et al.* 1991). One of the strategies to eliminate the coliforms from the gastrointestinal tract is by maintaining a lower pH, which is unsuitable for the growth of this organism. This aim could be achieved by supplementation of BA and prebiotics such as chicory to broiler diets.

Prebiotics and organic acids maintained a better microbial environment in digestive tract of birds by reducing the number of pathogenic microbes. It is accepted that the main direct beneficial effect attributable to prebiotics is to induce changes in the intestinal microbiota by selective stimulation of health promoting bacteria (Gibson *et al.* 1995). Van Immerseel *et al.* (2004) have indicated significantly reduced levels of *Salmonella* in the ceca of birds fed organic acids. Kwan and Ricke (1998) demonstrated that amongst the short chain fatty acids, butyrate has the highest bactericidal efficacy against the acid-intolerant species such as *E. coli* and *Salmonella*. Rebole *et al.* (2010) have shown increases in the cecal number of *Bifidobacteria* and *Lactobacilli* due to the effect of the dietary inulin, but this prebiotic caused neither an increase in the concentration of total SCFAs (acetic plus propionic and n-butyric acids) nor a decrease in the pH of digesta. However, they suggested that dietary inulin altered the fermentation patterns of cecal content as evidenced by the significant increase in the concentration of n-butyric acid and the n-butyric acid:acetic acid molar ratio in parallel with the increase in the concentration of d-lactic acid. This positive effect of inulin on the content of n-butyric acid in the cecal digesta was also observed by Rehman *et al.* (2007).

The n-butyrate provides energy for the growth of mucosal epithelium and can be involved directly or indirectly in various mechanisms regulating cellular differentiation, intestinal permeability, and gene expression (Mroz *et al.* 2005). Sharma *et al.* (1995) suggested the beneficial effects of BA on crypt cell growth in rats that might reflect changes in the gut microflora, which is known to be a major modulator of epithelial cell activity.

Chicory as a source of inulin in industry contributes to animal well-being in various ways. Inulin has been reported as a highly fermentable fiber (Franck and Bosscher, 2009), highly fermentable dietary fibers are characterized by being readily fermented by enteric bacteria, producing SCFAs especially butyrate, which are the end products of fermentation of polysaccharides by the colonic flora. These fatty acids are used as an energy source by intestinal epithelial cells.

The levels of SCFAs are quite low in the intestine and caeca of young chicks (Van der Wielen, 2006).

Table 2 Effects of different dietary treatments on performance of broiler chickens

Item	CON	CRSP15	CRSP30	CRSP45	SEM	P-value
Body weight gain (g)						
1-21 days	379.2[b]	324.8[b]	321.3[b]	452.0[a]	8.02	0.0002
22-46 days	981.5[b]	1215.9[b]	1054.7[b]	1375.3[a]	19.32	0.0001
1-46 days	1360.67[b]	1540.67[a]	1375.33[b]	1612.83[a]	25.24	0.0001
Feed intake (g/bird)						
1-21 days	707.2	757.0	708.0	810.3	0.16	0.5203
22-46 days	3097.8	3645.0	3653.0	4027.7	59.68	0.8915
1-46 days	3805.0	4402.0	4361.0	4838.0	65.90	0.9522
Feed conversion ratio						
1-21 days	2.86[a]	2.32[ab]	2.20[ab]	1.79[b]	0.17	0.0011
22-46 days	3.10	2.90	2.50	2.40	0.12	0.4268
1-46 days	2.80	2.70	2.10	2.00	0.89	0.5622

CON: basal diet without chicory root and stem powder (CRSP) and butyric acid (BA) (control group); CRSP15: basal diet + 15 g/kg CRSP + 0.3% BA; CRSP30: basal diet + 30 g/kg CRSP + 0.3% BA and CRSP45: basal diet + 45 g/kg CRSP + 0.3% BA.
The means within the same row with at least one common letter, do not have significant difference (P>0.05).
SEM: standard error of the means.

Table 3 Effects of different dietary treatments on carcass traits, abdominal fat and some internal organs (% of live body weight)

Item	CON	CRSP15	CRSP30	CRSP45	SEM	P-value
Carcass yield[1]	52.58[c]	59.69[b]	56.13[b]	65.97[a]	1.87	0.0004
Drumstick	14.17[b]	15.88[ab]	14.21[b]	17.32[a]	0.57	0.0003
Breast	14.67[b]	14.52[b]	13.99[b]	16.76[a]	0.64	0.0001
Liver	1.99[b]	1.98[b]	2.45[a]	2.16[a]	0.04	0.0001
Heart	0.54	0.51	0.52	0.52	0.01	0.6501
Spleen	0.12	0.10	0.11	0.11	0.06	0.9223
Gizzard	1.77	1.92	1.96	2.06	0.08	0.3528
Abdominal fat	1.52	1.45	1.51	1.26	0.30	0.0512
Gastric intestinal (GIT) (cm)	7.50	6.60	6.70	8.20	0.52	0.1180

[1] Carcass yield, without head either feet.
CON: basal diet without chicory root and stem powder (CRSP) and butyric acid (BA) (control group); CRSP15: basal diet + 15 g/kg CRSP + 0.3% BA; CRSP30: basal diet + 30 g/kg CRSP + 0.3% BA and CRSP45: basal diet + 45 g/kg CRSP + 0.3% BA.
The means within the same row with at least one common letter, do not have significant difference (P>0.05).
SEM: standard error of the means.

Table 4 Some blood metabolite (mg/dL) at day 46 of broiler chickens fed different dietary treatments

Item	CON	CRSP15	CRSP30	CRSP45	SEM	P-value
Glucose	187.9	176.7	196.2	193.9	6.16	0.1464
cholesterol	118.33	117.33	118.33	121.33	5.86	0.9801
Triglyceride	71.75	82.08	81.92	60.25	7.35	0.1340
HDL-c	50.33	49.46	52.43	52.80	2.30	0.4387
LDL-c	54.33	52.23	50.86	51.15	0.30	0.9569

CON: basal diet without chicory root and stem powder (CRSP) and butyric acid (BA) (control group); CRSP15: basal diet + 15 g/kg CRSP + 0.3% BA; CRSP30: basal diet + 30 g/kg CRSP + 0.3% BA and CRSP45: basal diet + 45 g/kg CRSP + 0.3% BA.
HDL: high density lipoprotein and LDL: low density lipoprotein.
The means within the same row with at least one common letter, do not have significant difference (P>0.05).
SEM: standard error of the means.

Thus it could be suggested here that young chicks are therefore the best candidate for diet supplementation of organic acid especially butyric acid because of its both bactericidal and stimulant of villi growth property. Liu et al. (2011) demonstrated that dietary supplementation of CRSP at 1% and 3% levels promoted digestion and absorption through the histomorphological changes of villi and jejunum parameters. Izadi et al. (2013) attributed improved performance of broiler fed diet supplemented with chicory root powder to the enhancements of length, number and surface area of intestinal villi that are paralleled with an increased digestive and absorptive capacity of the jejunum.

Notably, several studies have shown significant enlargement of villus height or crypt depth in the small intestine in monogastric animals that have ingested various dietary NSP sources (Rehman et al. 2007). Moreover, beneficial effects of CRSP can be attributed to reduced activity of urease producing bacteria in the gut. Yeo et al. (1997) have indicated that dietary prebiotic suppressed the growth of bacteria that produce urease. They suggested that this effect may be responsible for the increased weight gain during the first 3 wk of feeding in chicks fed the diet supplemented with prebiotic. Suppressing urease activity and ammonia production can be beneficial for improving animal health and enhancing growth, because ammonia locally produced by ureolysis in the intestinal mucosa can exert a significant damage to the surface cells.

Another important finding of the present study was the improvement in dressing percentage by supplementation of chicory powder plus butyrate to broilers diet. This result is in consistency with Panda et al. (2009), Yusrizal and Chen (2003b), Ammerman et al. (1989). Abdominal fat weight was lower numerically in the group supplemented with CRSP45 in comparison with the control group. It was shown that abdominal fat content in broiler chickens is reduced statistically by dietary supplementation of BA (Panda et al. 2009). Generally, it can inferred that CRSP45 diet by improving intestinal microenvironment, reducing endogenous nitrogen losses, and lowering the secretion of immune mediators, yields higher carcass percentages in broiler chickens (Panda et al. 2009; Yusrizal and Chen, 2003b).

The length of the GI tract was not statistically different (P>0.05) among the trial groups but it was numerically higher in the CRSP45 diet (P>0.05). This result is in disagreement with Elrayeh et al. (2012) who indicated that supplementation of 0.7% inulin to broiler diet increased significantly length value for total intestine. Some studies support the idea of using prebiotics for increasing the length of the intestinal villus, which affects the length of the intestine, as well. It has been reported that the longer the intestine is better for nutrient absorption, which in turn causes an increase in BWG (Yusrizal and Chen, 2003b).

Blood serum biochemical values were not significantly different among the groups that are in agreement with the results of Safamehr et al. (2013). Moreover, the results of the current study are in disagreement with the findings of Yusrizal and Chen (2003b) and Elrayeh and Yildiz (2012) and Mirza agazadehh et al. (2015) who observed that adding chicory-based fructans to feeds decreased the serum cholesterol, VLDA and TG level in broilers.

CONCLUSION

In conclusion, supplementing CRSP plus BA in the broiler diet enhanced BWG in all feeding periods and FCR in the rearing period (1-21 d). Improving digestion and absorption during the rearing period is detrimental to the health of broilers. In this study serum biochemical parameters and slaughter characteristic were not affected by dietary treatment excluding carcass yield and relative weight of liver.

ACKNOWLEDGEMENT

The authors would like to thank the Islamic Azad University, Shabestar Branch for financial support.

REFERENCES

Ammerman E., Quarles C. and Twining P.V. (1989). Evaluation of fructooligosaccharides on performance and carcass yield of male broilers. Poult. Sci. 68, 167-175.

Bach Knudsen K.E. (2001). The nutritional significance of dietary fibre analysis. Anim. Feed. Sci. Technol. 90, 3-20.

Calnek B.W., Barnes H.J., Beard C.W., Reid W.M. and Yoder HW. (1991). Diseases of Poultry. Iowa State University Press, Ames, USA.

Dahiya J.P., Wilkie D.C. and Van Kessel A.G. (2002). Drew Potential strategies for controlling necrotic enteritis in broiler chickens in post-antibiotic era. Anim. Feed Sci. Technol. 129, 60-88.

Elrayeh A.S. and Yildiz G. (2012). Effects of inulin and β-glucan supplementation in broiler diets on growth performance, serum cholesterol, intestinal length, and immune system. Turkish J. Vet. Anim. Sci. 36(4), 388-394.

Flickinger E.A., Van Loo J. and Fahey G.C. (2003). Nutritional responses to the presence of inulin and oligofructose in the diets of domesticated animals: a review. Crit. Rev. Food Sci. Nutr. 43, 19-60.

Franck A. and Bosscher D. (2009). Inulin. Pp: 41-60 in Fiber Ingredients, S.S. Cho and Samuel P. Eds. CRC Press, Boca Raton, Florida.

Gibson G.R., Beatty E.R., Wang X. and Cummings J.H. (1995). Selective stimulation of Bifidobacteria in the human colon by oligofructose and inulin. Gastroenterology. 108, 975-982.

Gonzalez-Alvarado J.M., Jimenez-Moreno E., Valencia D.G., Lazaro R. and Mateos G.G. (2008). Effects of fiber source and heat processing of the cereal on the development and pH of the gastrointestinal tract of broilers fed diets based on corn or rice. Poult. Sci. 87, 1779-1795.

Ivarsson E., Frankow-Lindberg B.E., Andersson H.K. and Lindberg J.E. (2011). Growth performance, digestibility and faecal coliform bacteria in weaned piglets fed a cereal-based diet including either chicory (Cichorium intybus) or ribwort (Plantago lanceolata) forage. Animal. 5, 558-564.

Izadi H., Arshami J., Golian A. and Raji M.R. (2013). Effects of chicory root powder on growth performance and histomorphometry of jejunum in broiler chicks. Vet. Res. Forum. 4(3), 169-174.

Knarreborg A., Simon M.A., Engberg R.M., Jensen B.B. and Tannock G.W. (2002). Effects of dietary fat source and subtherapeutic levels of antibiotic on the bacterial community in the ileum of broiler chickens at various ages. Appl. Environ. Microbiol. 68, 5918-5924.

Kwan Y.M. and Ricke, S.C. (1998). Induction of acid resistance of Salmonella typhimurium by exposure to short chain fatty acids. Appl. Environ. Microbiol. 64, 3458-3463.

Lesson S., Namkung H., Antongiovanni M. and Lee E.H. (2005). Effect of butyric acid on the performance and carcass yield of broiler chickens. Poult. Sci. 84, 1418-1422.

Liu H.Y., Ivarsson E., Jonsson L., Holm L., Lundh T. and Lindberg J.E. (2011). Growth performance, digestibility and gut development of broiler chickens on diets with inclusion of chicory (*Cichorium intybus*). *Poult. Sci.* **90**, 815-823.

Looft T., Johnson T.A., Allen H.K., Bayles D.O., Alt D.P., Stedtfeld R.D., Sul W.J., Stedtfeld T.M., Chai B. and Cole J.R. (2012). In-feed antibiotic effects on the swine intestinal microbiome. *Proc. Natl. Acad. Sci. USA.* **109**, 1691-1696.

Lunn J. and Buttriss J.L. (2007). Carbohydrates and dietary fiber. *Nutr. Bull.* **32**, 21-64.

Mirza Aghazadeh A. and Nabiyar E. (2015). The effect of chicory root powder on growth performance and some blood parameters of broilers fed wheat-based diets. *J. Appl. Anim. Res.* **43(4)**, 384-389.

Montagne L., Pluske J.R. and Hampson D.J. (2003). A review of interactions between dietary fibre and the intestinal mucosa, and their consequences on digestive health in young non-ruminant animals. *Anim. Feed Sci. Technol.* **108**, 95-117.

Mroz Z., Koopmans S.J., Bannink A., Partanen K., Krasucki M., Overland M. and Radcliffe S. (2005). Carboxylic acids as bioregulators and gut growth promoters in non-ruminants. Pp. 81-133 in Biology of Nutrition in Growing Animals. Biology of Growing Animal Series. R. Mosenthin, J. Zentek and T. Zewrowska, eds. Elsevier, Amsterdam, Netherlands.

Panda K., Rama Rao S.V., Raju L.N., Shyam G. and Sunder M.V. (2009). Effect of butyric acid on performance, gastrointestinal tract health and carcass characteristics in broiler chickens. *Asian-Australas J. Anim. Sci.* **22(7)**, 1026-1031.

Patterson J.K., Yasuda K., Welch R.M., Miller D.D. and Lei X.G. (2010). Supplemental dietary inulin of variable chain lengths alters intestinal bacterial populations in young pigs. *J. Nutr.* **140**, 2158-2161.

Pryde S.E., Duncan S.H., Hold G.L., Stewart C.S. and Flint H.J. (2002). The microbiology of butyrate formation in the human colon. *FEMS. Microbiol. Lett.* **217**, 133-139.

Rebole A., Ortiz L.T., Rodriguez M.L., Alzueta C., Trevino J. and Velasco S. (2010). Effects of inulin and enzyme complex, individually or in combination, on growth performance, intestinal microflora, cecal fermentation characteristics, and jejunal histomorphology in broiler chickens fed a wheat and barley-based diet. *Poult. Sci.* **89**, 276-286.

Rehman H., Bohm J. and Zentek J. (2006). Effects of diets with sucrose and inulin on the microbial fermentation in the gastrointestinal tract of broilers. Pp. 155 in Proc. Soc. Nutr. Physiol. Gottingen, Germany.

Rehman H., Rosenkran C. and Bohm J. (2007). Dietary inulin affects the morphology but not the sodium-dependent glucose and glutamine transport in the jejunum of broilers. *Poult. Sci.* **86**, 118-122.

Safamehr A., Fallah F. and Nobakht A. (2013). Growth performance and biochemical parametersof broiler chickens on diets consist of Chicory (*Cichorium intybus*) and Nettle (*Urtica dioica*) with or without multi-enzyme. *Iranian J. Appl. Anim. Sci.* **3(1)**, 131-137.

SAS Institute. (2004). SAS®/STAT Software, Release 9.1. SAS Institute, Inc., Cary, NC. USA.

Sharma R., Schumarcher U., Ronaasen V. and coates M. (1995). Rat intestinal mucosal responses to a microbial flora and different diets. *Gut.* **36**, 206-214.

Van der Wielen P.W., Biesterveld J.S., Notermans S., Hofstra H., Urlings B.A. and vanKapen F. (2006). Role of volatile fatty acids in development of the cecal microflora in broiler chickens during growth. *Appl. Environ. Microbiol.* **66**, 2536-2540.

Van Immerseel F., Fievez V., Buck J., Pasmans F., Martel A., Haesebrouck F. and Ducatelle R. (2004). Microencapsulated short-chain fatty acids in feed modify colonization and invasion early after infection with *Salmonella enteritidis* in young chickens. *Poult. Sci.* **83**, 69-74.

Varotto S., Lucchin M. and Parrin P. (2000). Immature embryos culture in Italian red chicory (*Cichorium intybus*). *Plant. Cell. Tiss. Org. Cult.* **62**, 75-77.

Xu Z.R., Hu C.H., Xia M.S., Zhan X.A. and Wang M.Q. (2003). Effects of dietary fructooligosaccharide on digestive enzyme activities, intestinal microflora and morphology of male broilers. *Poult. Sci.* **82**, 1030-1036.

Yeo J. and Kim K.Y. (1997). Effect of feeding diets containing an antibiotic, a probiotic or yucca extract on growth and intestinal urease activity in broiler chicks. *Poult. Sci.* **76**, 381-385.

Yusrizal Y. and Chen T.C. (2003a). Effects of adding chicory in feed on broiler growth performance, serum cholesterol and intestinal length. *Int. J. Poult. Sci.* **2**, 214-219.

Yusrizal Y. and Chen T.C. (2003b). Effect of adding chicory fructans in feed on faecal and intestinal microflora and excretory volatile ammonia. *Int. J. Poult. Sci.* **2**, 188-194.

13

Effect of Vegetable Oil Source and L-Carnitine Supplements on Growth Performance, Carcass Characteristics and Blood Biochemical Parameters of Japanese Quails (Coturnix japonica)

A. Abedpour[1], S.M.A. Jalali[1*] and F. Kheiri[1]

[1] Department of Animal Science, Faculty of Agriculture and Natural Resources, Shahrekord Branch, Islamic Azad University, Shahrekord, Iran

*Correspondence E-mail: sma.jalali@iaushk.ac.ir

ABSTRACT

An experiment was conducted to study the effects of soybean, linseed and sunflower oil (various sources of fatty acids) with and without L-carnitine supplements (0 and 50 mg kg^{-1}) on performance, blood biochemical parameters and carcass traits of Japanese quail. One hundred and ninety-two of 7-day old female Japanese quail were randomly assigned to 6 dietary treatments with 4 replicates and fed in the duration of 28 days. A 3×2 factorial arrangement (three oil sources and two levels of L-carnitine) was used in a completely randomized design with 8 birds per cage. Feed intake (FI), body weight gain (BWG) and feed conversion ratio (FCR) of birds were measured during the experiment. Moreover, 2 birds from each cage (replicate) were randomly selected at the end of the experiment and then the concentration of cholesterol, albumin and total protein of blood sera were measured. Results showed that BWG and FCR were affected by dietary treatments and BWG of quails significantly increased by the addition of L-carnitine to linseed oil treatment (P<0.05). Different types of vegetable oil and L-carnitine supplementation had no significant (P>0.05) effect on liver and gizzard of quails. Linseed and sunflower oil increased heart weight of quail. Addition of L-carnitine to diet, containing sunflower oil reduced the relative weight of liver and heart. Sunflower oil reduced the concentration of total protein and albumin in blood serum of birds. L-carnitine supplement increased the concentration of total protein and globulin in female blood serum. Data of the present experiment showed that source of dietary oil affected the performance of Japanese quail and also use of different plant oil sources with L-carnitine supplementation affected the blood biochemical parameter of female Japanese quail.

KEY WORDS blood metabolites, carcass characteristics, Japanese quail, L-carnitine, oil source.

INTRODUCTION

In the intensive feeding system of poultry production, oil and fats have commonly been used as energy sources. Some of the advantages of including these oil and fats in diets are decreasing of dust, supplying of essential fatty acids and fat-soluble vitamins and also helps to produce lower heat increment as compared to carbohydrates and proteins (Nobakht *et al.* 2011). Soybean, sunflower and canola oil are some of the important vegetable oils, which are commonly used in poultry diets (Burlikowska *et al.* 2010). The efficient utilization of oil and fat as sources of energy in poultry diets depends on their fatty acids composition (Shahriar *et al.* 2007; Burlikowska *et al.* 2010). It has been reported that by using of linseed, sunflower and olive oil instead of poultry fat, at 1.5, 2 and 3% of diets, improved the body weight and body weight gain (BWG) of quail (El-Yamany *et al.* 2008). Additionally, dietary supplementation of a mixture of canola and soybean oil (2%+2%) in broilers diets improved the FCR (Nobakht *et al.* 2011). Since oil sources have different fatty acid compositions, it seems that using different sources of oils in diets of Japanese quails may have a synergistic effect.

L-carnitine is a non-essential nutrient (Harpaz, 2005) and is extensively utilized as animal feed byproducts (Arslan, 2006). The L-carnitine content of plant products is low, especially in poultry diets which are mainly composed of corn and soybean meal. Methionine and lysine are the precursors of L-carnitine biosynthesis and usually the most important limiting amino acids in poultry nutrition. When diets are not supplemented with these two amino acids, chickens may not be able to synthesize adequate amounts of L-carnitine (Arslan, 2006). L-carnitine is required for the transfer of long-chain fatty acids from the cytosol to the mitochondrial matrix during lipid catabolism (β-oxidation); therefore, it plays a vital role in fat combustion and energy production (Jalali Haji-Abadi *et al.* 2010). Studies with broiler chickens have revealed that supplementing dietary L-carnitine increased the BWG, improved FCR and reduced abdominal fat content in broiler chickens (Rabie *et al.* 1997a; Rabie *et al.* 1997b; Rabie and Szilagyi, 1998). In contrary to this, other studies showed that dietary L-carnitine supplementation did not affect growth performance of Japanese quail (Sarica *et al.* 2005) and broiler chicks (Corduk *et al.* 2007; Kheiri *et al.* 2011).

The contradictory response of birds to dietary supplementation of L-carnitine may be related to different fatty acid composition of oil, which is used in poultry diets. Therefore, this study was conducted to determine the effect of dietary plant oil source, such as soybean, linseed and sunflower oil as a different fatty acid composition and L-carnitine supplementation on growth performance and blood biochemical parameters of Japanese quails.

MATERIALS AND METHODS

Birds and dietary treatment

One hundred and ninety-two 7 day-old female Japanese quails (*Coturnix japonica*) were randomly distributed to 6 dietary treatments with 4 replicates (cages) and 8 birds in each cage base on a 3×2 factorial arrangement (three oil sources and two level of L-carnitine) in a completely randomized design. Birds were fed *ad libitum* during 28 days of experiment. The dietary treatments contained two levels of L-carnitine (0 and 50 mg/kg) and three sources of plant oils (soybean, linseed and sunflower). The basal diet was balanced on the basis of corn and soybean meal and formulated to meet nutrient requirements provided by NRC

(1994). Each of the dietary treatments group contained 24% crude protein and 2900 kcal ME/kg (Table 1). To supply metabolizable energy for experimental diets, the oil level of all diet was 4% which is supplied by soybean, linseed and sunflower oil.

Table 1 Ingredients and composition of experimental diets

Dietary composition	Without L-carnitine	With L-carnitine
Ingredients (g kg^{-1})		
Basal diet*	947.4	947.4
Oil**	40.0	40.0
Washed sand	12.55	12.475
L-carnitine premix***	0.00	0.125
Premix free L-carnitine****	0.05	0.00
Nutrient composition		
ME(kcal/kg)	2900	2900
Crud protein (%)	24	24
Crud fat (EE) (%)	6.60	6.60
Calcium (%)	0.802	0.802
Available phosphorus (%)	0.305	0.305
Lysine (%)	1.37	1.37
Arginine (%)	1.57	1.57
Methionine (%)	0.405	0.405
Methionine + cysteine (%)	0.778	0.778

* 94.74% basal diet contain: Ground corn: 44.9%; Soybean meal 37% (44% CP); Fish meal: 5%; Wheat bran: 6%; CaCO₃: 1.07%; Dicalcium phosphate: 0.07% and Salt: 0.20%; 0.5%-vitamin and mineral premix (provided per kilogram of diet): vitamin A: 7700 IU; vitamin D₃: 3300 IU; vitamin E: 6.6 IU; vitamin K₃: 0.55 mg; Thiamine: 1.5 mg; Riboflavin: 4.4 mg; Pantothenic acid: 22 mg; Niacin: 5.5 mg; Pyridoxine: 3 mg; Choline chloride: 275 mg; Folic acid: 1.1 mg; Biotin: 0.055 mg; vitamin B₁₂: 0.088 mg; Antioxidant: 1 mg; Manganese: 66 mg; Zinc: 66 mg; Iron: 33 mg; Copper 8.8 mg; Iodine 0.9 mg and Selenium, 0.3 mg.
** According to experimental diet contain soybean, linseed and sunflower oil.
*** Content: 60% L-carnitine L-tartrate (40% pure L-carnitine).
**** Content: lactose, starch and cellulose microcrystal.

Fatty acid analysis of oils

To determine fatty acids composition of oils, fatty acid methyl esters of oils were analyzed according to the method described by Christie (1990). Fatty acid methyl esters were separated and quantified by gas chromatography (Agilent model 6890, USA) which is equipped with a Flame ionization detector (FID) and a SGE BPX70 column. Nitrogen was used as a carrier gas at a flow of 0.9 mL/min. The injector and detector temperatures were 260 and 300 °C, respectively. Individual methyl esters were identified by comparison with known mix fatty acid methyl standards and quantified by comparing their peak area with that of the external standard (Methyl Erucate). Fatty acid composition of soybean, linseed, and sunflower oil are presented in Table 2.

Performance and carcass traits

The body weight gain (BWG), feed intake (FI) and feed conversion ratio (FCR) were measured during the whole period of the experiment (1-5 week old). At 35 days of age, two birds were randomly selected from each replicate (cage) then, weighed, slaughtered and weight of some internal organs such as liver, heart and gizzard was determined and expressed as percentage of live body weight.

Table 2 Composition of fatty acids in soybean, linseed and sunflower oils

Fatty acids (%)	Soybean oil	Linseed oil	Sunflower oil
C14:0	0.21	0.11	0.16
C15:0	0.03	0.02	-
C15:1	0.03	0.02	-
C16:0	14.85	4.45	5.24
C16:1	0.17	0.34	0.08
C17:0	0.11	0.08	0.05
C17:1	0.07	0.05	0.04
C18:0	4.36	1.16	4.78
C18:1	25.20	15.76	16.06
C18:2 (ω-6)	47.87	19.18	70.9
C20:0	0.32	0.68	0.38
C18:3 (ω-3)	6.21	55.31	0.39
C20:1	0.19	0.4	0.19
C22:0	0.35	0.13	0.98
C24:0	0.03	0.61	0.36
C24:1	-	1.70	0.39
Saturated	20.26	7.24	11.95
Mono unsaturated	25.66	18.27	16.76
Poly unsaturated	54.08	74.49	71.29
ω-3/ω-6	0.129	2.884	0.0055
Unsaturated/saturated	3.936	12.812	7.368

Blood biochemical parameters

Blood samples (5 cc) were collected from quails during slaughtering from jugular vein for analysis of biochemical parameters. Serum was separated by centrifuge (2000×g for 10 minutes) and was kept in the freezer at -20 °C until assay. Total protein, albumin, and cholesterol of blood sera were measured by spectrophotometer enzymatic methods (by Pars Azmoon commercial kits) and globulin was calculated by the differences between concentration of total protein and albumin.

Statistical analysis

The data obtained from the experiment were analyzed using the general linear model (GLM) procedure in the SAS software (SAS, 2004). Significant differences between treatment means were determined using Tukey's HSD test at a probability of (P<0.05).

RESULTS AND DISCUSSION

The fatty acid composition is an important criterion to evaluate the use of fat in the intensive feeding of poultry (Burlikowska et al. 2010). Fatty acid composition of soybean, linseed, and sunflower oil are presented in Table 2.

Soybean oil has the highest mono-unsaturated fatty acids (especially oleic acid, C18:1) and also the lowest ratio of unsaturated to saturated fatty acids (3.93) in comparison to other oils. The highest linolenic acids (C18:3, ω-3), highest ratio of unsaturated to saturated and also greatest ω-3 to ω-6 fatty acids ratio were seen in linseed oil. It has been reported that using linseed and rapeseed oils, possible to increase the ω-3 fatty acids content in the form of linolenic acid, which is the precursor of the whole ω-3 family (El Yamany, 2008).

Sunflower oil has the highest fatty acid percent of linoleic acid (C18:2, ω-6), as omega 6 fatty acids and therefore this plant oil has the lowest ω-3 to ω-6 fatty acid ratio in comparison to other experimental oils. Base on fatty acid composition of soybean, linseed and sunflower oil are source of omega 9, 3 and 6 fatty acids, respectively.

The effect of dietary oil source, L-carnitine supplement and its interaction on growth performance of Japanese quail is shown in Table 3. The FI of birds was neither affected by oil source, nor by L-carnitine supplementation. Diets which contained soybean oil led to the highest BWG and the lowest FCR quails while L-carnitine supplement had no effect on BWG and FCR.

There was an interaction between oil sources and L-carnitine on BWG of quails as addition of L-carnitine to the diets contained linseed oil improved the BWG of birds. Feeding quails with linseed oil along with L-carnitine supplement led to the lowest FCR of the birds.

The main effects showed that soybean oil improved BWG and FCR of quails. This may be due to the optimum ratio of unsaturated to saturated fatty acids of soybean oil. The important factor affecting the amount of fats metabolizable energy is their digestibility and it is dependent on the length of carbon chain and the degree of saturation of fatty acids (Leeson and Summers, 1997), Moreover, it has been reported that the optimum ratio of unsaturated to saturated fatty acids for maximizing fat digestibility and metabolizable energy value of fat is around 3 to 1 (Lesson and Atteh, 1995).

Also, soybean oil had higher content of oleic acid (C18:1) compared to the other oils (Table 3), in which this fatty acid plays a direct role in the absorption of saturated fatty acids in the lumen and mucosa cells and facilitate their absorption (Leeson and Atteh, 1995). Therefore, soybean oil may improve digestibility, metabolizable energy, BWG and FCR of quails. Scaife et al. (1994) fed female broiler chicks by different lipids sources; such as beef tallow, soybean oil, canola oil, marine fish oil or a mixture of these oils and observed improved live body weight using soybean oil while beef tallow fat had the poorest FCR. Dietary L-carnitine supplementation individually had not any effect on FI, BWG and FCR of quails.

Table 3 Effects of dietary oil source and L-carnitine supplementation on feed intake, body weight gain (g.bird^{-1}.day^{-1}) and feed conversion ratio of female Japanese quails

Main factors	Feed intake	Body weight gain	Feed conversion ratio
Oil			
Soybean oil	21.60	5.857[a]	3.690[b]
Linseed oil	21.43	5.706[b]	3.760[ab]
Sunflower oil	21.85	5.696[b]	3.840[a]
L-carnitine level (mg/kg)			
0	21.50	5.712	3.768
50	21.76	5.794	3.759
Treatments			
Soybean oil	21.48	5.832[a]	3.684[b]
Linseed oil	21.35	5.571[b]	3.835[ab]
Sunflower oil	21.68	5.733[ab]	3.782[ab]
Soybean oil + L-carnitine	21.73	5.882[a]	3.696[ab]
Linseed oil + L-carnitine	21.51	5.842[a]	3.683[b]
Sunflower + L-carnitine	22.04	5.658[ab]	3.896[a]
PSEM	0.195	0.061	0.06
Source of variation			
Oil	0.1283	0.0376	0.0442
L-carnitine	0.1283	0.1265	0.8555
Oil × L-carnitine	0.1565	0.0428	0.0452

The means within the same column with at least one common letter, do not have significant difference (P>0.05).
PSEM: pooled standard error of mean.

This finding is in agreement with other researchers (Corduk *et al.* 2007; Lien and Horng, 2001; Kheiri *et al.* 2011) who, reported that L-cranitine supplementation did not have any effect on growth performance of broiler chicks. Sarica *et al.* (2005) demonstrated that dietary L-carnitine supplementation at 0, 30, 40 and 50 ppm did not affect FI, growth and FCR of Japanese quails. Also in another experiment, Sarica *et al.* (2007) showed that using 50 ppm L-carnitine in diet of Japanese quails containing 1% fish oil or sunflower oil had no effect on growth performance. These two reports are in contrast to our results. On the other hand, results of Parsaeimehr *et al.* (2012) showed that using L-carnitine (300 ppm) in diets containing 5% animal fat improved BWG and FCR of broiler chicks compared with diets containing 5% soybean oil. Sayed *et al.* (2001) have demonstrated that addition of L-carnitine (50 ppm) to diet contained 2 and 4% sunflower oil increased FI, BWG and FCR compared to the control group.

In the present study, growth performance of quails improved by using L-carnitine when the diet contained linseed oil as ω-3 fatty acids, and this may be related to improved oxidation of fatty acids in linseed oil by L-carnitine supplement. Therefore, the growth response of birds to L-carnitine supplementation was affected by dietary oil source.

On the other hands, results of Sadeghzadeh *et al.* (2014) showed that optimum production of chicks was occurred by supplementation of 150 ppm L-carnitine and 115% of methionine requirement of female broiler chicks recommended by NRC when the diet contained 3% soybean oil.

According to results of this research, response of quails to L-carnitine supplement may be related to dietary composition such as amino acids level especially methionine and lysine; which are precursors of endogenous biosynthesis of L-carnitine (Arslan, 2006). In addition, type of diet and level of oil also might be effective.

The results indicated that different types of plant oils and L-carnitine supplementation had no significant (P<0.05) effect on body weight, relative weights of liver and gizzard of quails (Table 4). Linseed and sunflower oil significantly increased heart weight and its relative weight of quail (P<0.05). Addition of L-carnitine to the diet contained sunflower oil reduced relative weight of liver and heart. Results of Nobakht *et al.* (2011) showed that sunflower, soybean, and canola oil as well as its combination in chicks' diet had no effect on relative weight of liver in birds. Also, Lien and Horng (2001) showed that supplementation of 160 ppm L-carnitine to broilers diet had no effect on liver weight.

The results of Rabie and Szilagyi (1998) showed that relative weights of liver, gizzard and heart of chicks were not affected by feeding 50 ppm L-carnitine. Other researchers also showed that the relative weight of liver, gizzard, and heart of quails were not affected by supplementation of 0, 30, 40 and 50 ppm L-carnitine (Sarica *et al.* 2005). The absolute and relative heart weights of quails were significantly depressed by carnitine supplemented diets contained sunflower oil compared to other dietary treatments. In the same way, the liver relative weight was also reduced significantly (P<0.05).

Table 4 Effects of dietary oil source and L-carnitine supplementation on live body and some internal organ weight and its relative weight of them to live body of female quails

Main factors	Weight (g)				Relative weight (%)		
	Live body	Liver	Heart	Gizzard	Liver	Heart	Gizzard
Oil							
Soybean oil	227.70	4.717	1.200[b]	4.317	2.102	0.5415[b]	1.902
Linseed oil	214.20	4.517	1.600[ab]	4.900	2.133	0.7482[a]	2.306
Sunflower oil	235.10	5.167	1.650[a]	5.000	2.201	0.7068[a]	2.140
L-carnitine level (mg/kg)							
0	228.20	5.155	1.544	4.989	2.270	0.6784	2.189
50	219.80	4.444	1.422	4.489	2.022	0.6526	2.044
Treatments							
Soybean oil	234.90	5.167	1.200	5.000	2.200[ab]	0.5097[c]	2.124
Linseed oil	228.00	4.633	1.600	4.967	2.066[ab]	0.6974[abc]	2.198
Sunflower oil	221.70	5.667	1.833	5.000	2.543[a]	0.8280[a]	2.244
Soybean oil + L-carnitine	210.50	4.267	1.200	3.633	2.005[ab]	0.5733[bc]	1.680
Linseed oil + L-carnitine	200.50	4.400	1.600	4.833	2.201[ab]	0.7990[ab]	2.415
Sunflower + L-carnitine	248.50	4.667	1.467	5.000	1.860[b]	0.5855[bc]	2.036
PSEM	14.546	0.505	0.161	0.570	0.169	0.05	0.23
Source of variation							
Oil	0.3824	0.4445	0.0312	0.4569	0.8377	0.0033	0.2481
L-carnitine	0.4935	0.1105	0.3709	0.3043	0.0969	0.5396	0.4548
Oil × L-carnitine	0.1535	0.7183	0.4460	0.3044	0.0494	0.0092	0.3767

The means within the same column with at least one common letter, do not have significant difference (P>0.05).
PSEM: pooled standard error of mean.

These results were in accordance with previous studies in which Koksal et al. (2011) have observed reduction of heart and liver weights in broiler diets supplemented with 100 ppm carnitine. The reduced weights of the heart and the liver at a lesser extend recorded in L-carnitine supplemented birds which may be related to increasing lipid utilization (Koksal et al. 2011).

The results indicated that experimental dietary treatments had significant effect on some blood biochemical parameters of quails (Table 5) (P<0.05). Sunflower oil, in comparison with the other plant oils, reduced the blood level of total protein, albumin and globulin while soybean oil decreased cholesterol level in sera of quails. Dietary L-carnitine supplementation increased the concentration of total protein and globulin, but reduced the cholesterol in the blood sera of female quails.

The lowest concentration of total protein and globulin in quail's serum were found in quails fed on sunflower oil without L-carnitine supplement (P<0.05). On the other hand, the lowest concentration of blood cholesterol was observed in birds that fed on diets containing soybean oil plus L-carnitine.

Supplementations of L-carnitine to the diet contained linseed oil significantly increased globulin and reduced cholesterol level in the serum of quails (Table 5) (P<0.05). Reducing the blood cholesterol of quails using soybean oil may be related to inhibition of hepatic 3-hydroxy-3-methylglutaryl-CoA reductase (HMG-CoA reductase) activity, a key enzyme which is regulating cholesterol synthesis (Lee et al. 2003) and or stimulate bile formation and biliary lipid secretion, particularly cholesterol output in bile by lecithin in soybean oil (LeBlanc et al. 2003).

L-carnitine supplementation decreased blood cholesterol of quail which is in agreement with those of Rezaei et al. (2007) and Parsaeimehr et al. (2012). L-carnitine may increase fatty acid oxidation and thus reduced blood cholesterol levels in quails.

On the other hand, the results of Parizadian et al. (2011) indicated that quails fed with diet contained L-carnitine supplementation (250 ppm), had lower blood cholesterol compared with the control group.

Sayed et al. (2001) also, have demonstrated that addition of L-carnitine (50 ppm) to diet contained 2 and 4% of sunflower oil decreased serum cholesterol compared to the control group.

The plasma cholesterol-lowering effects of L-carnitine would be associated with several possible processes, including increase of cholesterol turnover due to increased conversion of cholesterol to bile acids and biliary excretion or due to modified repartition of whole body cholesterol (Arslan, 2006).

There were some interactions between blood cholesterol level and dietary plant oil as well as L-carnitine supplement. This may be related to fatty acid composition of oils. Sunflower oil reduced protein, albumin and globulin concentration but L-carnitine increased them.

Table 5 Effect of treatments on blood biochemical parameters in female quails

Treatments	Total protein (g.dL^{-1})	Albumin (g.dL^{-1})	Globulin (g.dL^{-1})	Cholesterol (mg.dL^{-1})
Main factors				
Oil				
Soybean oil	3.483^a	1.752^a	1.732^{ab}	125.0^b
Linseed oil	3.433^a	1.478^{ab}	1.955^a	187.8^a
Sunflower oil	2.417^b	1.198^b	1.218^b	188.8^a
L-carnitine level (mg/kg)				
0	2.833^b	1.401	1.432^b	190.6^a
50	3.389^a	1.551	1.838^a	143.8^b
Treatments				
Soybean oil	3.067^{ab}	1.677	1.390^{bc}	164.0^{abc}
Linseed oil	3.267^{ab}	1.483	1.783^{ab}	246.7^a
Sunflower oil	2.167^b	1.043	1.123^c	161.3^{abc}
Soybean oil + L-carnitine	3.900^a	1.827	2.073^a	86.0^c
Linseed oil + L-carnitine	3.600^{ab}	1.473	2.127^a	129.0^{bc}
Sunflower + L-carnitine	2.667^{ab}	1.353	1.313^{bc}	216.3^{ab}
PSEM	0.314	0.196	0.196	22.173
Source of variation				
Oil	0.0082	0.047	0.0081	0.0208
L-carnitine	0.051	0.367	0.0265	0.0237
Oil × L-carnitine	0.0241	0.136	0.0078	0.0054

The means within the same column with at least one common letter, do not have significant difference (P>0.05).
PSEM: pooled standard error of mean.

The higher concentration level of total protein and globulin by feeding L-carnitine may be related to its protein sparing action (Jalali Haji-Abadi *et al.* 2010) and reduced using of amino acid precursor (lysine and methionine) for L-carnitine biosynthesis.

CONCLUSION

This investigation proved that 50 ppm L-carnitine could improve growth performance of quails which the diet containing 4% linseed oil. The present study also showed that soybean oil improved the FCR of Japanese quail and total protein, albumin and globulin of blood female birds reduced by sunflower oil. Overall, results of this experiment showed that growth performance and blood biochemical response of Japanese quail to dietary supplementation of L-carnitine related to dietary oil source.

REFERENCES

Arslan C. (2006). L-carnitine and its use as a feed additive in poultry feeding a review. *Rev. Med. Vet.* **157,** 134-142.

Burlikowska K., Piotrowska A. and Szymeczko R. (2010). Effect of dietary fat type on performance, biochemical indices and fatty acids profile in the blood serum of broiler chickens. *J. Anim. Feed Sci.* **19,** 440-451.

Christie W.W. (1990). Gas Chromatography and Lipids, a Practical Guide. Oily Press, Bridgwater, UK.

Corduk M., Ceylan N. and Ildiz F. (2007). Effects of dietary energy density and L-carnitine supplementation on growth performance, carcass traits and blood parameters of broiler chickens. *South African J. Anim. Sci.* **37,** 65-73.

El-Yamany A.T., El-Allawy H.M.H., Abd El-Samee L.D. and El-Ghamry A.A. (2008). Evaluation of using different levels and sources of oil in growing Japanese quail diets. *American-Eurasian J. Agric. Environ. Sci.* **3,** 577-582.

Harpaz S. (2005). L-carnitine and its attributed functions in fish culture and nutrition. A review. *Aquaculture.* **249,** 3-21.

Jalali Haji-Abadi S.M.A., Mahboobi-Soofiani N., Sadeghi A.A., Chamani M. and Riazi G. (2010). Effects of supplemental dietary L-carnitine and ractopamine on the performance of juvenile rainbow trout (*Oncorhynchus mykiss*). *Aqua. Res.* **41,** 1582-1591.

Kheiri F., Pourreza J., Ebrahimnezhad Y. and Jalali Haji-Abadi S.M.A. (2011). Effect of supplemental ractopamine and L-carnitine on growth performance, blood biochemical parameters and carcass traits of male broiler chicks. *African J. Biotechnol.* **68,** 15450-15455.

Koksal B.H., Kucukersan M.K. and Cakin K. (2011). Effects of L-carnitine and / or inulin supplementation in energy depressed diets on growth performance, carcass traits, visceral organs and some blood biochemical parameters in broilers. *Rev. Med. Vet.* **162,** 519-525.

LeBlanc M.J., Brunet S., Bouchard G., Lamireau T., Yousef I.M.,

Gavino V., Lévy E. and Tuchweber B. (2003). Effects of dietary soybean lecithin on plasma lipid transport and hepatic cholesterol metabolism in rats. *J. Nutr. Biochem.* **14**, 40-48.

Lee K.W., Everts H., Kappert H.J., Frehner M., Losa R. and Beynen A.C. (2003). Effects of dietary essential oil components on growth performance, digestive enzymes and lipid metabolism in female broiler chickens. *Br. Poult. Sci.* **44**, 450-457.

Leeson S. and Atteh J.O. (1995). Utilization of fats and fatty acids by turkey poults. *Poult. Sci.* **74**, 2003-2010.

Leeson S. and Summers J.D. (1997). Commercial Poultry Nutrition. University Books, Guelph, Canada.

Lien T.F. and Horng Y.M. (2001). The effect of supplementary dietary L-carnitine on the growth performance, serum components, carcass traits and enzyme activities in relation to fatty acid β-oxidation of broiler chickens. *Br. Poult. Sci.* **42**, 92-95.

Nobakht A., Tabatbaei S. and Khodaei S. (2011). Effects of different sources and levels of vegetable oils on performance, carcass traits and accumulation of vitamin e in breast meat of broilers. *Curr. Res. J. Biol. Sci.* **3**, 601-605.

NRC. (1994). Nutrient Requirements of Poultry, 9th Rev. Ed. National Academy Press, Washington, DC., USA.

Parizadian B., Shams-shargh M. and Zerehdaran S. (2011). Study the effects of different levels of energy and L-carnitine on meat quality and serum lipids of Japanese quail. *Asian J. Anim. Vet. Adv.* **6**, 944-952.

Parsaeimehr K., Farhoomand P. and Najafi R. (2012). The effects of L-carnitine with animal fat on performance, carcass characteristics and some blood parameters of broiler chickens. *Ann. Biol. Res.* **3**, 3663-3666.

Rabie M.H. and Szilagyi M. (1998). Effects of L-carnitine supplementation of diets differing in energy levels on performance, abdominal fat content and yield and composition of edible meat of broilers. *Br. J. Nut.* **80**, 391-400.

Rabie M.H., Szilagyi M. and Gippert T. (1997a). Effects of dietary L-carnitine supplementation and protein level on performance and degree of meatness and fatness of broilers. *Acta Biol. Hung.* **48**, 221-239.

Rabie M.H., Szilagyi M., Gippert T., Votisky E. and Gerendai D. (1997b). Influence of dietary L-carnitine on performance and carcass quality of broiler chickens. *Acta Biol. Hung.* **48**, 241-252.

Rezaei M., Attar A., Ghodratnama A. and Kermanshahi H. (2007). Study the effects of different levels of fat and L-carnitine on performance, carcass characteristics and serum composition of broiler chicks. *Pakistan J. of Biol. Sci.* **10**, 1970-1976.

Sadeghzadeh S.S., Yazdian M.R. and Nasr J. (2014). The effects of different levels of L-carnitine and methionine on performance and blood metabolites in female broiler. *Res. Opin. Anim. Vet. Sci.* **4**, 427-431.

Sarica S., Corduk M. and Kilinc K. (2005). The effect of dietary L-carnitine supplementation on growth performance, carcass traits and composition of edible meat in Japanese quail (*Coturnix coturnix japonica*). *J. Appl. Poult. Res.* **14**, 709-715.

Sarica S., Corduk M., Ensoy U., Basmacioglu H. and Karatas U. (2007). Effects of dietary supplementation of L-carnitine on performance, carcass and meat characteristics of quails. *South African J. Anim. Sci.* **37**, 189-201.

SAS Institute. (2004). SAS®/STAT Software, Release 9.1. SAS Institute, Inc., Cary, NC. USA.

Sayed A.N., Shoeib H.K. and Abdel-Raheem H.A. (2001). Effect of dietary L-carnitine on the performance of broiler chickens fed on different levels of fat. *Assiut Vet. Med. J.* **45**, 37-47.

Scaife J.R., Moyo J., Galbraith H., Michie W. and Campbell V. (1994). Effect of different dietary supplemental fats and oils on the tissue fatty acid composition and growth of female broilers. *Br. Poult. Sci.* **35**, 107-118.

Shahriar H.A., Rezaei A., Lak A. and Ahmadzadeh A. (2007). Effect of dietary fat sources on blood and tissue biochemical factors of broiler. *J. Anim. Vet. Adv.* **6**, 1304-1307.

14

Genetic Analysis of Egg Quality Traits in Bovan Nera Black Laying Hen under Sparse Egg Production Periods

A. Sylvia John-Jaja[1], A.R. Abdullah[1] and C. Samuel Nwokolo[2*]

[1] Department of Animal Science, College of Agriculture, Babcock University, Ilshan Remo, Nigeria
[2] Department of Physics, Faculty of Science, University of Calabar, Calabar, Nigeria

*Correspondence E-mail: nwokolosc@stud.unical.edu.ng

ABSTRACT

The present research was designed to examine the genetic analysis of egg weight, egg yolk weight and egg albumen weight of Bovan Neva Black laying hens at 25, 51 and 72 weeks. For this purpose, thirty birds were selected from the layer flock in the Babcock University Teaching and Research Farm. They were individually housed in separate labeled battery cage. A total of thirty eggs were collected daily from the birds continuously in five days of egg production, at each age of 25, 51 and 72 weeks. The total number of eggs collected at each age were 150. Data collected for egg weight, egg yolk weight and egg albumen weight were used to evaluate the descriptive statistics, influence of age and Pearson correlation coefficient on different age groups. The mean values of the traits revealed a consistent increase in egg weight 55.02-63.29 g, egg yolk weight 13.14-19.39 g and egg albumen weight 35.52-39.21 g by aging. A significant positive genetic correlation was obtained among traits with linear regression equations at different age groups. Restricted maximum likelihood (REML) of Wombat software was used to obtain the repeatability and heritability estimates. From the results, it was revealed that all the traits recorded high estimates of heritability and repeatability while egg weight is more heritable and repeatable than egg yolk weight and egg albumen weight indicating that fewer records would be required to adequately characterize the inherent production ability of each trait as laying age progressed.

KEY WORDS Bovan Nera Black, egg, egg quality traits, heritability, repeatability.

INTRODUCTION

Poultry farming occupies an important place in the economy of Nigeria. It contributes immensely to three major sectors (petroleum, mining, and agriculture) in Nigeria's economy and has evolved from subsistence farming to an extremely profit driven commercial enterprise. Poultry population in Nigeria was put at 114.3 million comprising 82.4 million chickens, 11% of which was commercially raised, and 31.9 million other poultry including pigeons, ducks, guinea fowls and turkey. This could be attributed to high productivity, high feed conversion efficiency, im-

proved fertility, hatchability, growth rate, egg yield and meat quality through genetics and breeding within a short time and without a huge investment when compared with other livestock breeding. Poultry raised for meat and eggs are important sources of edible animal protein. Poultry meat accounts for 30% of global meat consumption. The worldwide average per capital consumption of poultry meat has nearly quadrupled since the 1960's (FAO, 2009). This transformation could be attributed to the wide spread of the necessity of animal protein intake per day of an average Nigerian and estimating genetic parameters especially heritability estimates of internal and external egg quality traits of exotic and local hen thereby improving the egg quality traits considering the economic need to increase edible animal protein so as to match the protein requirement of the teaming population through genetic breeding.

Egg quality traits are those that affect its acceptability by consumers. Thus, to maintain the superiority in the total egg quality, routine genetic and breeding experimentation should be carried out continuously through genetic parameters for a number of chicken traits, particularly the improved commercial breeds so as to select the best performers with respect to important economic traits through concentrating and enhancing the manifestation of the gene controlling these traits.

The improved stock will be conserved and multiplied for productive purposes. In this way, commercial chicken will boost the Nigeria poultry industry, thus providing a buffer against shortages of poultry products. Therefore, the aim of this study was to determine the genetic analysis of egg quality traits in Bovan Nera Black laying hen under sparse egg production periods in the tropics.

MATERIALS AND METHODS

The experimental site, Ilara, is located at the Teaching and Research Farm of Animal Science Department, Babcock University. Ilara is situated between Latitude 6.867 °N and Longitude 3.717 °E with an altitude of 235.2 meters above sea level in tropical rainforest belt of Nigeria. It has an annual rainfall of 1200 mm, 65% mean relative humidity and 21.40 °C mean temperature. The research lasted for 54 weeks.

Day-old pullets were randomly selected and purchased from the base population of Nera Black hens and kept on litter till 18 weeks before they are moved to the battery cage. The chicks were protected from cold during the first four weeks of brooding. During lay, the birds were fed twice daily and water was administered accordingly with feed composed of 16% crude protein, other constituents of their feed includes vitamins, minerals and amino acid. Water was provided. At inception, birds were quarantined separately for 7 days and dewormed. . The birds were routinely vaccinated at various stages of development against diseases such as Newcastle, Gumboro and Coccidiosis. Thirty eggs were collected daily in the morning at 8 a.m., afternoon at 2 p.m. and then the final collection was made in the evening at 6:00 p.m. for five days. One hundred and fifty eggs were collected from three age groups (25, 51 and 72 weeks) and 450 for the overall ages of the birds for egg yolk weight and albumen weight.

Two experimental designs were adopted in the course of this study via completely randomized design (CRD) and visual appraisal. The CRD was used to select healthy layers after quarantine and vaccination while visual appraisal was employed to select a total of thirty layers capable of laying 5-6 eggs weekly; and rest for 1-2 days after monitoring their laying cycle and patterns between 21-24 weeks.

The egg weight was measured using a 0.09 sensitive digital scale. The yolk carefully separated from the albumen and placed in a petri dish for weighing. Simultaneously, the associated albumen is weighed.

The descriptive statistics, linear regression and correlation were obtained using statistical analytical system program (SAS, 1999) while heritability and repeatability estimates for egg quality traits were obtained employing a tool for mixed model analysis in quantitative genetics by restricted maximum likelihood (REML) of as follows:

$$Y_{ijkl} = \mu + age_i + bird_j + bird \times age_k + e_{ijkl}$$

Y_{ijkl}: observation on the ith egg quality traits of the ith birds.
μ: mean of the population.
age_i: random effect of age.
$bird_j$: random effect of bird.
$age_i + bird_j$: random effect of age and bird.
e_{ijkl}: residual effect.

RESULTS AND DISCUSSION

Egg production

The descriptive statistics of egg weight, egg albumen weight and egg yolk weight varied for one age group to another as presented in Tables 1, 2 and 3. This could be attributed to the genetic potential, prevalent environment factor influencing each trait and age of the layers as age is a major factor that determines, to a great extent the growth and physiological development of the traits. The least egg albumen weight value 35.52 g was registered at 25 weeks while at 72 weeks the birds reported a maximum value of 39.21 g.

There was a consistent increase in egg albumen weight as the age of the hen advanced. These values are comparable with the report in the literature. Islam and Dutta (2010) registered 36.10 g at the 48 weeks for Rhode Island Red; Tadesse et al. (2015) reported 33.19 g and 34.54 g for Isa Bovan and Brown at 32 weeks; 31.53 g, 31.19 g and 33.18 g for three pure lines of white leghorns respectively at 40 weeks; Rath et al. (2015) recorded 35.76 g at 50 weeks for white leghorns.

However, lower values 19.71 g and 23.77 g reported for Onagadori and white leghorns, respectively at 20-34 weeks by Goto et al. (2015); Begli et al. (2010) reported 22.66 g at 30 weeks for Iranian fowl; 25.14 g at 32 weeks for Koekoek breed by Tadesse et al. (2015); Islam and Dutta (2010) recorded 18.51 g at 48 weeks for Fayoumi breed; Khalil et al. (2013) reported 24.2 g and 25.6 g for Golden Montazah and white leghorn, respectively at 60 weeks.

Table 1 Descriptive statistics of egg weight

Egg weight (g)	N	Mean	Standard error	Coefficient of variation
Age of bird				
25	150	55.02	0.40	8.83
51	150	62.20	0.47	9.17
72	150	63.29	0.45	8.77

N: number of observation per age group.

Table 2 Descriptive statistics of egg yolk weight

Egg weight (g)	N	Mean	Standard error	Coefficient of variation
Age of bird				
25	150	13.14	0.10	8.95
51	150	15.37	0.41	9.19
72	150	15.97	0.14	18.72

N: number of observation per age group.

Table 3 Descriptive statistics of egg albumen weight

Egg weight (g)	N	Mean	Standard error	Coefficient of variation
Age of bird				
25	150	35.52	0.39	13.52
51	150	36.09	0.47	14.67
72	150	39.21	0.49	16.57

N: number of observation per age group.

Several researchers reported higher values of egg albumen weight. Akintola *et al.* (2011) recorded 45.89 g, 45.42 g, 45.77 g and 45.80 for graded dosage levels of orabolin with 0 ug, 10 ug, 20 ug and 30 ug respectively at 69 weeks. Minelli *et al.* (2007) registered 41.5 g at 28-32 weeks, 42.9 g at 47-50 g and 43.8 g at 70-73 weeks for laying phases of commercial breed. This variation could be attributed to the breed differences, the age of the layers, management procedures and environmental temperature.

An increasing trend was observed for the standard error of egg albumen weight for different age groups. The trait recorded 0.39 g at 25 weeks, 0.47 g at 51 weeks, 0.49 g at 72 weeks with 0.27 g for the overall mean of the birds. These values are in agreement with the report of other researchers. Goto *et al.* (2015) recorded 0.31 g for Onagadori breed and 0.47 g for white leghorns; Sreenivas *et al.* (2013) registered 0.39-0.43 g for different pure lines of white leghorns.

The coefficient of variation for the egg albumen weight ranged between 13.52-16.57% at 25-72 weeks. These values are within the range of 14.2% reported by Mube *et al.* (2014); 15.29% obtained by Begli *et al.* (2010).

The mean egg weight recorded 55.02 g at 25 weeks, 62.20 g at 51 weeks and 63.29 g at 72 weeks with a corresponding mean value of 60.17 g for the overall ages of the hen indicating an increasing trend. These results are similar to 57.78 g recorded by Rath *et al.* (2015) for egg weight at 50 weeks for white leghorns; 48.1-63.9 g registered for single comb, while leghorn at 25-65 weeks reported 50.01-53.89 g obtained for three pure lines and one control lines of white leghorns at 40 weeks by Sreenivas *et al.* (2013);

60.3-62.4 g recorded for white egg lines of Lohmann Tierzucht Gambh at 67-70 weeks andBrown egg line of Lohmann Tierzucht Gambh at 32-36 weeks by Blanco *et al.* (2014); 50.6-55.6 g obtained for white leghorn at 26-54 weeks by Sabri *et al.* (1999a); 45.67-51.33 g reported for white leghorn (IWN line) at 32-56 weeks by Paleja *et al.* (2008); 60.6 g, 60.3 g and 61.1 g registered for ATAK-S commercial layers hybrids at 52 weeks employing incandescent bulb, mini fluorescent and light-emitting diodes by Kamanli *et al.* (2015); 58.0-62.1 g reported for white leghorn at 35-65 weeks by Ledur *et al.* (2002); 62.0-67.3 g observed for commercial layers at 28-73 weeks by Minelli *et al.* (2007); 58.75 g, 60.27 g and 48.8 g obtained for Isu Brown, Bovan Brown and PotchetstroomKoekoek breeds at 32 weeks by Tadesse *et al.* (2015); 51.09-61.04 g observed for Vanavoija male line (PDI) at 32-60 weeks by Padhi *et al.* (2015); 63.9-65.2 g for commercial layers at 60-80 weeks by Molnar *et al.* (2016); 64.78 g, 63.46 g and 47.79 recorded for Isa Brown, Novan Brown, Koekoek respectively under intensive production system and 58.92 g, 59.32 g and 47.53 g reported for Isa Brown, Bovan Brown and Koekoek respectively under village production system by Tadesse *et al.* (2015); 51.9-55.6 g for Isa Brown under graded dosage levels of Ovabolin (0 ug, 10 ug, 20 ug and 30 ug) at 69 weeks by Akintola *et al.* (2011); 61.58 g for white leghorn group and 60.72 g for Rhode Island Red at 38 weeks by Lukanov *et al.* (2015); 56.6 g for young (22-29 weeks) and 68.6 g for old of Lohmann Brown laying hens, 66.4 g for young (36-73 weeks) and 71.6 g for old (64-71 weeks) of Cobb 500 broiler breeders by Tumova and Gous (2012); 53.30 g and 56.72 g for Block Olympia and H and

N Brown Nick breeds respectively between 36-46 weeks by (Ewa *et al.* 2005).

However, lower value 42.87 g was obtained for Iranian fowl at 30 weeks by Begli *et al.* (2010); 34.84 g for Onagadori breed and 41.01 g for white leghorn at 20-34 g for both breed, by Goto *et al.* (2015); 46.80 g and 39.83 g for Cobb 500 of Broiler and Fayoumi breeds at 48 weeks by Islam and Dutta (2010); 44.0 g and 45.7 g for Golden Montazah and white leghorn respectively at 120 weeks by Khalil *et al.* (2013); whereas, Petek *et al.* (2007) reported higher values of 74.11 g, 73.20 g and 69.70 g for commercial brown egg laying hens under effects of non-feed removal molting methods (non-molting control, Barley, and Alfalfa, respectively). The variation could be attributed to the breed differences, the age of the layers and environmental temperature as recommendation by (FAO, 1998).

There was a consistent increase in the standard error of egg weight at different age groups except at 72 weeks. At 25 weeks, 0.40 g was registered, 0.47 g at 51 weeks and 0.45 g at 72 weeks.

These value are similar to the report in the literature. Begli *et al.* (2010) recorded 0.17 g; Khalil *et al.* (2013) registered 0.10-0.14 g for golden monstazah and white leghorn breeds; 0.73-0.74 recorded at 60 and 80 weeks by Molnar *et al.* (2016); 0.10-0.16 g between 32-60 weeks reported by Padhi *et al.* (2015); 0.26 g at 50 weeks for white leghorn by Rath *et al.* (2015); 0.42-0.48 g for three strains of white leghorn recorded by (Sreenivas *et al.* 2013).

At week 25, the birds recorded the least coefficient of variation of egg weight value of 8.83% while at 51 weeks, the birds recorded the maximum value of 9.17% with a corresponding value of 10.81% for total age of the hen. These values are similar to the range of 8.9-9.98% reported by Mube *et al.* (2014) and 11.75% obtained by Begli *et al.* (2010); and 8.34% registered by (Zhang *et al.* 2005). The bird egg yolk weight follows a progressive increase as the age of the hen advanced.

Egg yolk weight recorded 13.14 g, 15.37 g and 19.39 g at 25, 51 and 72 weeks, respectively. These values are in agreement with 16.17 g at 50 weeks for white leghorns as reported by Rath *et al.* (2015); 14.16 g, 14.70 g and 15.58 g, respectively were recorded for three different strain of white leghorn at 40 weeks registered by Sreenivas *et al.* (2013); 14.5 g, 17.0 g and 17.1 g at 28-32 weeks, 47-50 weeks and 70-73 recorded for commercial layers recorded by Minelli *et al.* (2007); 14.88 and 16.40 g at 48 weeks for Fayonmi and Sanali breeds respectively Islam and Dutta (2010); 15.65 g, 15.07 g, 14.82 g and 14.94 g at 69 weeks for Brown laying birds using by Akintola *et al.* (2011); 16.69 g and 16.14 g for Isa brown breed employing intensive and village management system; 15.39 g and 15.97 g for Bovan Brown using intensive and village management

system, and 14.54 g and 15.94 g for Koekoek breed employing intensive and village management system at 32 weeks by Tadesse *et al.* (2015).

However, lower values 12.91 g was obtained for Iranian fowl at 30 weeks by Begli *et al.* (2010); 10.38 g for Onagadori breed and 11.61 g for white leghorn at 20-34 weeks by Goto *et al.* (2015); 9.60 g for broiler (Cob 500) and 11.20 g for Rhode Island Red at 48 weeks registered by Islam and Dutta (2010).

The discrepancies could be due to the breed differences, age of the layers, management system and environmental temperature. There was a successive increase in the standard error of egg yolk weight at different age groups that is, 0.10 g at 25 weeks, 0.14 g at 51 weeks and 0.15 g at 72 weeks. These results agree favourably with 0.05 g obtained by Rath *et al.* (2015); 0.12-0.13 g for different pure lines of white leghorns by Sreenivas *et al.* (2013); 0.05 g recorded by Begli *et al.* (2010); and 0.22-0.28 for Onagadori and white leghorn respectively registered by Goto *et al.* (2015). The same trend was observed for coefficient of variation of egg yolk weight that ranged 8.95-9.45% at 25, 52 and 72 weeks. These are similar to the report of 11.2% observed by Mube *et al.* (2014); 10.31% obtained by Begli *et al.* (2010); and 8.99% registered by Zhang *et al.* (2005). The influence of age on egg weight, egg albumen weight and egg yolk weight are appreciable as observable changes resulted from increase in age of the hen from 25-51, 51-72 and 25-72 weeks are presented in Table 4.

The rate of change in egg weight recorded 7.180, 1.090 and 8.272 between 25-51 weeks, 51-72 weeks and 25-75 weeks respectively; egg yolk weight registered rate of change of 2.230, 4.020 and 6.250 between 25-51 weeks, 51-72 weeks and 25-75 weeks respectively; and albumen weight reported 3.690, 3.120 and 0.570 between 25-51 weeks, 51-72 weeks and 25-75 weeks respectively. It could be observed that egg weight recorded the highest influence of age variance at 25-51 weeks and 25-72 weeks and the lowest value of 1.090 at 51-72 weeks compared to egg albumen weight and egg yolk weight at 25-51, 25-72 and 51-72 weeks. This variation indicates that age is major determinant in the growth and development of egg quality traits which tends to influence the external egg quality trait (egg weight) than the internal egg quality traits (egg albumen weight and egg yolk weight) in this study. This is similar to the report of Molnar *et al.* (2016), Abdallah *et al.* (1995), Sabri *et al.* (1999a) that egg weight increased with age and weighted regression analysis of periods indicated a linear increase.

Genetic correlation

Pearson correlation coefficient between egg quality traits at 25, 51 and 72 weeks are presented in Table 5.

Table 4 Regression equation showing the rate of change of egg quality traits at different age groups

Egg quantity traits	N	Age group	Regression equation	Rate of change
Egg weight	150	25-51	$EW_{51} = 7.180 + EW_{25}$	7.180
	150	51-72	$EW_{72} = 1.090 + EW_{51}$	1.090
	150	25-72	$EW_{72} = 8.272 + EW_{25}$	8.272
York Weight	150	25-51	$EW_{51} = 2.230 + EW_{25}$	2.230
	150	51-72	$EW_{72} = 4.020 + EW_{51}$	4.020
	150	25-72	$EW_{72} = 6.250 + EW_{25}$	6.250
Albumen Weight	150	25-51	$EW_{51} = 3.690 + EW_{25}$	3.690
	150	51-72	$EW_{72} = 3.120 + EW_{51}$	3.120
	150	25-72	$EW_{72} = 0.570 + EW_{25}$	0.570

N: number of observation per age group; EW: egg weight; EY: egg yolk weight and EAW: egg albumen weight.

Table 5 Pearson correlation coefficients of egg quality traits

Traits/age	EW	EAW	EYW
25 weeks			
EW	1	0.521^{**}	0.641^{**}
EAW	-	1	0.911^{**}
EYW	-	-	1
51 weeks			
EW	1	0.521^{**}	0.641^{**}
EAW	-	1	0.911^{**}
EYW	-	-	1
72 weeks			
EW	1	0.521^{**}	0.641^{**}
EAW	-	1	0.911^{**}
EYW	-	-	1

EW: egg weight; EAW: egg albumen weight and EYW: Egg yolk weight.
** ($P<0.005$).

A significant positive and similar high magnitude correlations were recorded between egg weight and egg yolk weight (0.641); egg weight and egg albumen weight (0.521) but egg yolk weight was affected mostly by the weight of the albumen (0.911) for different age groups. This could be attributed to similar variation in the additive variance and presume non-additive genetic plus permanent environment variance on the egg quality traits as a proportion of phenotypic variance at different age groups as age is a major determinant of the growth and development of egg quality traits. Additionally, the significant and positive high correlation between egg weight and the egg components indicates that egg weight can be used to predict egg yolk weight and egg albumen weight with a reasonable level of accuracy and prevision in Bovan Neva Black laying hen. Similar high positive significant correlation of egg weight with yolk weight and albumen weight in Bovan Neva Black laying hens at 25, 51 and 72 weeks recorded in this study have been previously reported in white leghorn and other commercial breeds in literature (Molnar et al. 2016; Sreenivas et al. 2013; Sharma et al. 2002; Jayalaxmi et al. 2002).

Hereditability estimates

The genetic potential for the periodic egg production of egg weight, egg albumen, and egg yolk weight varied from one age group to another primarily because of the variation of

the genetic response and the inherent transmitting ability of parent traits from one generation to another at each stage of egg production as presented in Table 6. High heritability estimates of 0.783-0.889, 0.426-0.798 and 0.433-0.811 for egg weight, egg yolk weight and egg albumen weight at 25-72 weeks, respectively were obtained indicating higher role of additive genetic variance in phenotypic expression for different age groups. These high estimates generally agreed with the report in the literature. Rath et al. (2015) recorded values of 0.443, 0.740 and 0.460 for heritability estimates of egg weight, egg yolk weight and egg albumen weight respectively at 50 weeks for white leghorns using half-sib correlation analysis adopted to multifarious species and evaluated using PROC VAMCOMP of restricted maximum likelihood (REML) method of Statistical Package for Social Sciences. Sabri et al. (1999a) registered 0.457 at 26-30 weeks and 0.501 at 50-54 weeks for heritability estimates of egg weight employing white leghorn evaluated using mixed model least-squares and maximum likelihood method. Blanco et al. (2014) reported heritability estimates of 0.49 for egg weight at 32-36 weeks for male white egg lines of Lohmann selected leghorn and 0.65 for egg weight at 32-36 weeks for male Brown egg lines of Lohmann Brown breed using mixed procedure of statistical analytical system for a half-sib correlation analyses adopted to multifarious species.

Table 6 Estimates of heritability and repeatability for egg production at different age group

Traits	Weeks	Heritability	Repeatability
Egg Weight	25	0.783±0.01	0.843±0.18
	51	0.889±0.13	0.902±0.13
	72	0.792±0.19	0.880±0.09
York Weight	25	0.798±0.20	0.838±0.07
	51	0.426±0.14	0.666±0.09
	72	0.623±0.09	0.666±0.17
Albumen Weight	25	0.811±0.17	0.846±0.13
	51	0.433±0.08	0.712±0.14
	72	0.787±0.16	0.887±0.11

Begli et al. (2010) obtained heritability estimates of 0.45 for egg weight in Iranian fowl at 30 weeks using half-sib correlation analysis adopted to multifarious specie and employing restricted maximum likelihood of ASREML software for the estimation of variance components.

Paleja et al. (2008) obtained 0.778 at 28 weeks, 0.682 at 40 weeks and 0.542 at 56 weeks for egg weight using white leghorn breed employing least squares analysis of data with unequal subclass number. Akintola et al. (2011) recorded 0.50 for egg weight, 0.32 for egg yolk weight and 0.61 for egg albumen weight at 40-65 weeks for Iranian fowl employing restricted maximum likelihood (REML) procedure using the ASREML 9.0 software package.

However, Sreenivas et al. (2013) recorded low to moderate estimates of 0.180-0.255 for egg weight, 0.245-0.322 for egg yolk weight and 0.124-0.214 for egg albumen weight for IWH, IWI and IWK lines of white leghorns at 40 weeks using full-sib correlation method employing mixed model least squares and maximum likelihood compared to the high estimates obtained in this study. This could be attributed to the method of estimation, presumed non-additive genetic plus permanent environmental effects variance, and breed differences. Although egg weight, egg yolk weight and egg albumen recorded high estimates of heritability, observable variation was registered at different age groups. Egg weight was highly heritable at 25, 51 and 72 weeks whereas egg yolk weight and egg albumen weight were highly heritable at 25 and 72 weeks only. This suggested that the genetic potential of egg weight, egg yolk weight and egg albumen weight were well expressed at 25 and 72 weeks which minimized the effects of environmental factors thus enhanced the role of additive genetic variance in the phenotypic expression of these traits compared to report recorded at 51 weeks for egg albumen weight and egg yolk weight. At 51 weeks, there was a decline in the heritability estimates of egg production except egg weight that decline at 72 weeks suggesting that the genetic potential of egg weight and albumen weight are well expressed at peak egg production (25 weeks), which minimized the influence of environmental factors compared to later periods (51 weeks) of the laying year.

These agreed favourably with the report of (Gunder et al. 1989; Muir and Patterson, 1990; Hagger and Abplanalp, 1978; Sabri et al. 1999a; Sabri et al. 1991b) that obtained higher heritabilities at early stages of the laying year compared to later times of the laying year.

However, a robust heritability estimates were recorded for egg yolk weight and egg albumen weight at 72 weeks implying higher role of additive genetic variance in phenotypic expression of these traits compared to the influence of presumed non-additive genetic plus permanent environmental variance as a proportion of phenotypic variance thereby differing from the report in literature (Grunder et al. 1989; Muir and Patterson, 1990; Hagger and Abplanalp, 1978; Sabri et al. 1999a; Sabri et al. 1991b) employing white leghorn. This could be attributed to breed difference, environmental temperature, management procedures and method of estimation.

Repeatability estimates

The genetic potential, additive genetic variance and presumed non-additive genetic plus permanent environmental variance as a proportion of phenotypic variance from one age group to anther were the indications of the variations in the repeatability estimate observed for egg weight, egg yolk weight and egg albumen weight in this study as presented Table 6.

High repeatability estimates of 0.843-0.902, 0.666-0.838 and 0.712-0.887 for egg weight, egg yolk weight and egg albumen weight at 25-72 weeks respectively were recorded indicating higher role of additive genetic variance in the phenotypic expression compared to non-additive genetic plus permanent environmental variance as a proportion of phenotypic variance at 25-72 weeks. These high estimates are similar to the report in the literature. Toye et al. (2012) obtained egg weight of 0.46 for Black Harco breed and 0.50 for heavy Lohman Brown layers at 28 weeks employing mean square from the analysis of variance using statistical system and STAT programmed. Blanco et al. (2014) recorded repeatability estimates of egg weight of 0.75 for white egg line and 0.71 for Brown egg line between 67-70 weeks using mixed procedure of statistical analytical system for a half-sib correlation analysis adopted to multifarious species. Goto *et al.* (2015) reported estimates of 0.47 for egg weight, 0.48 for egg yolk weight and 0.51 for egg albumen weight for Onagadori breed; 0.42 for egg weight, 0.40 for egg yolk weight and 0.45 for egg albumen weight for white leghorn between 20-30 weeks employing one-way analysis of variance with Stat View for windows software. Udeh (2010) reported 0.44 for egg weight at 40 weeks in Black Olympia (strain 2) using one-way analysis of variance described by Becker (1984) for multifarious species and half-sib correlation. Ayorinde and Sado (1988) also reported repeatability estimates of 0.58-0.60 for egg weight from Hubbard layers; Ibe (1984) obtained 0.76 for egg weight in white leghorn hens; and obtained an estimate of 0.46-0.71 for egg weight in different lines of white leghorns.

It could be observed that at 51 weeks, there was a decline in the repeatability estimates of egg production except egg weight suggesting that the genetic potential of egg weight and albumen weight are well expressed at peak egg production (25 weeks) while egg weight experienced prolonged genetic potential between 25-51 weeks which minimized the influence of environmental factors compared to later periods (51 weeks for egg yolk weight and egg albumen weight, and 72 weeks for egg weight) of the laying year. This is in line with the report of that observed high repeatability estimates at peak egg production and declined estimates at later periods.

Comparison between heritability and repeatability estimates

Repeatability and heritability estimates for egg production varied slightly are shown in Table 6. It was observed that repeatability estimates were higher than estimates of heritability suggesting low influence of non-additive genetic and permanent environmental influence as a proportion of phenotypic variance on the traits. This trend was observed by Blanco *et al.* (2014) for estimates of repeatability and heritability in egg weight for white egg line and Brown egg line.

CONCLUSION

From the findings, it was observed that as the age of the laying hen increased, the magnitude of egg weight, egg yolk weight and egg albumen weight increased. The regression coefficient of the traits revealed that positive rate of change in trait from one age group to another. The Pearson correlation coefficient recorded a significant positive and the same high correlation between egg weight and other components and highest coefficient between egg yolk weight and egg albumen at different age groups. It was

equally observed that the heritability and repeatability estimates were generally high while repeatability estimates were slightly higher than heritability values indicating a slight influence of presuming non-additive plus permanent environment variance on the traits of different age groups. This indicates that adequate and proper mass selection, management procedures, experimental design and method of estimation employed in the research were genetically responsible for the improvement of the traits.

ACKNOWLEDGEMENT

Our thanks go to Babcock University, Nigeria for providing us a peaceable environment for carrying out the practical. We also wish to thank all authors cited in this paper for making the manuscript a success.

REFERENCES

Abdallah A.G., Harms R.H. and Russell G.B. (1995). Effect of age and resting on hens laying eggs with heavy or light shell weight. *J. Appl. Poult. Res.* **4**, 131-137.

Akintola O.A.I., Ajuogu P.K. and Aniebo A.O. (2011). Effects of Ovabolin (Ethylestrenol) on egg laying performance and egg quality traits in second year layers. *Adv. Agric. Biotechnol.* **1**, 49-58.

Ayorinde K.L. and Sado C. (1988). Repeatability estimates for some egg traits in a commercial laying strain. *Nigeria J. Anim. Prod.* **15**, 119-125.

Becker W.A. (1984). Manual of Qualitative Genetics. Fourth ed. Academic Enterprise, Washington, USA.

Begli H.E., Zerehdaran S., Hassani S. and Abbasi M.A. (2010). Heritability, genetic and phenotypic correlations of egg quality traits in Iranian fowl. *British Poult. Sci.* **51**, 740-748.

Blanco A.E., Ieken W., Ould-Ali D., Cavero D. and Schmutz M. (2014). Genetic parameters of egg quality traits on different pedigree layers with special focus on dynamics stiffness. *Poult. Sci.* **93**, 2457-2461.

Ewa V.U., Otuma M.O. and Omeje S.I. (2005). Interrelationship of external egg quality traits of four inbred line chicken stains. *Trop. J. Anim. Sci.* **8**, 23-30.

FAO. (1998). Food and Agriculture Organization of the United Nations (FAO), Rome, Italy.

FAO. (2009). Food and Agriculture Organization of the United Nations the State of Food Insecurity in the World.

Goto T., Shiraishi J., Bungo T. and Tsudzuki M. (2015). Characteristics of egg-related traits in the Onagoidori (Japanese Extremely Long Tail) breed of chickens. *Japan Poult. Sci. Assoc.* **52**, 81-91.

Grunder A.A., Hamilton R.M.G., Fairfull R.W. and Thompson B.K. (1989). Genetic parameters of egg shell quality traits and percentage of eggs remaining intact between oviposition grading. *Poult. Sci.* **68**, 46-54.

Hagger C. and Abplanalp H. (1978). Food consumption records for the genetic improvement of income over food costs in laying flocks of White Leghorns. *British Poult. Sci.* **19**, 651-667.

Ibe S.N. (1984). Repeatability of egg quality traits in white Leghorn chickens. *Nigeria J. Genet.* **4(1),** 12-19.

Islam M.S. and Dutta R.K. (2010). Egg quality traits of indigenous, exotic and crossbred chicken (*Gallu gallus domesticus*) in Rajshahi, Baugladesh. *J. Life Earth Sci.* **5,** 63-71.

Jayalaxmi P., Prasad V.I.K., Sree R.M.A. and Eswara R.C. (2002). Inheritance of various eggf quality traits in three strains of White Leghorns. *Indian Vet. J.* **78,** 820-823

Kamanli S., Durmus I., Demir S. and Tarim B. (2015). Effect of different light sources on performance and egg quality traits in laying hens. *European Poult. Sci.* **79,** 1-10.

Khalil M.H., Iraqi M.M. and El-Atrouny M.M. (2013). Effects on egg quality traits of crossing Egyptian Golden Montazah with white leghorn chickens. Livest. Res. Rural Dev. Available at: http://www./wd.org/lrrd25/6/kha/251103.htm.

Ledur M.C., Liljedahl L.E., Mcmillan I., Asselstine L. and Fairfull R.W. (2002). Genetic effects of aging on egg quality traits in the first laying cycle of White Leghorn strains and strain crosses. *Poult. Sci.* **79,** 296-304.

Lukanov H., Genchev A. and Pavlov A. (2015). Egg quality and shell colour characteristics of crosses between Araucan and Schijndelaar with highly productive white leghorn and Rhode strains. *Agric. Sci. Technol.* **7(3),** 366-371.

Minelli G., Sirri F., Folegatti E., Meluzzi A. and Franchini A. (2007). Egg quality traits of laying hens reared in organic and conventional systems. *Italian J. Anim. Sci.* **6,** 728-730.

Molnar A., Maertens B., Ampe B., Buyse J., Kempen I., Zoons J. and Delezie E. (2016). Changes in egg quality traits during the last phase of production: is there potential for an extended laying cycle? *British Poult. Sci.* **57(6),** 842-847.

Mube H.K., Kana J.R., Tadondjou C.D., Yemdjie D.D., Marijei Y. and Tegnia A. (2014). Laying performance and egg quality of local Barred Hens under improved conditions in Cameroon. *J. Appl. Biosci.* **74,** 6158-6163.

Muir W.M. and Patterson D.L. (1990). Genetic and environmental association of unconllectible egg production with shell quality, rate of lay and erratic timinig of oviposition in white leghorn hens. *Poult. Sci.* **69,** 509-516.

Padhi M.K., Chatterjee R.N., Rajkumar U., Haunshi S., Niranjan M., Panda A.K., Reddy M.R. and Bhanja S.K. (2015). Evaluation of Vanaraja male line (PDI) for different production and egg quality traits. *Indian J. Anim. Sci.* **85,** 634-642.

Paleja H.I., Savalia F.P., Patel A.B., Khanna K., Vataliya P.H. and Solanki J.V. (2008). Genetic parameter in White Leghon (IWN Line) chicken. *Indian J. Poult. Sci.* **43,** 151-157.

Petek M., Gezen S.S., Alpay F. and Cibik R. (2007). Effects of non-feed removal molting methods on egg quality traits in commercial brown egg laying hens in Turkey. *Trop. Anim. Health Prod.* **21,** 1-5.

Rath P.K., Mishra P.K., Mallick B.K. and Behura N.C. (2015). Evaluation of different egg quality traits and interpretation of their mode of inheritance in white leghorn. *Vet. World.* **8,** 449-451.

Sabri H.M., Wilson H.R., Harms R.H. and Wilcox C.J. (1999a). Genetic parameter for egg and related characteristics of white leghorn hens in a subtropical environment. *Genet. Mol. Biol.* **2,** 183-192.

Sabri H.M., Wilson C.J. Wilson H.R. and Harms R.H. (1991b). Comparison of energy utilization efficiency among six lines of White Leghorns. *Poult. Sci.* **70,** 229-233.

SAS Institute. (1999). SAS®/STAT Software, Release 6.11. SAS Institute, Inc., Cary, NC. USA.

Sharma P.K., Varma S.K., Sharma R.K. and Brijesh S. (2002). Genetic parameters of production and egg quality characters White Leghorn. *Indian J. Poult. Sci.* **37,** 181-182.

Sreenivas D., Prakash M.G., Mahender M. and Chatterjee R.N. (2013). Genetic analysis of egg quality traits in white leghorn chickens. *Vet. World.* **6,** 363-372.

Tadesse D., Esatu W., Girma M. and Dessie T. (2015). Comparative study on some egg quality traits of exotic chickens in different production systems in East Shewa, Ethiopia. *African J. Agric. Res.* **10,** 1016-1022.

Toye A.A., Sola-ojo F.E. and Ayorinde K.L. (2012). Egg production, egg weight and egg mass repeatability and genetic gain from use of multiple time-spaced records in Black Harco and Lohman Brown layers. *Centrepoint J.* **18(2),** 147-156.

Tumova E. and Goust R.M. (2012). Interaction of hen production type, age, and temperature on laying pattern and egg quality. *Poult. Sci.* **91,** 1269-1277.

Udeh I. (2010). Repeatability of egg number and egg weight in two strains of layer type chicken. *Int. J. Poult. Sci.* **9,** 675-682.

Zhang L.C., Ning Z.H., Xu G.Y., Chou Z. and Yang N. (2005). Heritability and genetic and phenotypic correlations of egg quality traits in brown-egg dwarf layers. *Poult. Sci.* **84,** 1209-1213.

15

Effect of Diets Formulated on the Basis of Four Critical Essential Amino Acids on Performance and Blood Biochemical Indices of Broiler Finisher Chickens Reared under Tropical Environment

E. Opoola[1*], S.O. Ogundipe[1], G.S. Bawa[1] and P.A. Onimisi[1]

[1] Department of Animal Science, Ahmadu Bello University, Zaria, Nigeria

*Correspondence E-mail: eopoola@abu.edu.ng

ABSTRACT

A study was conducted in a completely randomized design to evaluate the effect of diets formulated on the basis of four critical essential amino acids (lysine, methionine, tryptophan and threonine) on the crude protein requirement, carcass quality, nutrient digestibility, haematological and blood biochemical indices of broiler finisher chickens (28-56 days) reared under tropical environment. One hundred and eighty chickens were used in this experiment. There were four experimental diets each with three replicates (15 birds per replicate). The experimental diets were formulated with a gradual crude protein increase from 18 to 21% in 1% intervals. Diet 1, 2, 3 and 4 contained 18, 19, 20 and 21% dietary crude protein, respectively. Ileal digestible quantities of all essential amino acids (EAA) were almost equal in the diets and total amount of each EAA was maintained at or above NRC requirements. The performance of chickens fed 19% crude protein (CP) was similar to chickens fed 20 and 21% CP diets in terms of final weight, weight gain and feed conversion ratio. Feeding 18% CP with essential amino acids resulted in significantly (P<0.05) lower final weight, weight gain, average daily weight gain and poorer feed conversion ratio (FCR) than those fed diets higher crude protein diets. Generally, it was observed that chickens fed 19, 20 and 21% CP supplemented with balanced essential amino acids were statistically similar in terms of the carcass weight, dressing percentage, thighs, drumsticks, heart, lung and back weights compared to the chickens fed 18% CP supplemented with balanced essential amino acids. Chickens fed 21% CP fortified with balanced essential amino acids had the best values for apparent digestible crude protein, crude fibre (CF) and nitrogen free extract (NFE) compared with the chickens fed 18% CP with balanced essential amino acid. The observed means for most of the haematological and blood biochemical indices fell within the normal values for healthy chickens. It can be concluded that crude protein requirement of broiler finisher chickens (28-56 days) can be reduced to 19% with essential amino acids supplementation without having any adverse effect on growth, nutrient digestibility, carcass quality and haematological parameters of broiler finisher chickens reared under the tropical environment.

KEY WORDS broiler, carcass quality, essential amino acids, haematology, performance, tropical environment.

INTRODUCTION

It is well documented that nutrient requirements of poultry are significantly affected by seasons of rearing and ambient temperature. In the cool, hot and humid tropics, genetic potential of chickens, irrespective of the breed, might be affected adversely because of environmental constraints. Protein is generally considered as one of the major cost components of poultry diets (Firman and Boling, 1998). Therefore, placing more emphasis on crude protein requirement in broiler diet formulation becomes imperative to maximize the performance of broiler chickens as well as with economic considerations (Eits *et al.* 2004). Adding purified amino acids or amino acid precursors has been known for more than 50-years to allow for reduced levels of intact proteins to provide adequate levels of essential and non-essential amino acids (Pesti, 2009). NRC (1994) stated that poultry require a specific amount and balance of essential amino acids (EAA) and non-essential amino acids (NEAA), rather than the crude protein level *per se* to support growth, immune system and the whole body composition of broiler chickens. Some progress has been made to lower the crude protein (CP) content of broiler diets by providing all those amino acids considered to be critical. However, some results from several researchers still showed that reduction of CP has some negative effects on performance and meat quality.

This failure occurred even with provision of all requirements for those amino acids considered as essential. Failure to obtain optimum results in performance may be attributed to one or more of the following factors: lack of a nitrogen pool to synthesize nonessential amino acids (NEAA); inability of body capacity to meet all NEAA requirements especially Gly, Ser, Pro and Glu; decreased level of potassium or altered ionic balance; and imbalances among certain amino acids such as arginine to lysine, lysine to threonine (Namroud *et al.* 2008). The studies of Han *et al.* (1992); Dean *et al.* (2006) and Namroud *et al.* (2008) showed that standard dietary recommendations can be met by supplementing the lower protein diets with EAA and NEAA to meet the exact requirement. It is well known that the CP and amino acid (AA) status of a diet affects the carcass composition of broilers with increased carcass protein and decreased carcass fat accompanying increases in dietary protein or essential AA contents because protein has the ability to act as a lipotrophic agent (Neto *et al.* 2000; Si *et al.* 2001; Pesti, 2009).

Excess dietary CP results in a lean bird but reduces feed efficiency whereas less than optimal protein content increases fat retention (Buyse *et al.* 1992). Therefore, importance of using appropriate amounts of balanced dietary CP and AA for poultry is a high priority issue for several reasons including feed efficiency, cost and environmental concerns (Buyse *et al.* 1992).

In the tropics, it is still unclear as to what extent the amino acid supplementation could be used to replace the dietary protein in broiler rations as most of the previous studies of replacing dietary protein were conducted under the temperate condition.

Therefore, this study aims at evaluating the effect of diets formulated on the basis of four critical essential amino acids on the crude protein requirement that will enhance growth performance, nutrient digestibility, carcass quality, haematological and blood biochemical indices of broiler chickens reared under tropical environment.

MATERIALS AND METHODS

Experimental site

The experiment was carried out at the Poultry Unit of Kogi State Ministry of Agriculture, Kabba, located within the Southern Guinea Savannah Zone on latitude 7 °5′N, longitude 6 °4′E and altitude of 640 m above sea level. It has an annual rainfall of 1500 mm and rain starts between late April and early May to mid October. The dry season begins around the middle of November, with cool weather that ends in February. This is followed by relatively hot-dry weather between March and April just before the rain begins.

The minimum daily temperature is from 14 °C - 20°C during the cool season while the maximum daily temperature is from 19 °C - 40 °C during the hot season. The mean relative humidity during dry and wet seasons is 21% and 72%, respectively.

Experimental design and management of birds

One hundred and eighty mixed sex Arbor Acre broiler chickens were housed in a deep litter system and had free access to water and a common diet for 7 days. On d 7 they were randomly re-allocated to one of four groups on the basis of approximately equal weights with fifteen (15) chickens per replicate in a completely randomized design. The experimental birds were given *ad libitum* access to water and diet. The ambient temperature was gradually decreased from 30 to 20 °C over the period of 28 to 56 d of age. The chicks were vaccinated in the hatchery against Marek's disease followed by vaccination at 5 and 21 days against Gumboro disease and on the 8th day against Newcastle disease. Mash feed and water were supplied *ad libitum*.

Experimental diets

Four broiler finisher diets were formulated. Diets 1, 2, 3 and 4 contained 18, 19, 20 and 21% crude protein respectively (Table 1). The diets were formulated to be isocaloric. Diets were also formulated to meet the NRC (1994) nutrient requirements for essential amino acids. All the diets were chemically analyzed according to the standard of AOAC (1990) methods for their proximate compositions (Table 2). Diets were supplemented with complete vitamin and trace mineral premixes. The L-lysine HCl, DL-methionine, L-tryptophan and L-threonine used in the diets were feed grade (minimum 98% purity) and purchased from Ajinomoto Incorporation Japan.

Table 1 Composition of broiler finisher diets formulated on the basis of four critical essential amino acids on the crude protein requirement (5-8 weeks)

Ingredients	Treatments			
	18% CP	19% CP	20% CP	21% CP
Corn grain	65.78	62.76	58.01	53.28
Groundnut cake	14.00	16.56	20.50	24.50
Soya cake	8.00	8.00	8.00	8.00
Fish meal	3.00	3.00	3.00	3.00
Palm oil	3.50	4.10	4.97	5.87
Limestone	1.00	1.00	1.00	1.00
Bone meal	2.90	2.90	2.90	2.90
Common salt	0.30	0.30	0.30	0.30
Premix[1]	0.30	0.30	0.30	0.30
L-lysine HCl	0.49	0.45	0.40	0.34
DL-methionine	0.30	0.28	0.25	0.22
L-tryptophan	0.11	0.10	0.09	0.07
L-threonine	0.32	0.31	0.28	0.25
Total	**100**	**100**	**100**	**100**
Calculated analysis				
Metabolizable energy (kcal/kg)	3151	3152	3150	3151
Crude protein (%)	18.00	19.00	20.00	21.00
Ether extract (%)	7.53	7.56	7.65	7.79
Crude fibre (%)	3.00	3.07	3.12	3.25
Calcium (%)	1.33	1.33	1.34	1.35
Lysine (%)	1.15	1.15	1.15	1.15
Methionine (%)	0.60	0.57	0.60	0.60
Available P (%)	0.57	0.57	0.57	0.57
TSAA[2] (%)	0.85	0.85	0.85	0.85
Tryptophan (%)	0.21	0.21	0.21	0.21
Threonine (%)	0.83	0.83	0.83	0.83
Glycine (%)	1.16	1.26	1.42	1.52
Arginine (%)	1.17	1.25	1.40	1.54
Phenylalanine (%)	0.81	0.85	0.91	0.98
Leucine (%)	1.50	1.54	1.62	1.70
Isoleucine (%)	0.76	0.79	0.83	0.88
Valine (%)	0.82	0.85	0.90	0.95

[1] Biomix premix supplied per kg of diet: vitamin A: 10000 IU; vitamin D_3: 2000 IU; vitamin E: 23 mg; vitamin K: 2 mg: vitamin B_1: 1.8; vitamin B_2: 5.5 mg; Niacin: 27.5 mg; Pantothenic acid: 7.5 mg; vitamin B_{12}: 0.015 mg: Folic acid: 0.75 mg; Biotin: 0.06 mg; Choline chloride: 300 mg; Cobalt: 0.2 mg; Copper: 3 mg; Iodine: 1 mg; Iron: 20 mg; Manganese: 40 mg; Selenium: 0.2 mg; Zinc: 30 mg and Antioxidant: 1.25 mg.
[2] TSAA: total sulphur amino acid.

Table 2 Proximate composition of broiler finisher chickens fed diets formulated on the basis of four critical essential amino acids on crude protein requirement (5-8 weeks)

Parameters (%)	Level of crude protein (%)			
	18	19	20	21
Dry matter	90.14	89.49	92.84	90.90
Crude protein	17.01	18.22	18.99	20.64
Crude fibre	2.04	2.91	2.98	3.01
Ether extract	6.68	7.41	8.41	8.98
Ash	4.11	5.21	3.14	4.89
Nitrogen free extract	70.16	66.25	66.48	62.48

Parameters measured

The parameters measured included: final body weight and feed intake. From the primary data collected for feed intake and weight gain, data for feed conversion rate were generated. Mortality was checked twice daily; birds that died were weighed with the weight used to adjust the feed conversion.

Nutrient digestibility trial

A nutrient digestibility trial was carried out at the end of the experiment, using metabolic cages. This was done by randomly selecting 3 birds of approximately equal weight from each replicate. The birds were placed in alternate cages, with polythene bags attached beneath the cages. The birds were fed with the control diet for 7 days of the adjustment period and fasted for 24 hours with only water provided. Forty gram of the experimental diet was allocated to each bird by 8:00 a.m. daily.

Faecal samples were collected daily, separated from feed and other extraneous materials, weighed, bulked together and kept in a deep freezer. At the end of the 6th day, the birds were not fed. The remaining faeces were then collected by 8.00 a.m. on the 7th day. The total samples were thawed and weighed which they were thoroughly mixed together and oven dried for 72 hours at 65 °C. The dried samples were then weighed and ground, after which samples were taken for proximate analysis along with the sample of the feed fed at the Department of Animal Science Laboratory, Ahmadu Bello University, Zaria. Nutrient retention was determined for crude protein, crude fibre, ether extract, ash and nitrogen free extract.

Nutrient retention= ((nutrient intake-nutrient output)/(nutrient intake)) × 100

Carcass characteristics and whole-body analyses

At the end of the experimental period (d 56), three birds per replicate with weights closest to the mean body weight of the replicate were used for the carcass study. The birds were slaughtered by cervical dislocation after being kept off-feed for twenty four hours (with free access to water). The birds were weighed, de-feathered and eviscerated. Weights of the conventional cut up parts (breast muscle, thigh muscle, abdominal fat) and the data on organ weights (i.e., liver, intestine, kidney, spleen and gizzard) were recorded at this stage.

Haematological and blood serum evaluation

The collection of blood was carried out on d 56 of the experiment. Two mL of blood samples was collected from three birds per replicate via the wing veins and put into ethylenediaminetetraacetic acid (EDTA) treated Bijou bottles (1 mg/mL) for haematological assay. Blood samples were analyzed within one hour of collection. The packed cell volume (PCV) was determined by microhaematocrit method (Schalm et al. 1975), haemoglobin concentration (Hb) was measured spectrophotometrically by cyanomethaemoglobin method (Schalm et al. 1975) using a SP6-500UV spectrophotometer (Pye UNICAM England), red blood cells and white blood cells counts were estimated using haemocytometer (Schalm et al. 1975), the mean cell haemoglobin, mean cell volume and mean cell haemoglobin count were calculated from Hb, PCV and red blood cell (RBC) (Jain, 1986).

Also, two millilitres of blood from three birds per replicate were allowed to clot and then centrifuged and serum was separated and stored at -20 °C until analyzed for serum parameters. Serum total protein was determined by the Kjedahl method as described by Kohn and Allen (1995). Albumin was determined using the bromocresol green (BCG) method as described by Peters et al. (1982). Aspartate amino-transferase (AST) and alanine amino-transferase (ALT), alkaline phosphatase (ALP) and creatine kinase (CK) were determined using spectrophotometric methods described by Rej and Holder (1983). Cholesterol was determined according to the method of Roschlan et al. (1974) while urea was determined as described by Kaplan and Szabo (1979).

Statistical analyses

All data obtained were statistically analyzed using the general linear models (GLM) procedure of SAS software, (SAS, 2001) for the analysis of variance. Duncan's multiple range tests were used to determine differences among treatment means. Means were considered different at (P<0.05).

General linear model

$Y_{ij} = \mu + K_{i} + e_{ij}$

Where:

Y_{ij}: observation of the ith level of crude protein as shown by broilers performance.

μ: overall mean.

K_{i}: i^{th} effect of crude protein.

e_{ij}: random error.

RESULTS AND DISCUSSION

Dietary treatment had effects on final body weight, weight gain, average daily weight gain, feed intake, average daily feed intake and feed conversion ratio. Broiler chickens fed 19% CP, 20% CP and 21% CP had similar weight gains (Table 3).

Table 3 Crude protein requirements and performance of broiler finisher chickens fed diets formulated on the basis of four critical essential amino acids (5-8 weeks)

Parameters	Crude protein level (%)				SEM
	18	19	20	21	
Initial weight (g)	1401.83	1402.45	1402.81	1402.02	1.02
Final weight(g)	2857.33[b]	3090.16[a]	3248.81[a]	3263.49[a]	51.49
Weight gain (g)	1455.50[b]	1687.71[a]	1846.00[a]	1861.48[a]	51.32
Ave daily gain (g)	69.31[b]	80.37[a]	87.91[a]	88.64[a]	2.44
Feed intake (g)	4012.20[b]	4012.90[b]	4274.80[a]	4212.50[a]	77.75
Feed intake (g/b/d)	191.08[b]	191.90[b]	203.56[a]	200.60[a]	3.70
Feed conversion ratio	2.77[b]	2.38[a]	2.32[a]	2.27[a]	0.09
Mortality (%)	0.00	0.00	0.00	1.75	0.88

The means within the same row with at least one common letter, do not have significant difference (P>0.05).
SEM: standard error of the means.

This result is similar to the research conducted by Han *et al.* (1992), Moran and Bushong (1992) and Moran (1994). They reported that there were no adverse effects on weight gain of low crude protein diets supplemented with essential amino acids during the first six weeks of age. The increase in weight gain observed for chickens fed diets high in crude protein with the four most critical essential amino acids may be due to the efficient protein and amino acid utilizations, sufficient body capacity to meet all NEAA requirements and adequate nitrogen pool to synthesize NEAA. Chickens fed 18% crude protein with essential amino acids had the least weight gain.

This result is similar to the report of Bregendahl *et al.* (2002), who reported a significant depression in body weight when the crude protein in diets supplemented with essential amino acids were reduced from 23 to 18%. The poor performance with low crude protein (18%) supplemented with the four critical essential amino acids observed in this study could be associated to the differences in amino acids digestibility of diets because of the slight variation in the levels of the feed ingredients, insufficient body capacity to meet all NEAA requirement and inadequate nitrogen pool to synthesize NEAA.

The significant effect observed for feed intake disagreed with the findings of Bregendahl *et al.* (2002) and Elmutaz *et al.* (2014), who reported no significant differences across the treatments fed low crude protein versus high crude protein but similar to the findings of Kidds *et al.* (2001) who reported an increase in feed intake of chickens fed diets containing 19% CP as compared to those fed diets containing 22.5% CP with essential amino acids. The reason may be because of the changes in amino acid contents of low crude protein diets. However, it was reported by Han *et al.* (1992) that equal feed intakes may be expected if low crude protein diets with the same metabolizable energy are supplemented with limiting amino acid.

It is obvious in this present study that decreasing dietary CP below 19%, even with maintained EAA levels, retarded growth and feed intake and increased FCR.

This means that, in agreement with many researchers, crystalline amino acids are not able to completely re-place CP in diets (Si *et al.* 2004; Waldroup *et al.* 2005; Yamazaki *et al.* 2006). Furthermore, research conducted by Neto *et al.* (2000) revealed a 13% increase in FCR of birds fed 17% CP diets as compared to those fed 24% CP diets. The poor performance and high feed conversion ratio observed for birds fed 18% CP, even with added essential amino acids, might be attributed to low CP and imbalance of essential amino acids.

Dietary treatments had no significant effect on mortality, which shows that the diets up to the highest level tested did not have any adverse effect on the health of birds. Dietary treatments had significant effects on live weight, carcass weight, dressing percentage, back, gizzard, lungs, kidney, heart, breast weight, wings, thigh and drumstick (Table 4). Chickens fed 19, 20 and 21% CP with the supplementation of amino acids had the best results in most of the parameters measured for carcass characteristics. The results disagreed with the findings of Si *et al.* (2001) and Baker *et al.* (1993), who reported that lowering the dietary CP level may not affect carcass characteristics yield, breast meat and thigh yield of birds.

Moran and Stillborn (1996) found no effect on the carcass yield of broiler fed low CP diets adequate in essential amino acids (EAA). Chickens fed 18% CP with diets adequate in EAA had the least carcass yield; this report is similar to the findings of Kerr and Kidd (1999) who also reported a significant decrease in the carcass yield of the birds fed low CP diets supplemented with EAA. Deficiency of dietary protein is known to increase the fat deposition in broilers (Yamazaki *et al.* 2006). Many researchers have reported the effect of supplementing several amino acids such as Met + Cys (Bunchasak *et al.* 1996), Arg (Leclercq *et al.* 1994), Trp and Glu on decreasing carcass and abdominal fat and hepatic lipid content. In our study, inclusion of adequate Met, Lsy, Try and Thre did not result in any significant reduction in fat deposition in abdominal cavity and carcass across the treatments.

Table 4 Carcass characteristics of broiler finisher chickens fed diets formulated on the basis of four critical essential amino acids at each protein level (5-8 weeks)

Parameters	Crude protein levels (%)				SEM
	18	19	20	21	
Live weight (g)	2916.67[c]	3150.00[b]	3296.67[a]	3283.33[a]	0.03
Carcass weight (g)	2280.00[b]	2626.67[a]	2740.00[a]	2736.67[a]	0.03
Dressing (%)	78.21[b]	83.37[a]	83.13[a]	83.34[a]	1.26
Prime cuts and organ weights (% of dressed weight)					
Breast	16.75[b]	20.92[ab]	20.74[ab]	23.05[a]	1.43
Wings	5.94[c]	7.50[b]	7.56[b]	8.81[a]	0.29
Thigh	8.69[b]	11.58[a]	12.19[a]	12.21[a]	0.36
Drumsticks	5.80[b]	7.84[a]	7.28[a]	8.49[a]	0.41
Back	9.73[b]	11.09[a]	11.30[a]	11.90[a]	0.41
Liver	1.05	1.29	1.52	1.60	0.20
Heart	0.25[b]	0.52[a]	0.41[a]	0.48[a]	0.05
Kidney	0.36[b]	0.48[a]	0.46[ab]	0.46[ab]	0.03
Gizzard	1.07[b]	1.27[ab]	1.51[a]	1.59[a]	0.12
Abdominal fat	0.70	0.68	0.56	0.69	0.11
Spleen	0.08	0.10	0.11	0.08	0.02
Lung	0.29[b]	0.44[a]	0.43[a]	0.53[a]	0.09

The means within the same row with at least one common letter, do not have significant difference (P>0.05).
SEM: standard error of the means.

This result is in agreement with previous studies of Yamazaki et al. (2006) and Han et al. (1992), their argument was that NEAA biosynthesis was also a limiting factor in a low CP diet and this could affect broiler performance and body composition. However, abdominal fat pad was observed to be higher in birds fed 18% CP diet with adequate EAA.

This result is similar to the findings of Sterlings et al. (2002), Hai and Blaha (2000) and Neto et al. (2000) who observed that abdominal fat pad weight increased with low CP diets.

Dietary crude protein supplemented with essential amino acids had significant (P<0.05) effects on the digestibility of crude protein, crude fibre, ash and NFE (Table 5). Chickens fed 21% CP fortified with balanced essential amino acids had the best values for apparent digestible crude protein, crude fibre and NFE compared with the chickens fed 18% CP with balanced essential amino acid. This result agreed with the findings of Neto et al. (2000) and Ratriyanto et al. (2012) who reported that efficiency of nutrient utilization was better for chickens fed higher level of CP diets supplemented with essential amino acids. The reasons may be as a result of increased proventricular acid secretions, increased pancreatic and intestinal mucosa secretions at this phase of rearing.

Digestible ash however, was influenced by dietary treatment; the high value suggests that more minerals were available to the birds as none of the birds showed any mineral deficiency symptoms throughout the experimental period.

The evaluations of red blood cells, mean cell haemoglobin, mean cell volume, mean cell haemoglobin counts, urea, creatine, glucose, AST, ALT, ALP, cholesterol, triglycerides and total protein did not reveal any significant differences across the treatment groups (Table 6), although the observed means for all the parameters fell within the normal values for healthy chickens as reported by Jain (1993) and Suchy (2000). The non significant differences observed across the treatment groups on the plasma glucose is similar to the studies of Swennen et al. (2005) and Swennen et al. (2006). It indicated that carbohydrate metabolism was not affected by the diet. Chickens fed 18% CP with balanced essential amino acids had the highest value for PCV compared to those fed higher levels of CP. The reasons could as a result of amino acid imbalance from the variations in amount of the feed ingredient used and physiological status of chickens. The mean values for haemoglobin count of 8.67-9.67 g/dL obtained in this experiment are similar to the values of 11.30 ± 1.82 g/dL reported by Oladele and Ayo (1999). This means that the chickens in all the treatment groups were not anaemic. The high values of 10.70 - 16.07 × 10^6 / L for red blood cells in this study are contrary to the values of 1.58 - 3.82 × 10^6 / L reported by Mitruka and Rawnsley (1977). This implies that chickens were polycythaemic. Chickens fed diets containing 18% CP had the highest uric acid value of 250.33 (mg/dL) compared to other treatments with higher levels of CP. This result is contrary to the findings of Collins et al. (2003), Malheiros et al. (2003) who reported considerably lower uric acid levels in the plasma of low crude protein.

Table 5 Nutrient digestibility of broiler finisher chickens fed diets formulated on the basis of four critical essential amino acids at each protein level (5-8 weeks)

Parameters (%)	Crude protein levels (%)				SEM
	18	19	20	21	
Dry matter	88.18	87.33	84.56	87.41	1.87
Crude protein	64.57[d]	74.07[b]	70.88[c]	77.28[a]	1.98
Crude fibre	59.29[c]	67.93[b]	67.34[b]	75.31[a]	3.27
Ether extract	85.58	90.20	88.07	88.08	1.38
Ash	75.19[c]	87.53[a]	56.90[d]	82.35[b]	0.98
Nitrogen free extract	59.16[d]	70.33[b]	66.80[c]	78.15[a]	0.62

The means within the same row with at least one common letter, do not have significant difference (P>0.05).
SEM: standard error of the means.

Table 6 Haematological parameters and serum biochemical indices of broiler finisher chickens fed graded diets formulated on the basis of four critical essential amino acids at each protein level (5-8 weeks)

Parameters	Crude protein levels (%)				SEM
	18	19	20	21	
PCV (%)	32.33[a]	30.00[b]	30.00[b]	29.67[b]	0.84
Hb (g/dL)	9.67[a]	8.80[b]	8.67[b]	9.37[a]	0.45
RBC (×10⁶/L)	3.53	3.57	3.73	3.83	0.53
WBC (×10⁶/L)	16.07[a]	13.02[b]	10.70[c]	14.67[ab]	0.34
MCH (Pg)	27.46	26.42	23.33	24.47	2.87
MCV (fL)	91.75	88.92	80.56	77.75	7.89
MCHC (g/dL)	29.96	29.34	28.91	31.70	1.73
Urea (mmol/L)	4.03	3.43	3.53	3.37	0.31
Uric acid (mg/dL)	250.33[a]	47.67[b]	60.67[b]	85.67[b]	25.24
Creatine (mg/dL)	49.33	53.33	50.00	35.32	9.08
Glucose (mg/dL)	3.77	4.34	4.50	5.33	1.45
AST (UI/L)	38.67	47.33	32.00	34.33	8.38
ALT (UI/L)	23.67	21.00	25.33	29.33	5.64
ALP (UI/L)	45.33	53.33	66.33	65.67	8.88
Cholesterol (mg/dL)	3.10	3.20	3.73	2.47	0.66
Triglycerides (mg/dL)	2.30	2.33	1.53	2.20	0.52
Albumin (g/dL)	3.20[b]	4.23[a]	4.20[a]	3.12[b]	3.04
Total protein (g/dL)	3.20	3.80	4.07	3.67	0.70

PCV: packed cell volume; HC: haemoglobin count; RBC: red blood cells; WBC: white blood cells; MCH: mean cell haemoglobin; MCV: mean cell volume; MCHC: mean cell haemoglobin count; AST: aspartate amino-transferase; ALT: alanine amino-transferase and ALP: alkaline phosphatase.
The means within the same row with at least one common letter, do not have significant difference (P>0.05).
SEM: standard error of the means.

The high level of uric acid observed for chickens fed 18% CP could be attributed to the quality and quantity of the protein, indicating an imbalance of amino acids in the diet. The mean values of 3.12-4.23 g/dL of albumin are similar to the value reported by Jain (1993).

CONCLUSION

It was concluded that the crude protein requirement for broiler finisher chickens can be reduced from 21% to 19% CP as long as the essential amino acids and all other nutrients meet the requirements of broiler chickens without having any adverse effects on growth, carcass quality, nutrient digestibility, haematological and blood serum indices.

ACKNOWLEDGEMENT

The authors would like to acknowledge the management of Agro Allied Company, Kogi State Ministry of Agriculture for giving us access to their poultry unit. Our appreciation also goes to Mallam Yunusa of Haematology laboratory, Faculty of Veterinary Medicine, Ahmadu Bello University, Nigeria for the blood sample analysis and Mr Kwano of Animal Science Department, Ahmadu Bello University for the feed and faecal sample analyses.

REFERENCES

AOAC. (1990). Official Methods of Analysis. Vol. I. 15ᵗʰ Ed. Association of Official Analytical Chemists, Arlington, VA, USA.

Baker D.H., Becker D.H., Norton H.W., Jenson A.H. and Harmon, B.G. (1993). Lysine imbalance of corn protein in the growing pig. *J. Anim. Sci.* **28,** 23-26.

Bregendahl K., Sell J.L. and Zimmerman D.R. (2002). Effect of low protein diets on growth performance and body composi tion of broiler chicks. *Poult. Sci.* **81**, 1156-1167.

Bunchasak C., Tanaka K., Ohtani S. and Collado C.M. (1996). Effect of Met + Cys supplementation to low-protein diet on the growth performance and fat accumulation of broiler chicks at starter period. *Japanese J. Anim. Sci. Technol.* **67**, 956-966.

Buyse J., Decuypere E., Berghman L., Kuhn E.R. and Vandesande F. (1992). The effect of dietary protein content on episodic growth hormone secretion and on heat production of male broilers. *British Poult. Sci.* **33**, 1101-1109.

Collin A., Malheiros R.D., Moraes V.M.B., Van As P., Darras M., Taouis V.M., Decuypere E. and Buyse J. (2003). Effects of dietary macronutrient content on energy metabolism and uncoupling protein mRNA expression in broiler chickens. *British J. Nutr.* **90**, 261-269.

Dean D.W., Bidner T.D. and Southern L.L. (2006). Glycine supplementation to low protein, amino acid-supplemented diets supports optimal performance of broiler chicks. *Poult. Sci.* **85**, 288-296.

Eits R.M., Meijerhof R. and Santoma G. (2004). Economics determine optimal protein levels in broiler nutrition. *World Poult. J.* **20**, 21-22.

Elmutaz A.A., Mohamad F., Idrus Z., Abdoreza S.F. and Loh T.C. (2014). Amino acids fortification of low-crude diet for broiler under tropical climate: ideal essential amino acids profile. *Italian J. Anim. Sci.* **13**, 3166-3172.

Firman J.D. and Boling S.D. (1998). Ideal protein in turkeys. *Poult. Sci.* **77**, 105-110.

Hai D.T. and Blaha J. (2000). Effect of low-protein diets adequate in levels of essential amino acids on broiler chicken performance. *Czech J. Ani. Sci.* **45**, 429-436.

Han Y., Suzuki V., Parsons C.M. and Baker D.H. (1992). Amino acid fortification of a low protein corn and soybean meal diet for chicks. *Poult. Sci.* **71**, 1168-1178.

Jain N.C. (1986). Schalm's Veterinary Haematology. Lea and Febriger Publisher, Philadelphia, USA.

Jain N.C. (1993). Essentials of Veterinary Haematology. Lea and Febiger Publisher, Philadelphia, USA.

Kaplan A. and Szabo I.I. (1979). Clinical Chemistry Interpretation and Technique. Henry Kumpton Publisher, London, United Kingdom.

Kerr B.J. and Kidd M.T. (1999). Amino acid supplementation of low-protein broiler diets: glutamic acid and indispensable amino acid supplementation. *J. Appl. Poult. Res.* **8**, 298-309.

Kidd M.T., Gerard P.D., Heger J., Kerr B.J., Rowe D., Sistani K. and Burnham D.J. (2001). Threonine and crude protein responses in broiler chicks. *Anim. Feed Sci. Technol.* **94**, 57-64.

Kohn R.A. and Allen M.S. (1995). Enrichment of proteolysis activity relative to nitrogen in preparations from the rumen for *in vitro* studies. *Anim. Feed Sci. Technol.* **52**, 1-4.

Leclercq B., Chagneau A.M., Cochard T. and Khoury J. (1994). Parative responses of genetically lean and fat chickens to lysine, argentine and non-essential amino acid supply. *Br. Poult. Sci.* **35**, 687-696.

Malheiros R.D., Moraes V.M.B., Collin A., Janssens G.P.J., Decuypere E. and Buyse J. (2003). Dietary macronutrients, endocrine functioning and intermediary metabolism in broiler chickens. Pair wise substitutions between protein, fat and carbohydrate. *Nutr. Res.* **23**, 567-578.

Mitruka B.M. and Rawnsley H.M. (1977). Clinical Biochemical and Haematological Reference Values in Normal Experiment Animals. Masson Publishing USA Inc., New York.

Moran E.T. and Bushong R.D. (1992). Effects of reducing dietary crude protein to relieve litter nitrogen on broiler performance and processing yields. Pp. 466-470 in Proc. 19th World Poult. Sci. Assoc. Meet. Amsterdam, Netherlands.

Moran E.T. (1994). Significance of dietary crude protein to broiler carcass quality. Pp. 1-11 in Proc. Maryland Nutr. Conf. Feed Manufact. University of Maryland, USA.

Moran E.T. and Stilborn B. (1996). Effect of glutamic acid on broiler given sub-marginal crude protein with adequate essential amino acids using feeds high and low in potassium. *Poult. Sci.* **75**, 120-129.

Namroud N.F., Shivazad M. and Zaghari M. (2008). Effect of fortifying low crude protein diet with crystalline amino acids on performance, blood ammonia level and excreta characteristics of broiler chicks. *Poult. Sci.* **87**, 2250-2258.

Neto M.G., Pesti G.M. and Bakalli R.I. (2000). Influence of dietary protein level on the broiler chicken's response to methionine and betaine supplements. *Poult. Sci.* **79**, 1478-1484.

NRC. (1994). Nutrient Requirements of Poultry, 9th Rev. Ed. National Academy Press, Washington, DC., USA.

Oladele S.B. and Ayo J.O. (1999). Compaarative studies on the haematocrit, haemoglobin and total protein values of apparent healthy and clinically sick indigenous chicks in Zaria, Nigerian. Bull. Anim. Health Prod. **47**, 163-167.

Pesti G.M. (2009). Impact of dietary amino acid and crude protein levels in broiler feeds on biological performance. *J. Appl. Poult. Res.* **18(3)**, 477-486.

Peters T., Biamonte G.T. and Doumas B.T. (1982). Protein (total protein) in serum. Pp. 100-115 in Selected Methods of Clinical Chemistry. G.W.R. Faulkner and S. Mcites, Eds. Amercian Association of Clinical Chemistry, USA.

Ratriyanto A., Indreswari R., Sudiyono S. and Sofyan A. (2012). Potential use of betaine to substitute methionine in broiler diets. Pp. 135-138 in Proc. Natl Semin. Zootec. Indigen. Res. Dev. Diponegoro University, Indonesia.

Rej R. and Holder M. (1983). Aspartate aminotransferase. Pp. 416-433 in Methods of Enzymatic Analysis. H.U. Bergmeyer, J. Bergmeyer and M. Grass, Eds. Velag Chemie, Weinheim, Germany.

Roschlan P., Bernet E. and Guber W. (1974). Enzymatische bestimmung des gesanty cholesterium in serum. *J. Clin. Biochem.* **12**, 403-407.

SAS Institute. (2001). SAS®/STAT Software, Release 6.11. SAS Institute, Inc., Cary, NC. USA.

Schalm J.W., Jain N.C. and Carol E.J. (1975). Veterinary Haematology. Lea and Febriger Publisher, Philadelphia, USA.

Si J., Fritts C.A., Burnham D.J. and Waldroup P.W. (2001). Relationship of dietary lysine level to the concentration of all essential amino acids in broiler diets. *Poult. Sci.* **80**, 1472-1479.

Si J., Fritts C.A., Waldroup P.W. and Burnham D.J. (2004). Effects of tryptophan to large neutral amino acid ratios and overall amino acid levels on utilization of diets low in crude protein by broilers. *J. Appl. Poult. Res.* **13**, 570-578.

Sterling K.G., Costa E.F., Henry M.H., Pesti G.M. and Bakalli R.I. (2002). Responses of broiler chickens to cottonseed and soybean meal-based diets at several protein levels. *Poult. Sci.* **81,** 217-226.

Suchy I.K. (2000). Haematological studies in adolescent breeding cocks. *Act. Vet. Brno.* **69,** 189-194.

Swennen Q., Janssens G.P.J., Millet S., Vansant G., Decuypere E. and Buyse J. (2005). Effects of substitution between fat and protein on feed intake and its regulatory mechanisms in broiler chickens: endocrine functioning and intermediary metabolism. *Poult. Sci.* **84,** 1051-1057.

Swennen Q., Janssens G.P.J., Collin A., Bihan-Duval E.L., Verbeke K., Decuypere E. and Buyse J. (2006). Diet-induced thermogenesis and glucose oxidation in broiler chickens: influence of genotype and diet composition. *Poult. Sci.* **85,** 731-742.

Waldroup P.W., Jiang Q. and Fritts C.A. (2005). Effects of glycine and threonine supplementation on performance of broiler chicks fed diets low in crude protein. *Int. J. Poult. Sci.* **4,** 250-257.

Yamazaki M., Murakami H., Nakashima K., Abe H. and Takemasa M. (2006). Effect of excess essential amino acids in low protein diet on abdominal fat deposition and nitrogen excretion of the broiler chicks. *Japanese Poult. Sci.* **43,** 150-155.

16

The Effect of Formulation Diets Based on Digestible Amino Acids and Lysine Levels on Carcass and Chemical Composition of Broiler

J. Nasr[1*] and F. Kheiri[2]

[1] Department of Animal Science, Saveh Branch, Islamic Azad University, Saveh, Iran
[2] Department of Animal Science, Shahrekord Branch, Islamic Azad University, Shahrekord, Iran

*Correspondence E-mail: javadnasr@iau-saveh.ac.ir

ABSTRACT

This study was conducted to evaluate the carcass yield, abdominal fat deposition and chemical compositions of thigh and breast muscles of male Arian broilers fed three different lysine levels of 1) high lysine (110% NRC), 2) standard level suggested by National Research Council (NRC) and 3) low lysine (90% NRC) based on two expression ways of amino acid contents of feedstuffs, as total amino acids (TAA) or digestible amino acids (DAA). Three hundred one day old male broiler chickens were used in a completely randomized design with 6 treatments of 5 replicates (10 male broilers). All diets were iso-caloric and iso-nitrogenous. The results of this study showed that diet formulation based on DAA significantly influenced breast muscle yield and abdominal fat deposition. High lysine level (110% NRC) significantly improved carcass yield, breast and thigh muscle percentages and weights. The interaction between DAA × lysine levels of the feed affects carcass and breast muscle percentages of the broilers. Feeding the broilers with high level lysine containing diets (110% NRC) resulted in significantly high lysine content of the breast and thigh muscles. The results of this study suggest that diet formulation based on digestible amino acids with additional lysine at the level of 110% of NRC in starter and grower diets optimized body weight gain in Arain male broilers.

KEY WORDS Arian, broiler, carcass, digestible, lysine.

INTRODUCTION

The wide variation in the composition and digestibility of amino acids (AA) present in the poultry feedstuffs is of great concern when using these raw materials. The importance of utilizing the correct amount of balanced dietary protein and AA for poultry is a high priority issue for two reasons: first, the costs of protein and AA the most expensive nutrients in feeds and second, the environmental concerns about liberation of nitrogen compunds from poultry waste. The concentration of protein and AA in broiler diets have a large impact on breast muscle yield, feed/gain ratio, and feeding period of time required to gain the appropriate body weight for each type of market. Depending upon genetic strain and the market objectives for each broiler complex, a broiler integrator probably utilize several different protein and AA dietary programs (Acar *et al.* 1991). In commercial practice, formulating diets to adequate AA minimums is critical to optimize live production and meat yield of broiler chickens. Within the last 10 years, demand for breast fillets and value-added products has contributed to increasing market weights of broiler chickens. Market weight, product mix, live cost and genetic strain are factors that may govern AA supplementation levels. Amino acids are critical for muscle development (Tesseraud *et al.* 1996) and lysine (Lys) content of in breast muscle is relatively higher than those of other AA. Lysine represents approximately 7% of the protein in breast meat (Munks *et al.* 1945).

Dietary Lys inadequacy has been shown to reduce breast meat yield compared with other muscles (Tesseraud *et al.* 1996). Therefore, defining dietary AA needs for optimum growth and meat yield is of utmost importance. Lys, one of the key AAs in protein synthesis and muscle deposition has also been demonstrated to be involved in the synthesis of cytokines, proliferation of lymphocytes and thus in the optimum functioning of immune system in response to infection. An inadequate supply of Lys would reduce antibody response and cell-mediated immunity in chickens (Geraert and Mercier Adisseo, 2010). It is well known that protein and Lys and its interaction is considered as an important factor which affects performance and carcass quality of growing chicks and so, dietary requirement of protein is actually a requirement for the Lys contained in the protein. Essential amino acid recommendations for broilers by the NRC (1994) are largely based on results of the experiments conducted several decades ago (Todd and Roselina, 2014).

Therefore the objective of this study was to evaluate the three different lysine requirement levels, high lysine (HLys, 110% NRC), standard (SLys, NRC) and low lysine (LLys, 90% NRC) and two ways to express amino acids of feedstuffs, either as total amino acid (TAA) or digestible amino acid (DAA) with same protein and energy requirements recommended by NRC (5) effects on the carcass yeilds and chemical composition of thigh and breast muscles of Arian male broiler.

MATERIALS AND METHODS

An experiment with Arian male broilers was conducted from 1 to 6 weeks of age. At day 1, 300 male chicks were placed in 30 floor pens (10 chicks per pen and 0.1 m^2 floor space/chick). Water and feed were also supplied *ad libitum*. The basic chemical composition of the feed materials and breast and thigh muscles were determined according to AOAC (1990). The total amino acid values of the ingredients were assayed by high-pressure liquid chromatography analysis. In order to determine digestibility, the levels of TAA determined in the analysis were multiplied by their respective digestibility coefficients, as determined by Yaghobfar and Zahedifar (2003). A completely randomized experimental design was used, in a factorial scheme (2×3). The first factor was included three different lysine requirement levels, high lysine (110% NRC), standard (NRC) and low lysine (90% NRC) and the second factor was included two levels of total amino acids (TAA) and digestible amino acids (DAA) of feedstuffs. Therefore, the following treatments were applied:

1) diet with high lysine (HLys) requirement level (110% NRC), formulated based on TAA, 2) diet with standard lysine (SLys) requirement level (NRC), formulated based on TAA, 3) diet with low lysine (LLys) requirement level (90% NRC), formulated based on TAA, 4) diet with high lysine (HLys) requirement level (110% NRC), formulated based on DAA, 5) diet with standard lysine (SLys) requirement level (NRC), formulated based on DAA and 6) diet with low lysine (LLys) requirement level (90% NRC), formulated based on DAA.

Diets were formulated according to NRC (1994) recommendations to contain 23 and 19% crude protein (CP) and 3100 and 3200 kcal ME/kg in starter and grower diets, respectively. The diets were formulated as iso-energetic and iso-nitrogenic (Tables 1 and 2). Body weights (BW) was obtained at 42 d of age. Mortality was zero in treatments. At the end of the experimental period (at 42 day of age), in order to evaluate carcass quality, 4 birds from each treatments with body weights as close as to the average body weight of the experimental unit were slaughtered per repetition. These birds were weighed, eviscerated, and weighed again after removal of the head and feet to obtain carcass weight. The breast and thigh muscles, liver and abdominal fat weights were also determined. Yields of the breast and thigh muscles, carcass were determined in relation to body weight, and expressed as percentage of body weight (%). Data were analyzed by factorial analysis of variance (ANOVA) using the general linear models procedure of SAS (2004). Where significance occurred, means were compared with the Duncan multiple range test at 0.05 levels. The data were expressed as means with SEM.

RESULTS AND DISCUSSION

The results of carcass characteristics and chemical composition of thigh and breast muscles are given in Tables 3 and 4. The diet formulated based on DAA did not influence carcass, thigh percentage muscle and liver weight (Table 3). These results are consistent with Rostagno *et al.* (1995), who found no differences in carcass yield with the similar formulation.

The diet formulated on DAA basis significantly (P<0.05) influenced breast muscle percentage and abdominal fat deposition (Table 3). The excess AA intake is possibly related to the energy/protein ratio of the diet, and consequently, to carcass composition. In the present study, the diets formulated with the same amino acids (iso-nitrogenic) and same energy contents (iso-energetic) but different AAs containing feedstuffs and lysine levels. Therefore, the broilers given the diet formulated based on DAA, received more AA than those given the diets prepared based on TAA basis.

Table 1 Composition of the experimental diets in starter (0-21 day) period

Amino acids of feed	Total			Digestible		
Lysine levels	110%	Standard	90%	110%	Standard	90%
	1	2	3	4	5	6
Ingredients						
Corn grain	55.47	56.27	57.32	54.01	55.01	55.39
Soybean meal (48%)	35.56	35.01	33.94	37	36.21	36.01
Soybean oil	3	3	3	2.8	2.8	2.8
Fish meal	2	2	2	2	2	2
Oyster shells	1.88	1.88	1.88	1.88	1.88	1.88
Dicalcium phosphate	1	1	1	1	1	1
Common salt	0.2	0.2	0.2	0.2	0.2	0.2
Vitamin and mineral premix[1]	0. 5	0. 5	0. 5	0. 5	0. 5	0. 5
DL-methionine	0.11	0.09	0.1	0.11	0.1	0.1
L-lysine HCl	0.28	0.05	0.05	0.5	0.3	0.12
Nutrients contents						
Apparent metabolizable energy corrected for nitrogen (AMEn) (Mcal/kg)	3.10	3.10	3.10	3.10	3.10	3.10
Protein (%)	23.00	23.00	23.00	23.00	23.00	23.00
Ether extract (%)	5	5	5	5	5	5
Linoleic acid (%)	2.5	2.5	2.5	2.5	2.5	2.5
Calcium (%)	1.00	1.00	1.00	1.00	1.00	1.00
Avail phosphorus (%)	0.5	0.5	0.5	0.5	0.5	0.5
Sodium (%)	0.16	0.16	0.16	0.16	0.16	0.16
Lysine (%)	1.25	1.14	1.03	1.25	1.14	1.03
Methionine (%)	0.51	0.51	0.51	0.51	0.51	0.51
TSAA (%)	1.11	1.11	1.11	1.11	1.11	1.11

[1] Provides per kg of diet: vitamin A: 7000 IU; vitamin D_3: 1400 IU; vitamin E: 16.65 mg; vitamin K: 1.5 mg; vitamin B_1: 0.6 mg; vitamin B_2: 2.36 mg; vitamin B_6: 0.6 mg; vitamin B_{12}: 0.013 mg; Biotin: 0.15 mg; Choline: 1.54 g; Pantothenic acid: 9.32 mg; Niacin: 30.12 mg; Folic acid: 1.42 mg; Selenium: 0.65 mg; Iodine: 0.35 mg; Iron: 57.72 mg; Copper: 12.30 mg; Zinc: 141.48 mg and Manganese: 173 mg.
TSAA: total sulfur amino acids.

Table 2 Composition of the experimental diets in grower (22-42 day) period

Amino acids of feed	Total			Digestible		
Lysine levels	110%	Standard	90%	110%	Standard	90%
	1	2	3	4	5	6
Ingredients						
Corn grain	65.6	65.9	66.78	64.97	65.87	67.6
Soybean meal (48%)	28.72	28.2	27.14	28.93	28.1	27.01
Soybean oil	3	3.24	3.22	3.27	3.4	3
Oyster shells	1.6	1.7	1.8	1.5	1.5	1.4
Dicalcium phosphate	0.2	0.2	0.2	0.23	0.23	0.23
Common salt	0.2	0.2	0.2	0.2	0.2	0.2
Vitamin and mineral premix[1]	0. 5	0. 5	0. 5	0. 5	0. 5	0. 5
DL-methionine	0.1	0.06	0.11	0.2	0.1	0.06
L-lysine HCl	0.08	0	0.05	0.2	0.1	0
Nutients contents						
AMEn (Mcal/kg)	3.2	3.2	3.2	3.2	3.2	3.2
Protein (%)	19	19	19	19	19	19
Ether extract (%)	5	5	5	5	5	5
Linoleic acid (%)	2.5	3	2.5	2.5	3	2.5
Calcium (%)	0.09	0.09	0.09	0.09	0.09	0.09
Avail phosphorus (%)	0.42	0.42	0.42	0.42	0.42	0.42
Sodium (%)	0.16	0.16	0.16	0.16	0.16	0.16
Lysine (%)	1.1	1	0.9	1.1	1	0.9
Methionine (%)	0.42	0.42	0.42	0.42	0.42	0.42
TSAA (%)	0.88	0.88	0.88	0.88	0.88	0.88

[1] Provides per kg of diet: vitamin A: 7000 IU; vitamin D_3: 1400 IU; vitamin E: 16.65 mg; vitamin K: 1.5 mg; vitamin B_1: 0.6 mg; vitamin B_2: 2.36 mg; vitamin B_6: 0.6 mg; vitamin B_{12}: 0.013 mg; Biotin: 0.15 mg; Choline: 1.54 g; Pantothenic acid: 9.32 mg; Niacin: 30.12 mg; Folic acid: 1.42 mg; Selenium: 0.65 mg; Iodine: 0.35 mg; Iron: 57.72 mg; Copper: 12.30 mg; Zinc: 141.48 mg and Manganese: 173 mg.
TSAA: total sulfur amino acids.

Table 3 Effects of amino acids of feedstuffs and lysine requirement levels on carcass composition at 42 d

Amino acids×lysine levels[*]	Carcass yeild (%)	Carcass weight (g)	Breast (%)	Breast (g)	Thigh (%)	Thigh (g)	Abdominal fat (g)	Liver (g)
1- TAA × 110% NRC	66.63[a]	1237[b]	21.59[a]	394[b]	22.62	420[b]	17.05[b]	53.85[ab]
2- TAA × Standard NRC	66.18[a]	1215[b]	21.19[a]	396[b]	22.79	417[b]	15.10[b]	53.63[ab]
3- TAA × 90% NRC	60.94[b]	1113[c]	18.88[b]	345[b]	21.88	400[b]	19.10[b]	42.55[b]
4- DAA × 110% NRC	65.18[a]	1463[a]	21.92[a]	472[a]	23.30	524[b]	40.33[a]	60.40[a]
5- DAA × standard NRC	66.35[a]	1227[b]	21.27[a]	400[b]	23.71	438[ab]	12.43[b]	49.35[ab]
6- DAA × 90% NRC	63.03[ab]	1223[b]	20.78[ab]	402[b]	21.63	421[b]	33.10[a]	55.75[ab]
P-value	0.017	0.036	0.087	0.031	0.332	0.039	0.002	0.146
SEM	1.175	52.93	0.381	22.71	0.719	29.35	4.577	4.422
Amino acids of feedstuffs								
Total	64.58	1188[b]	20.56[b]	378[b]	22.43	412	17.08[b]	50.01
Digestible	64.85	1304[a]	21.79[a]	426[a]	22.88	461	28.62[a]	55.17
P-value	0.780	0.037	0.042	0.023	0.452	0.056	0.006	0.170
SEM	0.678	78.32	0.293	13.11	0.415	16.94	2.643	2.553
Lysine requiremnet levels								
110% NRC	65.91[a]	1350[a]	21.76[a]	433[a]	22.96	472	28.69[a]	57.13
Standard	66.26[a]	1221[ab]	21.24[ab]	398[ab]	23.25	427	13.76[b]	51.49
90% NRC	61.98[b]	1168[b]	19.83[b]	374[b]	21.75	410	26.10[a]	49.15
P-value	0.003	0.046	0.041	0.045	0.116	0.126	0.010	0.207
SEM	0.831	63.45	0.181	16.06	0.508	20.75	3.237	3.127

[*] Treatment 1: diet with 110% NRC (1994) lysine requirement level, formulated based on total amino acids (TAA); Treatment 2: diet with standard NRC (1994) lysine requirement level, formulated based on TAA; Treatment 3: diet with 90% NRC (1994) lysine requirement level, formulated based on TAA; Treatment 4: diet with 110% NRC (1994) lysine requirement level, formulated based on digestible amino acids (DAA); Treatment 5: diet with standard NRC (1994) lysine requirement level, formulated based on DAA and Treatment 6: diet with 90% NRC (1994) lysine requirement level, formulated based on DAA.
The means within the same row with at least one common letter, do not have significant difference (P>0.05).
SEM: standard error of the means.

If crude protein ratio is increased in a given broiler feed, significant decrease might occur in the energy/protein ratio, which results in carcasses with high fat content (Rosebrough and Steele, 1985).

Protein accretion at post-hatch period in boiler chicks is a result from either increasing protein synthesis or decreasing protein degradation. Diets with low Lys levels can limit breast muscle growth early in the development by reducing synthesis of muscle proteins and RNA content (Tesseraud et al. 1996). Significant effects of the dietary Lys levels both on the thigh muscle percentage and liver weight were not observed in this study. The results were in accordance with the previous studies (Kidd et al. 1998; Nasr and Kheiri, 2011; Bernal et al. 2014) which demonstrated that Lys requirement for growing chicks is higher than that of NRC (1994) recommendation for maximum growth. It is also confirmed that increasing dietary Lys level increases breast meat yield.

Dietary Lys levels significantly influenced the breast muscle percentage (Table 3) and the weights of carcass, breast and thigh muscle (Figure 2). HLys and LLys levels resulted in higher abdominal fat deposition. This result is supported by the finding that the abdominal fat weight significantly (P=0.002) increased both with the increasing or decreasing in Lys levels.

The increase in abdominal fat weight is probably resulted from the imbalanced AA content of the diets. This study showed that increasing Lys level (110% NRC) in diet increased breast muscle yield significantly, as shown in the previous studies (Gorman and Balnave, 1995; Kidd et al. 1998; Nasr and Kheiri, 2011; Bernal et al. 2014). Feeding broilers with HLys containing diets throughout the feeding period optimizes breast meat yield (Kidd et al. 1998; Nasr and Kheiri, 2011; Vieira and Angel, 2012; Bernal et al. 2014), although it may not always be economically justified. However, evidence in the literature suggests that feeding diets with high Lys level during the starter period impacts subsequent breast meat yield (Kerr et al. 1999). The concentration of dietary Lys can significantly influence the breast meat yield for the following reasons: breast muscle contains a high concentration of Lys (Table 4), which represents a larger portion of carcass meat. Although the breast muscle development is also affected by sex, age, breed and genetics strain, previous studies (Acar et al. 1991; Tesseraud et al. 1996; Gorman and Balnave, 1995) also demonstrated the increasing effect of additional feed Lys on the breast meat growth.

The weight of breast and thigh muscles of the treatment 4 (HLys-DAA) was significantly (P=0.031 and P=0.039) higher than those of the other treatments (Figure 1).

Table 4 Effects of amino acids of feedstuffs and lysine requirement levels on chemical composition of breast and thigh meat at 42 d

Amino acids×lys levels	Breast meat			Thigh meat		
	Lys (%)	Lipid (%)	CP (%)	Lys (%)	Lipid (%)	CP (%)
1- TAA × 110% NRC	5.83[a]	5.53	57.91	5.05[a]	9.86	60.59
2- TAA × Standard NRC	4.43[b]	3.04	49.26	3.83[b]	9.60	52.22
3- TAA × 90% NRC	3.92[b]	6.49	45.64	3.05[b]	11.39	48.74
4- DAA × 110% NRC	5.89[a]	4.77	59.34	5.40[a]	9.30	64.66
5- DAA × standard NRC	4.29[b]	3.66	52.17	3.92[b]	9.06	53.11
6- DAA × 90% NRC	4.09[b]	3.76	45.10	3.55[a]	11.59	49.82
P-value	0.009	0.295	0.289	0.002	0.517	0.122
SEM	0.424	1.128	5.211	0.362	1.162	4.452
Amino Acids of Feedstuffs						
Total	4.73	5.021	50.94	3.98	10.28	53.85
Digestible	4.76	4.061	52.20	4.29	9.980	55.86
P-value	0.93	0.311	0.769	0.30	0.755	0.587
SEM	0.245	0.651	3.009	0.209	0.671	2.569
Lysine requiremnet levels						
110% NRC	5.86[a]	5.149	58.63	5.23[a]	9.579	62.63[a]
Standard	4.36[b]	3.35	50.72	3.88[ab]	9.33	52.67[b]
90% NRC	4.00[b]	5.13	45.37	3.30[b]	11.49	49.28[b]
P-value	0.001	0.215	0.061	0.000	0.155	0.021
SEM	0.299	0.798	3.685	0.256	0.822	3.148

Treatment 1: diet with 110% NRC (1994) lysine requirement level, formulated based on total amino acids (TAA); Treatment 2: diet with standard NRC (1994) lysine requirement level, formulated based on TAA; Treatment 3: diet with 90% NRC (1994) lysine requirement level, formulated based on TAA; Treatment 4: diet with 110% NRC (1994) lysine requirement level, formulated based on digestible amino acids (DAA); Treatment 5: diet with standard NRC (1994) lysine requirement level, formulated based on DAA and Treatment 6: diet with 90% NRC (1994) lysine requirement level, formulated based on DAA.
The means within the same row with at least one common letter, do not have significant difference (P>0.05).
CP: crude protein.
SEM: standard error of the means.

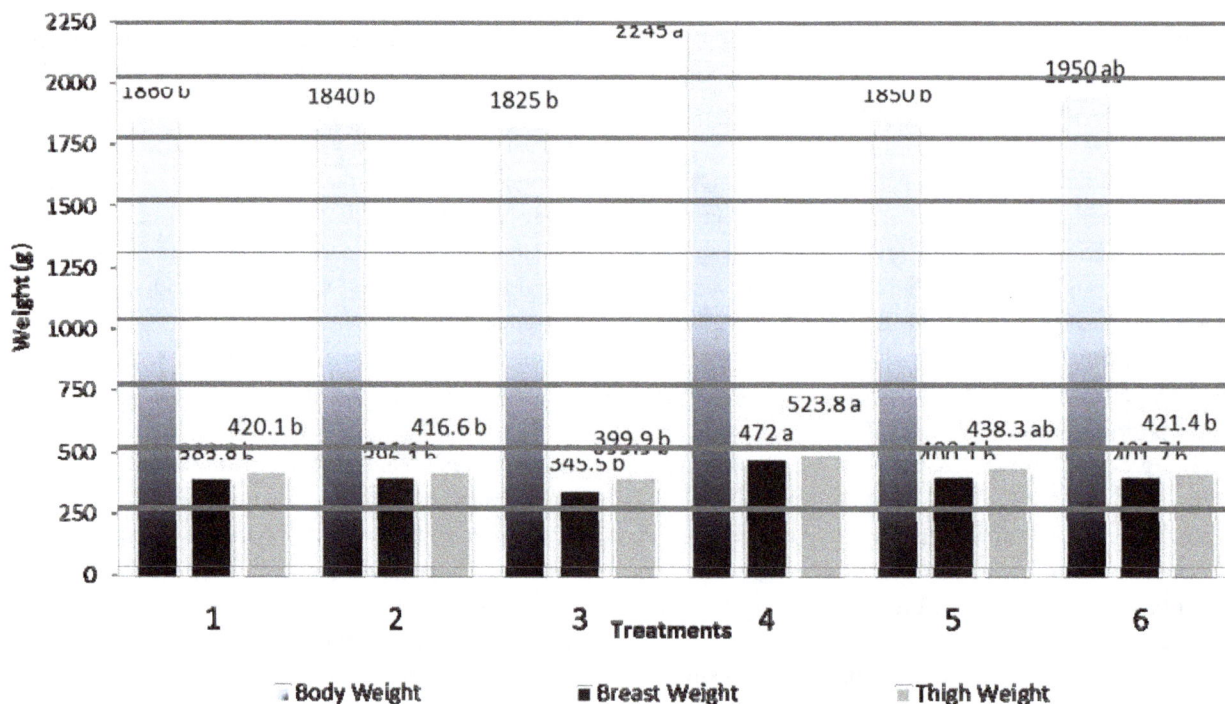

Figure 1 Effects of treatments on mean body, breast and thigh weight (at 42 day)
Treatment 1: diet with 110% NRC (1994) lysine requirement level, formulated based on total amino acids (TAA); Treatment 2: diet with standard NRC (1994) lysine requirement level, formulated based on TAA; Treatment 3: diet with 90% NRC (1994) lysine requirement level, formulated based on TAA; Treatment 4: diet with 110% NRC (1994) lysine requirement level, formulated based on digestible amino acids (DAA); Treatment 5: diet with standard NRC (1994) lysine requirement level, formulated based on DAA and Treatment 6: diet with 90% NRC (1994) lysine requirement level, formulated based on DAA

Previous reserchers (Eits *et al.* 2003; Dozier *et al.* 2007; Bernal *et al.* 2014) also showed that feeding broilers with HLys-AA diets increases breast meat yield. Results of the present study showed that the HLys-DAA had higher efficiency for tissue protein accretion and growth. This result is possibly related to the better availability of AA for synthesis of muscle proteins (Table 3). Diets formulated with HLys levels promoted a better conversion of DAA into carcass and breast and thigh meat yield (Figure 1). However, HLys-DAA groups had excess abdominal fat. This suggests that the excess AA intake caused imbalance and excess fat deposition. The combination of DAA and standard Lys possibly resulted in less AA availability for fat deposition. Leclercq (1998) stated that the Lys level should be at the highest level in the feed for minimizing abdominal fat percentage during maximizing breast meat yield and body weight gain.

Acar *et al.* (1991) found significant interactions between genotype and feed lysine level affecting abdominal fat, breast fillet yield, and breast meat yield. Moreover, dietary protein level has also been found to affect the lysine requirement (expressed as a percentage of the diet) of chicks (Hurwitz *et al.* 1998; Sterling *et al.* 2006).

Differences in the results among the previous researchers in relation to the dietary AA concentrations have been attributed to (Acar *et al.* 1993) differences between broiler strains.

The response to dietary AA/CP density (Smith *et al.* 1998; Sterling *et al.* 2006) and dietary Lys (Acar *et al.* 1991; Pesti *et al.* 1995) differs among the strain sources. A high-yielding strain was shown to contain more breast muscle, total RNA and protein on a weight basis and total DNA content over a low-yielding strain (Corzo *et al.* 2005). Muscle growth is largely related to the number of nuclei or total DNA (Kang *et al.* 1985).

Hence, strains exhibiting rapid muscle growth should have been supplied with high dietary AA needs for muscle accretion. Thus, HLys-DAA had significantly ($P<0.05$) higher in carcass, breast and thigh weights than those of the other groups, in this study.

The diet formulated based on AA and lysine levels did not affect protein and lipid contents of breast and thigh muscles.

The overall Lys content in carcass of broilers increased as the chicks aged because of relative increase of breast meat percentage (Tables 3 and 4).

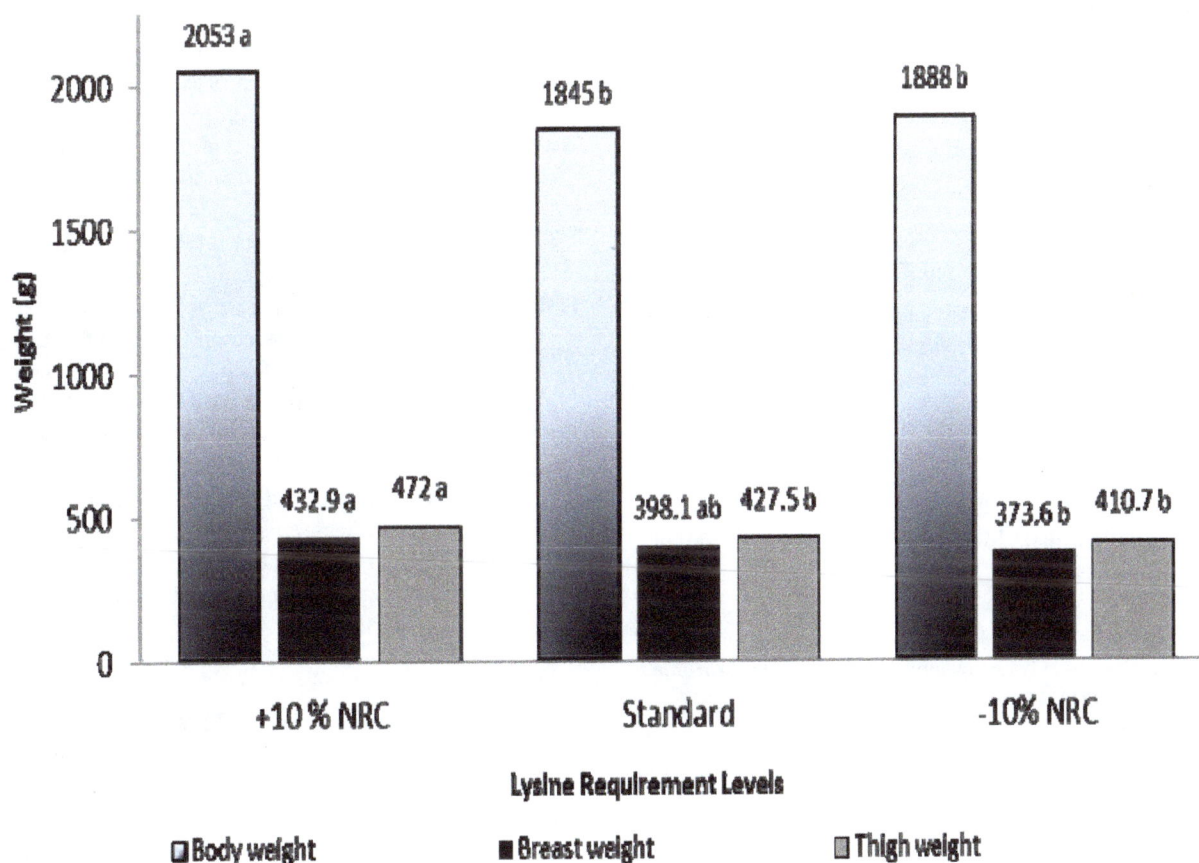

Figure 2 Effects of lysine requirement levels (% of NRC recommendation) on mean body, breast and thigh weight (at 42 day)

The amino acid requirements for different genetic broiler strains will be partially dependent upon the amino acid content of each body components (i.e. breast, thigh and drum) and the extent of changes of the percentages of carcass components.

Nevertheless, the Lys contents of the breast and thigh muscles were 4.742% and 4.133% respectively, which were lower than those of Ross Strain.

CONCLUSION

1- Feeding broilers with HLys diets (110% NRC) increases carcass and breast percentage by 4.4% and 1.81%, the levels was higher than the broilers received LLys diets.
2- HLys-DAA positively affects the carcass and breast muscle percentages of broilers. The interaction between DAA and HLys allows Arian broilers to full express the genetic potential for growth. Formulating broiler diets based on DAA gives a better prediction of dietary protein quality and bird performance than total amino acids.
3- Feeding broilers with high Lys containing diets (110% NRC) significantly increases amount of Lys in breast and thigh meats.

ACKNOWLEDGEMENT

This research was partially supported by Saveh Branch, Islamic Azad University. We thank our colleagues from Saveh Branch, Islamic Azad University who provided insight and expertise that greatly assisted the research.

REFERENCES

Acar N., Moran Jr.E.T. and Bilgili S.F. (1991). Live performance and carcass yield of male broilers from two commercial strain crosses receiving rations containing lysine below and above the established requirement between six and eight weeks of age. *Poult. Sci.* **70**, 2315-2321.

Acar N., Moran Jr.ET. and Mulvaney D.R. (1993). Breast muscle development of commercial broilers from hatching to twelve weeks of age. *Poult. Sci.* **72**, 317-325.

AOAC. (1990). Official Methods of Analysis. Vol. I. 15th Ed. Association of Official Analytical Chemists, Arlington, VA, USA.

Bernal L.E.P., Tavernari F.C., Rostagno H.S. and Albino L.F.T. (2014). Digestible lysine requirements of broilers. *Brazilian J. Poult. Sci.* **16(1)**, 49-55.

Corzo A., Kidd M.T., Dozier W.A., Walsh T.J. and Peak S.D. (2005). Impact of dietary amino acid density on broilers grown for the small bird market. *Japan Poult. Sci.* **42**, 329-336.

Dozier W.A., Corzo A., Kidd M.T. and Branton S.L. (2007). Dietary apparent metabolizable energy and amino acid density effects on growth and carcass traits of heavy broilers. *J. Appl. Poult. Res.* **16**, 192-205.

Eits R.M., Kwakkel R.P., Verstegen M.W. and Emmans G.C. (2003). Responses of broiler chickens to dietary protein: effects of early life protein nutrition on later responses. *British Poult. Sci.* **44**, 398-409.

Geraert P.A. and Mercier Adisseo Y. (2010). Amino Acids: Beyond the Building Blocks! ADISSEO France SAS, 10 Place du Général de Gaulle, 92160 ANTONY, France.

Gorman I. and Balnave D. (1995). The effect of dietary lysine and methionine on the growth characteristics and breast meat yield of Australian broiler chickens. *Australian J. Agric. Res.* **46**, 1569-1577.

Hurwitz S., Sklan D., Talpaz H. and Plavnik I. (1998). The effect of dietary protein level on the lysine and arginine requirements of growing chickens. *Poult. Sci.* **77**, 689-696.

Kang C.W., Sunde M.L. and Swick R.W. (1985). Characteristics of growth and protein turnover in skeletal muscle of turkey poults. *Poult. Sci.* **64**, 380-387.

Kerr B.J., Kidd M.T., Halpin K.M., McWard G.W. and Quarles C.L. (1999). Lysine level increases live performance and breast yield in male broilers and breast yield in male broilers. *J. Appl. Poult. Res.* **8**, 381-390.

Kidd M.T., Kerr B.J., Halpin K.M., McWard G.W. and Quarles C.L. (1998). Lysine levels in starter and grower finisher diets affect broiler performance and carcass traits. *J. Appl. Poult. Res.* **7**, 351-358.

Leclercq B. (1998). Specific effects of lysine on broiler production: comparison with threonine and valine. *Poult. Sci.* **77**, 118-123.

Munks B., Robinson A., Beach E.F. and Williams H.H. (1945). Amino acids in the production of chicken egg and muscle. *Poult. Sci.* **24**, 459-464.

Nasr J. and Kheiri F. (2011). Effect of different lysine levels on Arian broiler performance. *Italian J. Anim. Sci.* **10(3)**, 170-174.

NRC. (1994). Nutrient Requirements of Poultry, 9th Rev. Ed. National Academy Press, Washington, DC., USA.

Pesti G.M., Leclercq B.A., Chagneau M. and Cochard T. (1994). Comparative responses of genetically lean and fat chickens to lysine arginine and non-essential amino acid supply. II. Plasma amino acid responses. *British Poult. Sci.* **35**, 697-707.

Rosebrough R.W. and Steele N.C. (1985). Energy and protein relationships in the broiler. 1. Effect of protein levels and feeding regimens on growth body composition and *in vitro* lipogenesis of broiler chicks. *Poult. Sci.* **64**, 119-126.

Rostagno H.S., Pupa J.M.R. and Pack M. (1995). Diet formulation for broiler based on total versus digestible amino acids. *J. Appl. Poult. Res.* **4(1)**, 293-299.

SAS Institute. (2004). SAS®/STAT Software, Release 9.1. SAS Institute, Inc., Cary, NC. USA.

Smith E.R., Pesti G.M., Bakalli R.I., Ware G.O. and Menten J.F.M. (1998). Further studies on the influence of genotype and dietary protein on the performance of broilers. *Poult. Sci.* **77**, 1678-1687.

Sterling K.G., Pesti G.M. and Bakalli R.I. (2006). Performance of different broiler genotypes fed diets with varying levels of dietary crude protein and lysine. *Poult. Sci.* **85**, 1045-1054.

Tesseraud S., Peresson R. and Chagneau A.M. (1996). Dietary lysine deficiency greatly affects muscle and liver protein turnover in growing chickens. *British J. Nutr.* **75,** 853-865.

Todd J.A. and Roselina A. (2014). Nutrient requirements of poultry publication: history and need for an update. *J. Appl. Poult. Res.* **23,** 567-575.

Vieira S.L. and Angel C.R. (2012). Optimizing broiler performance using different amino acid density diets: what are the limits? *J. Appl. Poult. Res.* **21,** 149-155.

Yaghobfar A. and Zahedifar M. (2003). Endogenous losses of energy and amino acids in birds and their effect on true metabolisable energy values and availability of amino acids in maize. *British Poult. Sci.* **44,** 719-725.

17

Effect of Clinoptilolite Coated with Silver Nanoparticles on Meat Quality Attributes of Broiler Chickens during Frozen Storage

S.R. Hashemi[1*], D. Davoodi[2] and B. Dastar[1]

[1] Department of Animal Science, Gorgan University of Agricultural Science and Natural Resources, Gorgan, Iran
[2] Department of Nanotechnology, Agricultural Biotechnology Research Institute of Iran (ABRII), Karaj, Iran

*Correspondence E-mail: hashemi711@gau.ac.ir

ABSTRACT

This study was carried out to assess the effect of clinoptilolite coated with silver nanoparticles on meat quality attributes of broiler chickens during frozen storage. A total of 375 one-day-old broiler chicks were assigned in a completely randomized design to 1 of 5 experimental groups including: basal diet, basal diet supplemented with 1% clinoptilolite and basal diet supplemented with 1% clinoptilolite coated with either 25, 50 or 75 ppm nanosilver. On d 42, five birds per treatment were slaughtered and breast and thigh meat were kept 3 or 7 days at -17 °C before assessing meat quality attributes. The addition of nanosilver coated on clinoptilolite at all levels increased water-holding capacity (WHC) in thigh muscles after 7 days frozen storage. The lowest value of springiness and chewiness was for the breast muscle of broilers fed clinoptilolite coated with 25 ppm nanosilver diet. Adhesiveness, cohesiveness and gumminess value were not influenced by treatment (P>0.05). In conclusion, nanosilver coated on clinoptilolite can be used as potential feed additive in the broiler diet without negative implications on meat quality characteristics.

KEY WORDS broiler, clinoptilolite, frozen storage, meat quality, nanosilver, zeolite.

INTRODUCTION

Poultry meat quality attributes may be affected by several factors such as genotype, rearing conditions and feed additives which could affect muscle metabolism as well as chemical composition (Meluzzi et al. 2009). Nanotechnology is may be used for assurance of food safety in different food products (Kannan and Subbalaxmi, 2011). Food scientists predicts that nanotechnology will also have a significant impact on food products in a variety of ways both directly and indirectly (Sekhon, 2014). There are various types of nanoparticles such as Ag, Au and Zn. Recently, silver nanoparticles have been shown to have unique antibacterial properties and are widely used as antimicrobial agents and as replacement for antibiotic growth promoter in the poultry industry (Chauke and Siebrits, 2012; Hashemi et

al. 2014). Despite demonstrated beneficial effects of silver nanoparticles on digestive microbial biodiversity and function in animals (Li et al. 2006; Sawosz et al. 2004), other effects of silver nanoparticles on physiological status, such as digestive enzymatic activity, immunological status and intestinal structure and their effect on meat quality characteristics are unknown. The use of feed additives and utilization of exogenous compounds is a new nutritional strategies that may help the poultry industry improve the quality of meat (Khalafalla et al. 2011). By the exploitation of the use of nanoparticles, poultry products can produce high safety. In previous papers, we have shown that broiler diets supplemented zeolite coated with silver nanoparticles improved water-holding capacity of thigh muscle in broiler chickens (Hashemi et al. 2014). Appearance of broiler meat is an important quality criterion when making the decision to purchase and in final product satisfaction. After slaughter, poultry carcass has to be chilled and / or frozen to ensure a high quality and safe product. Chicken meat is kept cold during distribution to retail stores to prevent the growth of bacteria and to increase its shelf life. The storage time of broiler meat in the refrigerator and freezer can influence meat quality. The objective of this study is to evaluate the effect of clinoptilolite coated with silver nanoparticles on meat quality attributes of broiler chickens during frozen storage.

MATERIALS AND METHODS

A total of 375 one-day-old Ross 308 broilers (male and female) from a commercial hatchery, were randomly assigned to 5 experimental groups. Each group comprised five pens each of 15 birds.

Experimental diets were following:

1) basal diet (control, without nanosilver and clinoptilolite) (C).

2) basal diet supplemented with 1% clinoptilolite (Z).

3) basal diet supplemented with 1% clinoptilolite coated with 25 ppm nanosilver (ZN25).

4) basal diet supplemented with 1% clinoptilolite coated with 50 ppm nanosilver (ZN50).

5) basal diet supplemented with 1% clinoptilolite coated with 75 ppm nanosilver (ZN75).

Natural clinoptilolite (zeolite) used in this research was prepared from well-defined zeolitic stratigraphic units from Semnan province region, Iran. Zeolitic rock was pulverized and sieved to give a particle size of 1-2 mm and then washed with distilled water to remove all the soluble impurities and then and dried in the oven over night at 105 °C. The chemical formula of pure clinoptilolite was $(K_2, Na_2, Ca, Mg)_3 Al_6Si_{30}O_{72} \cdot 24H_2O$. Clinoptilolite coated with silver nanoparticles were prepared by Nano Nasb Pars Company (Tehran, Iran). Silver nanoparticles had a maximum diameter of 50 nm.

In order to study the chemical composition of the clinoptilolite sample were analyzed using X-ray Fluorescence (XRF) technique and the XRF data was collected on a PHILIPSPW1480 XRF spectrometer with Rh tube (Nikpey et al. 2013) (Table 1). Field emission scanning electron microscopy (FESEM, Mira, 3-XMU) was employed for the detailed study of morphology and additionally, Energy dispersive X-ray spectroscopy (EDX) and elemental mapping analyses were used to investigate the materials composition at Razi Metallurgical Research Center, Iran (Figure 1).

The silver (Ag) content of clinoptilolite was measured by the method as described by Kulthong et al. (2010). In brief, 0.2-0.3 g sample was weighed and then digested by a microwave digestion system in 5 mL of 14.4 M HNO_3 to dissolve all the silver content. The microwave irradiation cycles were 250 W for 5 min, 400 W for 5 min and 600 W for 5 min. The digested sample was then cooled and diluted up to 25 mL with deionized water to enable quantification of silver by a graphite furnace atomic absorption spectroscopy or GFAAS (Perkin Elmer Analyst 300, Waltham, MC).

Table 1 The summarized chemical composition of the used clinoptilolite by means of X-ray Fluorescence (XRF)[1] technique

Semnan natural clinoptilolite-rich tuffs	
Constituents	% by weight
SiO_2	68.95
Al_2O_3	11.14
Fe_2O_3	0.97
CaO	4.83
Na_2O	0.95
K_2O	0.90
MgO	0.79
TiO_2	0.201
MnO	0.011
P_2O_5	0.012
SO_3	0.068
L.O.I[2]	10.64
Si/Al	4.81
Ag[3]	< 5 ppm

[1] XRF PHILIPS PW1480 XRF spectrometer with Rh tube.
[2] LOI: loss on ignition.
[3] The silver content was measured by a graphite furnace atomic absorption spectroscopy (GFAAS).

Diets were formulated to meet broiler nutrient requirements according to the Ross 308 management guideline and proximate analyses confirmed formulated values for all critical nutrients in the diets fed (Table 2). The birds had free access to water and all diets were fed *ad libitum*. Birds had access to light according to a 23 L/1 D program. The basal diet ingredients and composition of the control diet are presented in Table 1. All experimental protocols were reviewed and approved by the Animal Care Committee of the Gorgan University of Agricultural Sciences and Natural Resources.

On d 42, five birds from each treatment, close to the mean weight of all birds in each pen were selected, slaughtered and were plucked, kept in a chiller for approximately 30-45 minutes until the internal temperature of the birds reach 2-3 °C and then, eviscerated and cut in parts and the breast muscles (pectoralis major) and right thigh. Then the samples were packaged in low-density polyethylene and stored in the freezer (-17±1 °C) until meat quality attributes (moisture, pH, oxidative stability, water-holding capacity, texture profile analysis and color) at d 3 and 7 after the slaughter.

For pH analyses portable pH-meter (Model pH 211; Hanna Instruments,Woonsocket, RI, USA) was used. A total of 10 g of sample was weighed and homogenized with 10 mL of distilled water.

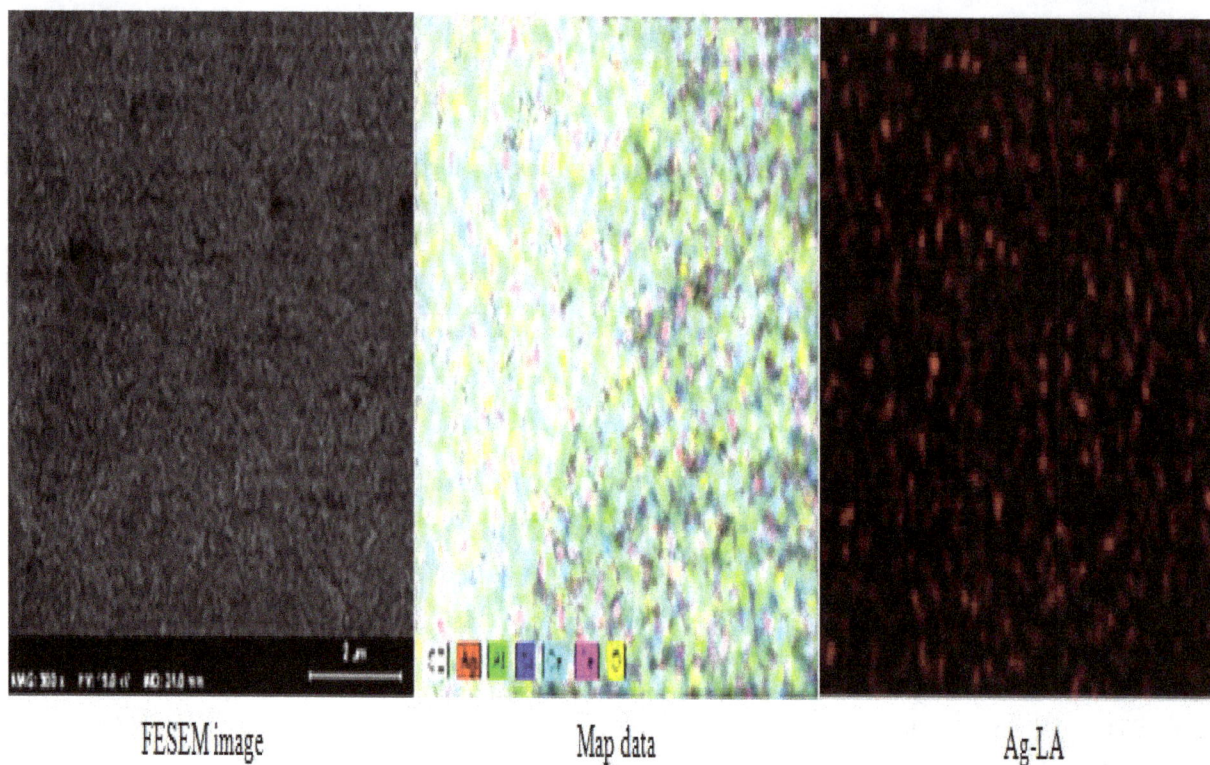

FESEM image   Map data   Ag-LA

Figure 1 Field emission scanning electron microscopy (FESEM) micrographs and energy dispersive X-ray spectroscopy (EDX) for elemental mapping analyses of the nanosilver coated on clinoptilolite

Table 2 Composition and analysis of the diets (g kg^{-1} as-fed)

Ingredients	Control diet		Exprimental diet	
	Starter (1-21)	Grower (22-42)	Starter (1-21)	Grower (22-42)
Yellow corn	53.70	59.96	51.6	57.84
Soybean meal (44%)	39.52	33.25	39.95	33.68
Soybean oil	3	3.41	3.69	4.11
Coated nanosilver on zeolite	0	0	1	1
Dicalcium phosphate	1.47	1.09	1.47	1.09
Limestone	1.19	1.29	1.18	1.28
Salt	0.43	0.32	0.43	0.32
Vitamin premix[1]	0.25	0.25	0.25	0.25
Mineral premix[1]	0.25	0.25	0.25	0.25
DL-methionine	0.13	0.05	0.13	0.05
L-lysine	0.06	0.13	0.05	0.13
Analysis (dry matter basis)				
Metabolizable energy (ME) (kcal kg^{-1})	2950	3050	2950	3050
Crude protein (CP) (%)	21.2	19.06	21.2	19.06
Calcium (%)	0.92	0.86	0.92	0.86
Phosphorus (%)	0.41	0.33	0.41	0.33
Sodium (%)	0.18	0.14	0.18	0.14
Lysine (%)	1.01	0.95	1.01	0.95
Methionine (%)	0.47	0.36	0.47	0.36
Cystine (%)	0.36	0.37	0.36	0.37
Arginine (%)	1.45	1.27	1.45	1.27
Threonine (%)	0.84	0.74	0.84	0.74

[1] Supplied per kilogram of diet: vitamin A: 1500 IU; Cholecalciferol: 200 IU; vitamin E: 10 IU; Riboflavin: 3.5 mg; Pantothenic acid: 10 mg; Niacin: 30 mg; Cobalamin: 10 µg; Choline chloride: 1000 mg; Biotin: 0.15 mg; Folic acid: 0.5 mg; Thiamine: 1.5 mg; Pyridoxine: 3.0 mg; Iron: 80 mg; Zinc: 40 mg; Manganese: 60 mg; Iodine: 0.18 mg; Copper: 8 mg and Selenium: 0.15 mg.

The probe of the pH meter was dipped into the solution and the pH values recorded according to the AOAC (1995). The pH meter was calibrated by measuring buffer solutions (pH=4 and pH=7) after every 5 observations. Moisture content of chicken thighs and breast muscle were determined according to the Association Official Analytical Chemists (AOAC, 1998), by drying about 10 g of the sample at 105 °C in the oven until a constant weight was recorded. The method used for determining the water-holding capacity (WHC) was a modification of the high speed centrifugation method (Jang et al. 2008). Oxidative stability in the meat samples is based on the reaction of one molecule malondialdehyde (MDA) with 2 molecules of thiobarbituric acid (TBA). The color of the final complex is pink and the absorbance of the complex is measured spectrophotometrically. MDA is a major degradation product of oxidation of polyunsaturated fatty acids. Evaluation of TBA was performed.

Texture profile analyses (adhesiveness, chewiness, springiness, hardness, gumminess and cohesiveness) were assessed using a texture analyzer (Brookfield, LFRA. 4500 Texture Analyser, USA) as described by Santhi and Kalaikannan (2014) and defined in our previous study (Hashemi et al. 2014). Briefly, samples were allowed to equilibrate at room temperature for 20 mins and then cut into uniformly sized cubes of 1 cm (width) × 1 cm (thickness) × 1 cm (length). Each sample was compressed twice to 80% of the original height using a compression probe (TA 11/1000, 20 mm). A crosshead speed of 10 mm/s was used. The values were recorded based on the software available in the instrument. Color analysis, the lightness (L*), redness (a*) and yellowness (b*) were measured using a colorimeter (Lovibond CAM-system 500). Briefly, after exiting samples from the freezer, thigh and breast were taken out of the bags and rinsed thoroughly with tap water. Breast and thigh skins were raised up carefully and kept at room temperatures for 30 minutes. Then, the surfaces of meat samples (thigh and breast) were photographed from the similar sections. A Hunter Lab spectrophotometer model Color Quest II was used, calibrated with white standard and gray standard. A completely randomized design with 5 treatments and 5 replicates of 15 birds was employed. Statistical analyses were performed using the GLM procedure of SAS software (SAS, 2004). Significant differences were further separated using Duncan's multiple range test. Statistical significance was considered at (P≤0.05).

RESULTS AND DISCUSSION

The effect of different treatment diets on selected quality characteristics of broiler breast meat and meat color intensity is shown in Table 3.

No significant differences were noted on selected quality and meat color intensity between the different treatment diets at 3 and 7 days frozen storage(P>0.05).

Table 4 presents the values of selected quality of broiler thigh meat and meat color intensity. The birds fed diets containing various levels of nanosilver coated on clinoptilolite (ZN25, ZN50 and ZN75) had lower MDA concentrations than the control group on d 3 after frozen storage (P<0.05). Thigh meat water-holding capacity (WHC) 7 days frozen storage was significantly affected by the dietary treatments.

Broilers fed diet containing 1% clinoptilolite and control diet had lower WHC value than the other groups. Also, on d 7 after refrigerated storage, MDA concentration of birds' thigh meat in ZN25 and ZN50 diets was lower than that of birds in the control group (P<0.05) but similar to that of birds in Z and ZN75 (P>0.05). However, no effect of diet on other thigh meat selected quality and meat color intensity was observed (P>0.05).

The texture profile of broiler chicken thigh meat is presented in Table 5. On d 3, there were no significant differences between treatments on hardness, adhesiveness, springiness, cohesiveness and gumminess (P>0.05), while chewiness value was influenced by treatment diets (P<0.05). The lowest value of chewiness was recorded by birds fed control diet. On d 7, the springiness, cohesiveness, gumminess and chewiness value were affected by treatment diets (P<0.05) and at the same time as hardness and adhesiveness were not influenced by dietary treatment (P>0.05). The effect of different experimental diets on texture profile analyses of broiler chicken breast muscle is offered in Table 6.

On d 3, adhesiveness and gumminess were not influenced by dietary treatment (P>0.05) while other traits were affected most strongly by experimental diets (P<0.05). Seven days after storage of broiler chicken breast in the freezer only springiness and chewiness value were affected by the dietary treatments (P>0.05). The lowest value of springiness and chewiness were measured in the broilers fed ZN25 diet. Hardness, adhesiveness, cohesiveness and gumminess value werenot influenced by treatment (P>0.05).

The present results showed that the addition of nanosilver particles coated on clinoptilolite at all levels increased WHC in thigh muscles 7 days after frozen storage. Many variables, such as broiler genotype, age, sex, nutrition, rearing system, carcass dressing and type of meat, can affect the nutritional value of meat. Dabes (2001) claimed that lower WHC indicated losses in the nutritional value through exudates that were released and this resulted in drier and tougher meat. Offer and Knight (1989) showed that muscle pH and protein denaturation are considered to be the main determinants of WHC in meat.

Table 3 Selected quality of broiler pectoralis major muscle fed different treatments diet

Selected quality traits	Treatments					SEM
	Control	Z	ZN25	ZN50	ZN75	
3 days storage at -17±1 °C						
Moisture (%)	74.11	73.25	73.79	74.81	73.62	4.1
pH	5.32	5.49	5.38	5.41	5.44	0.13
WHC (%)	75.23	73.92	75.81	77.90	74.93	3.8
TBA-RS (mg MDA/kg)	2.50	2.53	2.48	2.02	2.51	0.27
Color parameters						
L* (Lightness)	59.22	57.78	59.74	64.74	67.45	4.1
a* (Redness)	11.40	12.02	12.26	12.56	12.80	0.56
b* (Yellowness)	6.81	6.72	7.76	6.86	6.71	0.39
7 days storage at -17±1 °C						
Moisture (%)	67.61	68.23	67.96	67.10	68.22	3.8
pH	5.37	5.51	5.42	5.43	5.46	0.12
WHC (%)	63.93	65.92	64.21	64.18	63.81	4.3
TBA-RS (mg MDA/kg)	4.61	4.58	6.37	4.49	4.71	0.28
Color parameters						
L* (Lightness)	59.71	58.04	60.36	65.58	67.60	5.2
a* (Redness)	12.18	12.24	13.16	13.08	13.02	0.64
b* (Yellowness)	10.52	9.02	9.16	9.08	9.12	0.48

Z: basal diet supplemented with 1% zeolite.
ZN25, ZN50 and ZN75: basal diet supplemented with 1% zeolitecoated with 25, 50 and 75 ppm nanosilver respectively.
TBA-RS: thiobarbituric acid reactive substances and WHC: water holding capacity.
SEM: standard error of the means.

Table 4 Selected quality of broiler thigh muscles fed different treatments diet

Selected quality traits	Treatments					SEM
	Control	Z	ZN25	ZN50	ZN75	
3 days storage at -17±1 °C						
Moisture (%)	76.70	76.32	75.61	75.80	74.63	3.7
pH	5.57	5.63	5.42	5.38	5.39	0.18
WHC (%)	78.00	74.91	79.83	73.41	68.75	6.1
TBA-RS (mg MDA/kg)	2.35[a]	2.94[ab]	2.81[b]	2.84[b]	2.78[b]	0.12
Color parameters						
L* (Lightness)	54.36	58.28	57.28	62.73	61.18	4.2
a* (Redness)	13.56	13.52	14.12	14.08	14.21	0.39
b* (Yellowness)	8.84	8.12	7.92	7.78	7.76	0.21
7 days storage at -17±1 °C						
Moisture (%)	74.01	70.42	71.54	71.61	71.32	5.2
pH	5.61	5.64	5.43	5.37	5.42	0.13
WHC (%)	64.31[b]	60.76[b]	75.32[a]	77.13[a]	76.25[a]	3.4
TBA-RS (mg MDA/kg)	4.36[a]	3.98[ab]	3.22[b]	3.41[b]	3.82[ab]	0.18
Color parameters						
L* (Lightness)	56.36	58.73	58.12	64.52	65.06	3.9
a* (Redness)	13.52	13.94	14.74	14.96	14.34	0.21
b* (Yellowness)	10.44	9.04	9.82	9.61	10.04	0.50

Z: basal diet supplemented with 1% zeolite.
ZN25, ZN50 and ZN75: basal diet supplemented with 1% zeolitecoated with 25, 50 and 75 ppm nanosilver respectively.
TBA-RS: thiobarbituric acid reactive substances and WHC: water holding capacity.
The means within the same column with at least one common letter, do not have significant difference (P>0.05).
SEM: standard error of the means.

Hence, the decrease in net protein charge results in diminished WHC due to the availability of fewer charged protein sites for binding water which forces more of the immobilized water into the free water compartment (Bowker and Zhuang, 2015). It is demonstrated that early postmortem events including rate and extent of pH decline, proteolysis and even protein oxidation are key in influencing the ability of meat to retain moisture.

Effect of Clinoptilolite Coated with Silver Nanoparticles on Meat Quality Attributes of Broiler...

129

Table 5 Effect of treatments on texture profile analyses (TPA) of broiler chicken thigh muscle

Treatments	TPA					
	Hardness (g)	Adhesiveness	Springiness (cm)	Cohesiveness (ratio)[1]	Gumminess[2]	Chewiness[3]
3 days storage at -17±1 °C						
Control	1553.11	-35.44	1.23	0.41	636.73	783.19[c]
Z	1642.16	-37.73	1.84	0.50	821.02	1510.64[a]
ZN25	1481.34	-43.17	1.64	0.49	725.69	1190.13[b]
ZN50	1588.65	-42.19	1.53	0.53	841.64	1287.71[ab]
ZN75	1475.14	-45.56	1.62	0.56	826.11	1338.12[ab]
SEM	61.72	5.45	0.03	0.05	41.98	58.15
7 days storage at -17±1 °C						
Control	1267.11	-31.17	1.99[b]	0.71[a]	899.57[a]	1790.15[a]
Z	1512.32	-36.31	1.87[c]	0.66[a]	997.92[a]	1866.11[a]
ZN25	1235.46	-39.52	2.92[a]	0.47[b]	580.45[b]	1694.91[a]
ZN50	1465.12	-42.22	2.23[ab]	0.41[b]	600.65[b]	1339.44[b]
ZN75	1378.19	-42.35	2.50[a]	0.49[b]	675.27[b]	1668.05[a]
SEM	72.11	4.88	0.04	0.03	48.12	61.11

[1] Cohesiveness is: area under second curve / area under first curve.
[2] Gumminess was calculated as: hardness × cohesiveness.
[3] Chewiness was calculated as: hardness × springiness × cohesiveness.
Z: basal diet supplemented with 1% zeolite.
ZN25, ZN50 and ZN75: basal diet supplemented with 1% zeolitecoated with 25, 50 and 75 ppm nanosilver respectively.
The means within the same column with at least one common letter, do not have significant difference (P>0.05).
SEM: standard error of the means.

Table 6 Effect of treatments on texture profile analyses (TPA) of broiler chicken pectoralis major muscle

Treatments	TPA					
	Hardness (g)	Adhesiveness	Springiness (cm)	Cohesiveness (ratio)[1]	Gumminess[2]	Chewiness[3]
3 days storage at -17±1 °C						
Control	1160.12[b]	-43.25	1.44[a]	0.48[a]	556.81	801.79[a]
Z	1281.34[b]	-33.52	1.53[a]	0.51[a]	653.32	999.57[a]
ZN25	1231.56[b]	-36.79	1.12[c]	0.42[c]	529.33	592.57[b]
ZN50	1588.15[a]	-44.62	1.33[ab]	0.43[c]	682.84	908.16[a]
ZN75	1575.22[a]	-46.21	1.27[bc]	0.46[bc]	724.49	920.12[a]
SEM	51.90	6.25	0.03	0.01	46.11	52.11
7 days storage at -17±1 °C						
Control	1105.11	-34.10	1.68[a]	0.44	530.88	891.87[a]
Z	1220.56	-36.77	1.45[a]	0.57	695.40	1008.33[a]
ZN25	1227.22	-41.48	0.99[b]	0.50	613.52	607.36[b]
ZN50	1165.56	-46.67	1.40[a]	0.62	722.31	1011.22[a]
ZN75	1112.78	-41.71	1.33[a]	0.53	589.36	783.84[ab]
SEM	48.30	5.12	0.02	0.03	39.13	54.98

[1] Cohesiveness is: area under second curve / area under first curve.
[2] Gumminess was calculated as: hardness × cohesiveness.
[3] Chewiness was calculated as: hardness × springiness × cohesiveness.
Z: basal diet supplemented with 1% zeolite.
ZN25, ZN50 and ZN75: basal diet supplemented with 1% zeolitecoated with 25, 50 and 75 ppm nanosilver respectively.
The means within the same column with at least one common letter, do not have significant difference (P>0.05).
SEM: standard error of the means.

The water holding capacity of foods can be defined as the ability to hold its own and added water during the application of forces, pressing, centrifugation, or heating. Hermansson (1986) defined WHC as a physical property and is the ability of a food structure to prevent water from being released from the three-dimensional structure of the protein. Unacceptable water-holding capacity costs the meat industry millions of dollars annually (Huff-Lonerganand Lonergan, 2005). Much of the water in the meat is entrapped in the structures of the cell, within the

myofibrils, between the myofibrils and between the myofibrils and the cell membrane (sarcolemma), between muscle
cells and between muscle bundles (Honikel, 2004). Therefore, key changes in the intracellular architecture of the cell
influence the ability of muscle cells to retain water. As rigor
progresses, the space for water to be held in the myofibrils
is reduced and fluid can be forced into the extra-
myofibrillar spaces where it is more easily lost as drip. Lat-
eral shrinkage of the myofibrils occurring during rigor can
be transmitted to the entire cell if proteins that link myofibrils together and myofibrils to the cell membrane (such as
desmin) are not degraded. Limited degradation of cy-
toskeletal proteins may result in increased shrinking of the
overall muscle cell, which is ultimately translated into drip
loss. Recent evidence suggests that degradation of key cy-
toskeletal proteins by calpain proteinases has a role to play
in determining water-holding capacity (Huff-Lonerganand
Lonergan, 2005).

The present study also indicated that the addition of
nanosilver coated on clinoptilolite at all levels 3 and 7 days
after frozen storage of thigh muscles decreased malonaldehyde (MDA) production. Malonaldehyde, a product of lipid
peroxidation, which is used as an indicator of oxidative
damage to cells and tissues and in directly indicates the
degree of oxidative stress. Lipid oxidation is one of the
main factors limiting the quality and acceptability of meats
and meat products. Oxidative damage to lipids occurs in the
living animal because of an imbalance between the production of reactive oxygen species and the animal's defense
mechanisms (Jiang et al. 2007). It is reported that silver
nanoparticles showed an efficient cellular electron exchange mechanism, which arrest electron leakage, reducing
the reactive oxygen species (ROS) production and MDA
levels (Lu et al. 2002; Hatami and Ghorbanpour, 2013).
Lower oxidative damage is associated with an improved
ability of the sample to bind more water (Melody et al.
2004).

In this study, lower oxidative damage improved WHC
suggest an improved ability of the sample to bind water.
The observed relation between WHC, less oxidative damage and nanosilver particles coated on clinoptilolite in the
diet may be due to reducing proteolysis and protein oxidation and thus quality characteristics influenced by proteolysis such as water holding capacity. There are, however,
other possible explanations and there is insufficient evidence of effects of silver nanoparticles on meat quality.

CONCLUSION

Poultry research and food production are aimed at increasing the meat nutritional value without lowering the sensory
quality or consumer's acceptability. Based on the data gath

ered from this trial, it can be concluded the use of
nanosilver coated on clinoptiloliteas feed additive had a
beneficial effect on broiler meat. Texture profile analysis
showed that nanosilver coated on clinoptilolite had no any
negative effects on texture or sensory attributes of chicken
meat. Although more research is required in this area to
confirm this finding and to assess the safety of the meat
produced in this way. However, basic information about the
meat quality attributes and nanosilver and clinoptilolite as
feed additives in the chicken diets is poorly documented.

ACKNOWLEDGEMENT

This work was carried out with the support of the Vice
Presidency of Research and Technology at Gorgan University of Agricultural Science and Natural Resources. The
authors would like to extend their sincere appreciation for
funding of this research through the Research Group Project No.94-336-85.

REFERENCES

AOAC. (1995). Official Methods of Analysis. Vol. I. 16th Ed.
Association of Official Analytical Chemists, Arlington, VA,
USA.

AOAC. (1998). Official Methods of Analysis. Vol. I. 18th Ed.
Association of Official Analytical Chemists, Arlington, VA,
USA.

Bowker B.C. and Zhuang H. (2015). Relationship between water-
holding capacity and protein denaturation in broiler breast
meat. Poult. Sci. 94, 1657-1664.

Chauke N. and Siebrits F.K. (2012). Evaluation of silver nanoparticles as a possible coccidiostat in broiler production. South
African J. Anim. Sci. 42, 493-497.

Dabes A.C. (2001). Propriedades da carne fresca. Rev. Nac. Carne. 25, 32-40.

Hashemi S.R., Davoodi D., Dastar B., Bolandi N., Smaili M. and
Mastani R. (2014). Meat quality attributes of broiler chickens
fed diets supplemented with silver nanoparticles coated onzeo-
lite. Poult. Sci. J. 2, 183-193.

Hatami M. and Ghorbanpour M. (2013). Effect of nanosilver on
physiological performance of pelargonium plants exposed to
dark storage. J. Hort. Res. 21, 15-20.

Hermansson A.M. (1986). Water-and fat holding. Pp. 31-38 in
Functional Properties of Food Macromolecules. R. Mitchell,
and D.A. Ledward, Eds. Elsevier Applied Science Publishers,
London, UK.

Honikel K.O. (2004). Water-holding capacity of meat. Pp. 389-
400 in Muscle Developmentof Livestock Animals: Physiology, Genetics and Meat Quality. M.F. tePas, M.E. Everts and
H.P. Haagsman, Eds. CABI Publishing, Cambridge, USA.

Huff-Lonergan E. and Lonergan S.M. (2005). Mechanisms of
water-holding capacity of meat: the role of postmortem biochemical and structural changes. Meat Sci. 71, 194-204.

Jang A., Liu X.D., Shin M.H., Lee B.D., Lee S.K., Lee J.H. and Jo

C. (2008). Antioxidative potential of raw breast meat from broiler chicks fed a dietary medicinal herb extract mix. *Poult. Sci.* **87**, 2382-2389.

Jiang S.Q., Jiang Z.Y., Lin Y.C., Xi P.B. and Ma X.Y. (2007). Meat quality and antioxidative property of male broilers fed oxidized fish oil. *Asian-Australasian J. Anim. Sci.* **20**, 1252-1257.

Kannan N. and Subbalaxmi S. (2011). Green synthesis of silver nanoparticles using *Bacillus subtillus* IA751 and its antimicrobial activity. *Res. J. Nanosci. Nanotechnol.* **1**, 87-94.

Khalafalla F.A., Ali F.H.M., ZahranDalia A. and Mosa A.M.M.A. (2011). Influence of feed additives in quality of broiler carcasses. *J. World's Poult. Res.* **2**, 40-47.

Kulthong K., Srisung S., Boonpavanitchakul K., Kangwansu-pamonkon W. and Maniratanachote R. (2010). Determination ofsilver nanoparticle release from antibacterial fabrics into artificial sweat. *Part. Fibre. Toxicol.* **7**, 1-9.

Li Y., Leung P., Yao L., Song K.W. and Newton E. (2006). Antimicrobial effect of surgical masks coated with nanoparticles. *J. Hosp. Infect.* **62**, 58-63.

Lu C.M., Zhang C.Y., Wen J.Q. and Wu G.R. (2002). Research on the effect of nanometer materials on germination and growth enhancement of glycine max and its mechanism. *Soybean Sci.* **21**, 68-171.

Melody J.L., Lonergan S.M., Rowe L.J., Huiatt T.W., Mayes M.S. and Huff-Lonergan E. (2004). Early postmortem biochemical-factors influence tenderness and water-holding capacity of threeporcine muscles. *Anim. Sci.* **82**, 1195-1205.

Meluzzi A., Sirri F., Castellini C., Roncarati A., Melotti P. and Franchini A. (2009). Influence of genotype and feeding on chemical composition of organic chicken meat. *Italian J. Anim. Sci.* **8**, 766-768.

Nikpey A., Kazemian H., Safari-Varyani A., Rezaie M. and Sirati-Sabet M. (2013). Protective effct of microporous natural clinoptilolite on leadinduced learning and memory impairment in rats. *Health Scope.* **2**, 52-57.

Offer G., Knight P., Jeacocke R., Almond R., Cousins T., Elsey J., Parsons N., Sharp A., Starr R. and Purslow P. (1989). The structural basis of the water-holding, appearance, and toughness ofmeat and meat products. *Food Microstruct.* **8**, 151-170.

Santhi D. and Kalaikannan A. (2014). The effect of the addition of oat flour in low-fat chicken nuggets. *J. Nutr. Food Sci.* **4**, 260.

SAS Institute. (2004). SAS®/STAT Software, Release 9.1. SAS Institute, Inc., Cary, NC. USA.

Sawosz E., Binek M., Grodzik M., Zielinska M., Sysa P., Szmidt M., Niemec T., Sondi I. and Salopek-Sondi B. (2004). Silver nanoparticles as antimicrobial agent: a case study on *E. Colias* a model for Gram-negative bacteria. *J. Colloid Interf. Sci.* **275**, 177-182.

Sekhon B.S. (2014). Nanotechnology in agri-food production: an overview. *Nanotechnol. Sci. Appl.* **7**, 31-53.

18

Quantitative Trait Loci for some of Behavior and Performance Traits on Chromosome 4 of Japanese Quail

E. Rezvannejad[1*] and E. Nasirifar[1]

[1] Department of Biotechnology, Institute Science and High Technology and Environmental Science, Graduate University of Advanced Technology, Kerman, Iran

*Correspondence E-mail: e.rezvannejad@kgut.ac.ir

ABSTRACT

The current study was conducted to identify the quantitative trait locus (QTL) for the body weight at age 1, 7, 14, 21 and 28 days and daily gain at age 0-1, 1-2, 2-3 and 3-4 weeks, slighter carcass weight and tonic immobility in Japanese quail. Two divergently lines of wild and white Japanese quail which maintained in the Animal Science Research Center of the Shahid Bahonar University of Kerman was used. Birds in two lines were mated in reciprocal cross for generating F1 generation. The F2 generation was generated by random mating of F1 birds. The half sib and F2 model analysis of QTL express was used for QTL mapping the measured traits. Total F2 individuals (422 birds), their parents (34 F1 birds) and pure lines (16 birds) were genotyped by 2 informative microsatellite markers on chromosomes 4. On this chromosome, with the half sib model, no significant QTL at the 5%, 1% chromosome wide was detected. But, with the F2 model, a significant QTL at the 5% chromosome wide was identified for body weight (BW) at age 21 and 28 days, carcass weight and daily gain traits during 1-2, 2-3 and 3-4 weeks in 54cM position. The all of identified QTLs explain less of 4% of the phenotypic variance, because the traits of growth are complex traits affected by many loci influencing appetite, feed uptake, nutrient allocation, metabolic rate, physical activity and so forth

KEY WORDS body weight trait, daily gain trait, Japanese quail, QTL mapping, tonic immobility.

INTRODUCTION

Rearing of the domestic fowl in Iran and its dissemination throughout the country has an old history. Iran (Persia) was a great empire from the 5th century BC to approximately the 7th century AD and extended from India (Delhi) to the Black and the Mediterranean seas. At those times and later, in the middle ages, Persia was located at the crossroads of major ways for transporting goods, including the domestic fowl, from the East to the West, both by land and waterways. Numerous wars in the territory of Persia and adjacent countries during those periods could also facilitate the spreading of the chicken populations. Archaeological exca-vations confirmed the presence of the domestic fowl in the territory of Iran at the ancient times (Mohammadabadi *et al*. 2010). According to West and Zhou (1989), the chicken bones were found here in three regions: two findings in Tepe Yahya (southeastern Iran) dated at 3.900-3.800 BC and 1.000 BC, respectively, and one in Takht-e-Suleiman (northwestern Iran) dated at 1.000 BC. Persian merchants maintained strong trade ties with the Mediterranean area and sailed upstream the Volga river, reaching Nizhniy Novgorod in Russia. It is known that Persian chickens from the Guilan Province took part in the origin of the Russian Orloff breed (Mohammadabadi *et al*. 2010). Since 1981, many centers were organized for reproducing native and commercial poultry varieties. One of these commercial varities is Japanese quail. The Japonica quail belongs to the Phasianidae family. It is wildly reared for egg and meat production in several countries of Asia, Europe and America (Minvielle, 2004). Until recently, in contrast to chicken (*Gallus gallus*), only a few linkage groups have been known for the Japanese quail, and few genes have been mapped (Tsudzuki, 2008). Despite the large variety of traits that have been studied in the Japanese quail.

Therefore, improvement of economic traits in Japanese quails has been desirable response by traditional breeding programs. For example, the growth rate is improved based on phenotype, but advantages of molecular genetics are important in rapidly and useful changes, especially in quantitative traits such as meat quality (Tsudzuki, 2008). QTL refers to chromosomal regions that controlling quantitative traits. It may include one or a group of genes with weak or strong linkage. Many quantitative traits are controlled by a relatively large number of genes, these genes known as genes with a major effect. So, the QTL is the gene with major effect (Falconer and Mackay, 1996).

In chickens, the first report for QTL mapping appeared in 1998 on the basis of microsatellite DNA markers (Vallejo *et al.* 1998; Van kaam *et al.* 1998). Science then, about 200 QTLs has been discovered for economic traits of chickens (Abasht *et al.* 2006).

Contrastingly, in Japanese quail there are at present only a few reports concerning QTLs, although QTL mapping is possible with DNA markers in addition to the mapping of traits. Beaumont *et al.* (2005) tried to identify QTLs for body weight and 11 kinds of fear fullness-related traits with AFLP markers developed by Roussot *et al.* (2003). Furthermore, Minvielle *et al.* (2005) carried out QTL analysis using microsatellite markers developed by Kayang *et al.* (2004). Esmailizadeh *et al.* (2012) have recently detected significant QTL for body weight (weight at 3, 4, 5 and 6 weeks of age) on chromosome 1 and 3 in Japanese quail. The extend of the microsatellites and AFLP markers has made it possible to detect new genes. This increased the interest of the quail in biological research. Also, the genetic maps were assistant researches for QTL mapping with microsatellites and AFLP markers in F2 and back cross populations (Minvielle *et al.* 2005). Because of, the body weight and carcass traits are the main of economic traits in Japanese quail, detecting QTL for growth traits is very important. Although Japanese quail in Iran have studied by molecular techniques for different purpose (Sohrabi *et al.* 2012; Ori *et al.* 2014; Moradian *et al.* 2014; Mohammadifar *et al.* 2009; Moradian *et al.* 2015), but until now researchers have not identified QTL affecting body weight, growth rate and important behavior trait on chromosome 4 in Japanese quail.

Moreover low energy expenditure, in addition to increased energy intake, has been a major cause of future weight gain, and variations in energy expenditure may be one of the underlying sources of variation in body weight (Moazeni *et al.* 2016). Various determinants, including body composition, hormonal levels, activity of the sympathetic nervous system, and genetics are responsible for differences in metabolic rate among individuals (Moazeni *et al.* 2016). Hence, the objective of the present work was to identify QTL affecting body weight, growth rate and important behavior trait on chromosome 4 in Japanese quail and compassion of half sib and F2 model for detecting QTL.

MATERIALS AND METHODS

Experimental design

A F2 population was developed using two distinct Japanese quail strains, wild (meat type) and white (layer type). Eight pairs of white (S) and wild (W) birds were crossed reciprocally and 34 F1 birds (9 males and 25 females) were produced. The F1 birds including 17 SW and 17 WS progeny were generated by reciprocal crosses. The F1 birds were intercrossed randomly generating 422 F2 offspring (246 males and 176 females) in five continuing hatches. The F2 progeny was raised for 5 weeks on a floor covered with wood shavings in an environmentally controlled room with continuous artificial lighting and at a temperature was decreased gradually from 37 to 25 °C. The progeny received water and food *ad libitum*.

Phenotyping

Hatching weight, body weights at age 1, 2, 3 and 4 weeks and carcass slaughter weight were recorded on individuals of the F2 generation. The growth rate (average daily gain in body weight between consecutive ages) was derived based on the body weights. Tonic immobility was measured as the length of time during which a chick remained immobile by keeping it on its back for 10 s.

Genotyping

The 2 microsatellite markers on chromosome 4, based on the recognized microsatellite markers (Kayang *et al.* 2004) and their polymorphism information, were selected. Genomic DNA was isolated from blood samples using salting out method (Helms, 1990). Polymerase chain reactions for each marker were carried out separately in a reaction volume of 25 μL included 120 ng of template DNA, 2.5 μL 1X PCR reaction buffer, 0.5 μ*M* of each primer, 2 μ*M* of each deoxy nucleotide triphosphate (dNTP), 0.2 U of Taq polymerase and residual volume is deionizer water. The PCR products were electrophoresed in 6% polyacrylamide gels. A sample of bands (alleles) was shown in Figure 1.

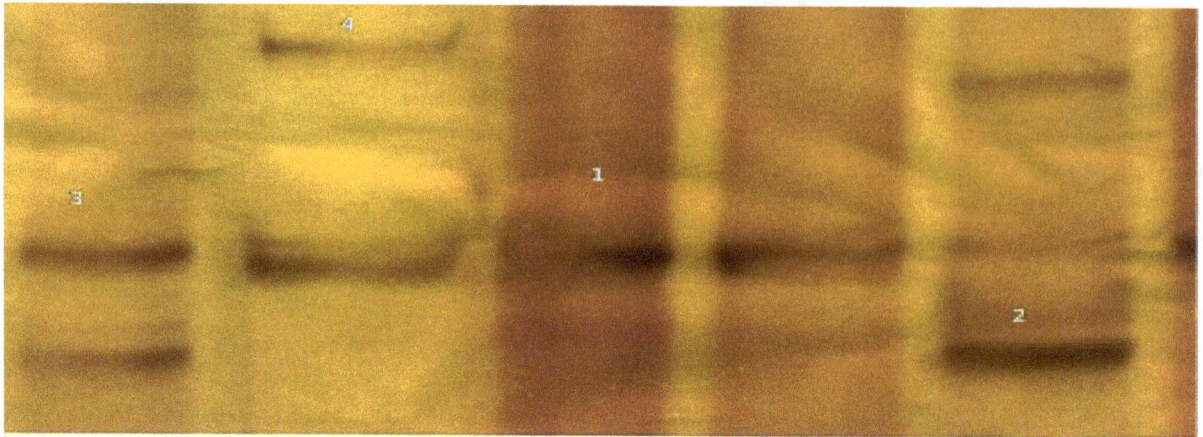

Figure 1 An example of the band scoring ("1" and "2" the homozygous genotype, "3" the heterozygous individuals and "4" is ladder)

Fragment sizes were analyzed with TOTAL LAB analysis software (http://www.totallab.com). All genotypes were checked twice. In all, parents (16 birds), F1 (34 birds) F2 (422 birds) were genotyped.

QTL analysis

Phenotypic data were analyzed by using SAS software (SAS, 2000). Means; standard deviation and coefficient of variation of traits were calculated. The QTL express software under half-sib and F2 models at http://qtl.cap.ed.ac.uk was used for QTL analyses. Data subjected to a model with sex and hatch as fixed effects in the model. The percentage difference in the residual sums of squares between the full and reduced model was calculated as the phenotypic variance, which that QTL could explain. Significant thresholds for analyses were calculated using a permutation test (Churchill and Doerge, 1994). A total of 1000 permutations were computed to determine the empirical distribution of the statistical test under the null hypothesis of no QTL associated with the part of genome under study. Three significance levels were used: suggestive, 5% and 1% chromosome-wide (Lander and Kruglyak, 1995).

RESULTS AND DISCUSSION

Phenotypic data

A summary of descriptive statistics of the studied traits (including number of observations, minimum and maximum values, means and standard deviations) for the F2 generation derived from a reciprocal intercross between the wild and white lines are presented in Table 1. The average daily gain ranged between 11.3 g (for ADG3-4) to 0.59 g (for ADG0-1). The effect of hatch was significant (P<0.05) for all of the traits studied while the sex of the bird had significant (P<0.05) effect on W3 and W4 and the related derived traits such as ADG2-3, ADG3-4.

The females were generally heavier than male and so the females were faster than the males in growth (P<0.05).

Half-sib analysis

The QTL for BW at hatch, 1 to 4 weeks of age and average daily gain for different consecutive ages and CW using the half-sib analysis model are presented in Table 2. The suggestive QTL were found for BW at age 3, 4 weeks and average daily gain at age 0-1, 1-2 weeks closed to marker CUJ0026 and for CW close to marker CUJ0074. F ratio for these QTLs were 1.01, 0.95, 1.24, 1.10 and 1.12.

In Figure 2, detected QTLs using the half-sib analysis model for studied traits with interval mapping model have been showed in chromosome 4.

F2 analysis

Information on suggestive and significant QTLs is presented in Table 3. Two main positions for detected QTLs identified on chromosome 4. The first is located close to marker CUJ0026 (22 cM), which contains QTLs associated with BW at age 2, 3 and 4 weeks f and average daily gain at age 1-2, 2-3 and 3-4 weeks. F ratios for suggestive and significant QTLs were 1.28, 2.25, 2.35, 2.79, 2.14 and 3.07, respectively. The second is located close to marker CUJ0074 (54cM) contains A significant QTL for CW with F ratio=2.82. Most additive effects of identified QTLs in this research were negative (Table 3). Only QTLs for body weight at age 1week, slaughter carcass weight and average daily gain 0-1week illustrated positive additive effects. Moreover, depending on the trait, dominance effects of QTLs were positive or negative. Detected QTLs explained 0.59-3.85% of the total phenotypic variance.

The detected QTLs for studied traits with interval mapping model on chromosome 4 was shown in Figure 2. Illustrated thresholds of QTLs on chromosome are suggestive (P<0.10) and significant (P<0.05).

Table 1 Statistics of the various traits for the F2 generation derived from a reciprocal intercross between the wild and white lines in studied Japanese quail

Trait[1]	Number	Mean (g)	SD (g)	CV (g)	Minimum (g)	Maximum (g)
BW0	422	6.8	0.69	10.1	4.8	9.5
BW1	419	23.1	4.76	20.6	11.3	41.40
BW2	420	47.6	9.30	19.8	21.0	74.80
BW3	420	83.0	13.55	16.6	15.9	124.10
BW4	417	120.9	17.96	15.1	63.1	168.60
CW	420	152.4	17.58	11.6	83.7	199.20
ADG0-1	419	2.3	0.65	27.9	0.59	4.86
ADG1-2	418	3.5	0.79	23.1	0.79	6.21
ADG2-3	418	5.0	0.91	18.5	1.73	8.03
ADG3-4	416	5.4	1.14	21.0	2.01	11.33
Tonic immobility	425	75.21	32.25	19.02	9.59	198.06

[1] Numbers following body weight (BW) indicating age in weeks.
CW: carcass weight and ADG: average daily gain and numbers following them indicating consecutive age.
SD: standard deviation and CV: coefficient of variation.

Table 2 Quantitative trait loci for studied traits using the half-sib analysis model on chromosome 4 in Japanese quail

BW[1]	Closed marker	Position (cM)	F ratio	QTL variance %[2]
BW0	CUJ0026	22	0.89	0.43
BW1	CUJ0026	22	0.58	1.02
BW2	CUJ0026	22	0.8	0.77
BW3	CUJ0026	22	1.01[†]	0.95
BW4	CUJ0026	22	0.95[†]	0.98
CW	CUJ0074	54	1.12[†]	0.53
ADG0-1	CUJ0026	22	1.24[†]	0.80
ADG1-2	CUJ0026	22	1.10[†]	1.30
ADG2-3	CUJ0026	22	0.68	0.95
ADG3-4	CUJ0026	22	0.7	0.82
Tonic immobility	CUJ0074	54	0.72	1.04

[1] Numbers following body weight (BW) indicating age in weeks.
[2] Proportion of total variance explained by the quantitative trait locus (QTL).
CW: carcass weight and ADG: average daily gain and numbers following them indicating consecutive age.
† Denotes suggestive linkage.
* (P<0.05) and ** (P<0.05).

In this study, QTL were detected for body weight, slaughter carcass weight and daily gain which are important traits in poultry breeding. This research adds new information from a chromosome-wide map for QTL in Japanese quail, and it is the first report on the detection of loci affecting important economic traits on chromosome 4 in Japanese quail. The length of chromosome 4 is 55cM with two detected microsatellite markers (Kayang et al. 2004).

Two models of analysis were used for the QTL detection: the half sib and the F2 model. The F2 analysis model identified more QTL compared with the half sib analysis model. The detected QTLs with the F2 analysis model, 6 out of 7 suggestive QTLs in the chromosome wide were significant, whereas for the QTL detected with the half sib model, all of

5 QTL identified in the chromosome wide were in suggestive level. In general, the F2 analysis model is claimed to be more powerful than half-sib analysis, but only if the alleles in founder lines are fixed (Weller, 2001).

The founder lines of the present experimental population were divergent Japanese quail strains (meat type wild and layer type white strains). Therefore, it is likely that the QTL alleles for BW were fixed in the founder population and F2 model were perfect to half sib model. The positive additive effects by F2 model indicate that alleles increasing body weight came from the meat type line and negative additive effects came from the layer line. Although, to our knowledge, there is no report on QTL for body weight related traits on Japanese quail chromosome 4.

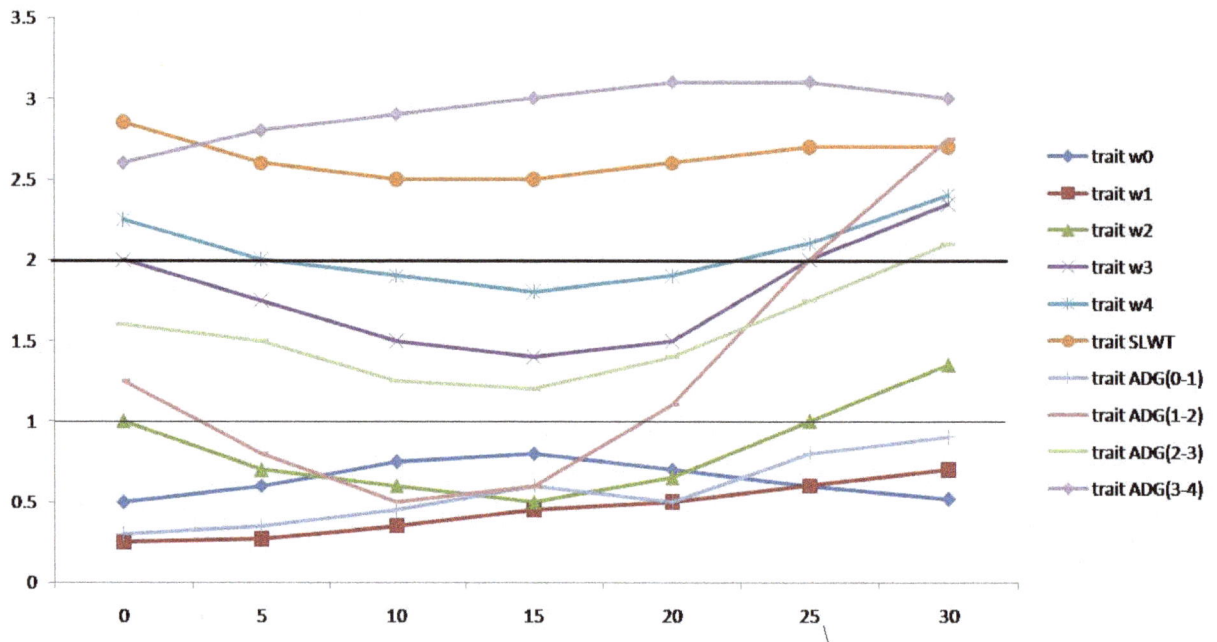

Figure 2 Interval mapping of QTL using an intercross between two Japanese quail strains on chromosome 4
The lower and upper horizontal lines represent 5 and 1% chromosome-wide significant levels of linkage, respectively

Table 3 Quantitative trait loci for studied traits using the F2 analysis model on chromosome 4 in Japanese quail

BW[1]	Closed marker	Position (cM)	F ratio	QTL effect			QTL variance %[2]
				(Additive effect±SE)	(Dominance effect±SE)	(Imprinting effect±SE)	
BW0	CUJ0026	22	0.69	-0.142±0.11	-0.04±0.16	0.03±0.15	0.60
BW1	CUJ0026	22	0.68	0.09±0.77	0.88±0.83	-0.11±1.17	0.92
BW2	CUJ0026	22	1.28[†]	-2.43±1.46	2.67±1.56	-1.25±2.22	1.07
BW3	CUJ0026	22	2.25[*]	-4.53±2.25	5.19±2.43	-4.18±3.43	2.56
BW4	CUJ0026	22	2.35[*]	-4.88±2.79	4.35±3.00	-8.10±4.30	3.85
CW	CUJ0074	54	2.89[*]	0.70±1.80	1.78±2.41	-8.41±3.03	3.39
ADG0-1	CUJ0026	22	0.92	0.02±0.10	0.13±0.11	-0.02±0.15	0.59
ADG1-2	CUJ0026	22	2.79[*]	-0.37±0.13	0.27±0.14	-0.10±0.20	0.87
ADG2-3	CUJ0026	22	2.14[*]	-0.31±0.16	0.27±0.17	-0.39±0.25	1.36
ADG3-4	CUJ0026	٢٢	3.07[*]	-0.12±0.18	-0.40±0.22	-0.66±0.24	1.35
Tonic immobility	CUJ0074	54	0.72	-0.133±0.09	-0.10±0.19	0.07±0.06	0.68

[1] Numbers following body weight (BW) indicating age in weeks.
[2] Proportion of total variance explained by the quantitative trait locus (QTL).
CW: carcass weight and ADG: average daily gain and numbers following them indicating consecutive age.
† Denotes suggestive linkage.
* (P<0.05) and ** (P<0.05).

However, there are some of QTL studies that detected QTL for bodyweight on chicken chromosome 4. Zhou *et al.* (2006) reported a QTL affecting body weight at age 2, 4 and 6 weeks f and growth rate at age 0-2, 2-4 and 4-6 weeks in chromosome 4 of chicken. In addition, De Koning *et al.* (2004), Sewalem *et al.* (2002), Van Kaam *et al.* (1999) and Van Kaam *et al.* (1998) confirmed a QTL on

chromoosome 4 affecting body weight and feed intake found in a commercial broiler line.

In this study, the interesting observation was that suggestive and significant detected QTLs in this intercross explained only 0.43–3.85% of the phenotypic variance for body weight, rate growth and carcass traits by using each of two models. Growth is a highly complex trait affected by many loci influencing appetite, feed uptake, nutrient allocation, body composition, physical activity and etc. This means that each locus affecting growth explains only a small fraction of the genetic variance.

CONCLUSION

In summary, commercial breeding programs of Japanese quail have become more complex and challenging because so many objective need to be simultaneously considered to reduce production costs, maintain health and improve product quality. Breeding goals must include increased growth rate, increased % breast weight, maintenance of good development and growth of the skeletal system and overall fitness. The relationships of these traits are complex and some of the traits are very difficult to measure. Therefore, molecular MAS may be required to improve genetic selection programs. The current study is the first step toward the fine mapping QTL affecting BW and average daily gain on chromosome 4. Following, valuable candidate genes may be found by combining results of fine mapping and the Japanese quail genome sequence, and further function study of the genetic background of growth and carcass traits of Japanese quail.

ACKNOWLEDGEMENT

We greatly acknowledge the management of Dufil Prima Food Limited located in Choba, Port Harcourt who helped us to identify other customers of the waste product with whom we jointly bought and shared the quantity of the waste needed for this experiment, since the company does not sell less than a tonne.

REFERENCES

Abasht B., Dekkers J.C.M. and Lamont S.J. (2006). Review of quantitative trait loci identified in the chicken. *Poult. Sci.* **85,** 2079-2096.

Beaumont C., Roussot O., Fève K., Plisson-Petit F., Leroux S., Pitel F., Faure J.M., Mills A.D., Guémené D., Leterrier C., Sellier N., Mignon-Grasteau S., Sellier N., Le Roy P., Perez-Enciso M. and Vignal A. (2005). Genetic analysis of tonic immobility in Japanese quail with AFLP markers. *Anim. Genet.* **36,** 401-407.

Churchill G.A. and Doerge R.W. (1994). Empirical threshold values for quantitative trait mapping. *Genetics.* **138,** 963-971.

De Koning D.J., Haley C.S., Windsor D., Hocking P.M., GriffinH. Morris A., Vincent J. and Burt D.W. (2004). Segregation of QTL for production traits in commercial meat-type chickens. *Genet. Res.* **83,** 211-220.

Esmailizadeh A.K., Baghizadeh A. and Ahmadizadeh M. (2012). Genetic mapping of quantitative trait loci affecting body weight on chromosome 1 in a commercial strain of Japanese quail. *Anim. Prod. Sci.* **52,** 64-68.

Falconer D.S. and Mackay T.F.C. (1996). Introduction to Quantitative Genetics. Longman, London, United Kingdom.

Helms C. (1990). Salting out procedure for human DNA extraction. In the Donis-Keller Lab- Lab Manual Home Page. Available at: http://hdklab.wustl.edu/lab_manual/dna/dna2.html.

Kayang B.B., Vignal A., Inoue-Murayama M., Miwa M., Monvoisin J.L., Ito S. and Minvielle F. (2004). A first generation micro satelite linkage map of the japaneas quail. *Anim. Genet.* **35,** 195-200.

Lander E.S. and Kruglyak L. (1995). Genetic dissection of complex traits: Guidelines for interpreting and reporting linkage results. *Nat. Genet.* **11,** 241-247.

Minvielle F. (2004). The future of Japanese quail for research and production.*World's Poult. Sci.* **60,** 500-507.

Minvielle F., Kayang B.B., Inoue-Murayama M., Miwa M., Vignal A., Gourichon D., Neau A., Monvoisin J.L. and Ito S. (2005). Microsatellite mapping of QTL affecting growth, feed consumption, egg production, tonic immobility and body temperature of Japanese quail. *BMC Genet.* **6,** 87-96.

Moazeni S.M., Mohammadabadi M.R., Sadeghi M., Moradi H.S., Esmailizadeh A.K. and Bordbar F. (2016). Association between UCP gene polymorphisms and growth, breeding value of growth and reproductive traits in Mazandaran indigenous chicken. *Open. J. Anim. Sci.* **6,** 1-8.

Mohammadabadi M.R., Nikbakhti M., Mirzaee H.R., Shandi M.A., Saghi D.A., Romanov M.N. and Moiseyeva I.G. (2010). Genetic variability in three native Iranian chicken populations of the Khorasan province based on microsatellite markers. *Rusian J. Genet.* **46,** 572-576.

Mohammadifar A., Amirnia S., Omrani H., Mirzaei H.R. and Mohammadabadi M.R. (2009). Analysis of genetic variation in quail population from Meybod Research Station using microsatellite markers. *Anim. Sci. J. (Pajouhesh and Sazandegi).* **22(1),** 72-79.

Moradian H., Esmailizadeh A.K., Sohrabi S. and Mohammadabadi M.R. (2015). Identification of quantitative trait loci associated with weight and percentage of internal organs on chromosome 1 in Japanese quail. *J. Agric. Biotechnol.* **6(4),** 143-158.

Moradian H., Esmailizadeh A.K., Sohrabi S.S., Nasirifar E., Askari N., Mohammadabadi M.R. and Baghizadeh A. (2014). Genetic analysis of an F2 intercross between two strains of Japanese quail provided evidence for quantitative trait loci affecting carcass composition and internal organs. *Mol. Biol. Rep.* **41(7),** 4455-4462.

Ori R.J., Esmailizadeh A.K., Charati H., Mohammadabadi M.R. and Sohrabi S.S. (2014). Identification of QTL for live weight and growth rate using DNA markers on chromosome 3 in an F2 population of Japanese quail. *Mol. Biol. Rep.* **41(2),** 1049-1057.

Roussot O., Fève K.Plisson-Petit F., Pitel F., Faure J.M., Beaumont C. and Vignal A. (2003). AFLP linkage map of the Japanese quail *Coturnix japonica. Gen. Sel. Evol.* **35,** 559-572.

SAS Institute. (1996). SAS®/STAT Software, Release 6.11. SAS Institute, Inc., Cary, NC. USA.

Sewalem A., Morrice D.M., Law A., Windsor D., Haley C.S., Ikeobi C.O., Burt D.W. and Hocking P.M. (2002). Mapping of quantitative trait loci forbody weight at three, six and nine weeks of age in a broilerlayer cross. *Poult. Sci.* **81,** 1775-1781.

Sohrabi S.S., Esmailizadeh A.K., Baghizadeh A., Moradian H., Mohammadabadi M.R., Askari N. and Nasirifar E. (2012). Quantitative trait loci underlying hatching weight and growth traits in an F2 intercross between two strains of Japanese quail. *Anim. Prod. Sci.* **52(11),** 1012-1018.

Tsudzuki M. (2008). Mutations of Japanese quail (*Coturnix japonica*) and recent advances of molecular genetics for this species. *J. Poult. Sci.* **45,** 159-179.

Vallejo R.L., Bacon L.D., Liu H.C., Witter R.L., Groenen M.A.M., Hillel J. and Cheng H.H. (1998). Genetic mapping of quantitative trait loci afecting susceptibility to Marek's disease virusinduced tumors in F intercross chickens. *Gene.* **148,** 349-360.

Van Kaam J.B., Groenen M., Bovenhuis H., Veenendaal A., Vereijken A.L. and van Arendonk J.A. (1999). Whole genome scan in chickens for quantitative trait loci affecting growth and feed efficiency. *Poult. Sci.* **78,** 15-23.

Van Kaam J.B., Van Arendonk J.A.M., Groenen M.A.M., Bovenhuis H., Vereijken A.L.J., Crooijmans R.P.M.A., Van der Poel J.J. and Veenendaal A. (1998). Whole genomes scan for quantitative trait loci affecting body weight in chickens using a three generation design. *Livest. Prod. Sci.* **54,** 133-150.

Weller J.I. (2001). Quantitative trait loci analysis in animals. CABI Publishing, London, United Kingdom.

West B. and Zhou B.X. (1989). Did chicken go north? New evidence for domestication. *World's Poul. Sci.* **45,** 205-218.

Zhou H., Deeb N., Evock-Clover C.M., Ashwell C.M. and Lamont S.J. (2006). Genome-wide linkage analysis to identify chromosomal regions affecting phenotypic traitsinthe chicken. I .Growth and average daily gain. *Poult. Sci.* **85,** 1700-1711.

19

Effects of Rosemary (*Rosmarinus officinalis*) Extract on Performance, Antioxidant Ability and Blood Gas Indices of Broiler Chickens Treated with Sodium Nitrate in Drinking Water

A.R. Akhavast[1] and M. Daneshyar[1*]

[1] Department of Animal Science, Faculty of Agriculture, Urmia University, Urmia, Iran

*Correspondence E-mail: m.daneshyar@urmia.ac.ir

ABSTRACT

A total of 220 broiler chicks in five groups were used to reveal the effects of different levels of 0.0, 1.5, 3.0 and 6.0 mL/L rosemary extract along with sodium nitrate (27.4 mg/L) in drinking water as compared to the control (without any supplement in water) on performance and antioxidant potential of treated broiler chickens. Body weight gain and feed conversion ratio were negatively affected by nitrate during finisher period and was compensated by all the rosemary levels. Both the blood uric acid and total antioxidant capacity were diminished by nitrate while consumption of 3.0 mL/L rosemary extract returned it to the control level. Nitrate decreased venous blood pO_2, and rosemary extract levels of 3.0 and 6.0 mL/L increased pO_2 to the same level as the control. Venous blood pCO_2 was affected in an opposite way to pO_2 by nitrate and rosemary extract. It is concluded that rosemary extract supplementation in drinking water can improve antioxidant ability and performance functions of broiler chickens treated with sodium nitrate.

KEY WORDS blood gas, body weight gain, creatinine, total antioxidant capacity, uric acid.

INTRODUCTION

Changes in the patterns of agricultural practice, food processing and industrialization have resulted in accumulation of nitrates and nitrites in the environment. Intensive farming practice increased the use of nitrogen-based fertilizers (Chow and Hong, 2002). However, the continuous use of nitrogenous fertilizers in agriculture is the major source of nitrates and nitrites which have been shown to be present in relatively high concentrations in a wide variety of food plants and drinking water (Walker, 1975). The main problem of nitrate consumption is methemoglobin production which cannot bind oxygen. Accumulation of methemoglobin (methemoglobinemia) occurs if this oxidation process overwhelms the protective reduction capacity of the cells (Jaffe, 1981). A further concern relating to the metabolism of dietary nitrate is the potential *in vivo* formation of N-nitroso compounds from nitrite. Carcinogenesis, hepatotoxicity, impairment of reproductive functions, endocrine disturbances, growth retardation, destruction of vitamin A, methaemoglobinaemia and impairment of certain defense mechanisms linked to the inflammatory response and tissue injury are the toxicological effects of nitrates and nitrites in different mammalian species (Slepchenko and Rhnelnitskh, 1988). Conversion of nitrate to nitrite potentiate the formation of N-nitroso compounds such as peroxynitrite that is the most damaging reactive nitrogen species (RNS). Peroxynitrite causes the oxidative stress, lipid peroxidation and damage to cellular components (Seven *et al.* 2010). Although the nitrates and nitrites effects have been investigated extensively in mammals, little is known about their possible negative effects in poultry. In an experiment on broiler chickens, 20 mg/L nitrate consumption in water decreased the growth rate (Barton et al. 1986). In the other experiment on slow growing native breed (Balady), dietary sodium nitrate (4.2 g/kgfeed) have retarded growth and caused methaemoglobinaemia (Atef et al. 1991). Oxidative damage, which is a consequence of excessive oxidative stress and/or insufficient antioxidant potential is the other consequence of nitrate or nitrite treating in animals (Safary and Daneshyar, 2012). Antioxidants play a major role in protecting cells from the actions of reducing chemical radicals and disrupting the process of lipid peroxidation (Yu, 1994). Today, there is increasing interest in the use of natural antioxidants such as rosemary (Rosmarinus officinalis) extracts, flavonoid and tocopherol for food preservation (Hras et al. 2000; Williams et al. 2004) because these natural antioxidants avoid undesirable health problems that may arise from the use of synthetic antioxidants such as butylated hydroxyanisole and butylated hydroxytoluene (Aruoma et al. 1992). Antioxidant effect of aromatic plants is due to the presence of hydroxyl groups in their phenolic compounds (Shahidi and Wanasundara, 1992). Rosemary, belonging to the Lamiaceae family, is well known for its antioxidative properties, is used for flavouring foods and beverages, and is also used in several pharmaceutical applications. These polyphenols also have important biological activities in vitro such as anti-tumor, chemo-preventive and anti-inflammatory activities (Shuang-Sheng and Rong-Liang, 2006; Cheung and Tai, 2007). Rosemary contains a variety of phenolic compounds, including carnosol, carnosic acid, rosmanol, 7-methyl-epirosmanol, isorosmanol, rosmadial and caffeic acid, with substantial in vitro antioxidant activity (Ibanez et al. 2003). Some experiments have indicated the potential antioxidant effect of rosemary and its components. Polat et al. (2011) investigated the dietary supplementation effects of rosemary plant (57, 86 and 115 g/kg) and its volatile oil (100, 150 and 200 mg/kg) in broilers. They showed that supplementation of 100 mg/kg rosemary volatile oils or 8.6 g/kg rosemary plant increases the plasma superoxide dismutase (SOD) activity. In the other study, feeding 5, 10 and 15 mL/L rosemary essential oil increased the serum serum superoxide dismutase (SOD) activity in broilers under oxidative stress (Yasar et al. 2011). The objective of this study was to evaluate the effects of hydroalcoholic extract of rosemary on broiler performance and antioxidant status chickens treated sodium nitrate in drinking water.

MATERIALS AND METHODS

Birds, diets and management

Two hundred and twenty one-day-old male broiler chicks (Ross 308) were obtained from a commercial hatchery.

All the birds were weighed on arrival and randomly divided among 25 pens (1 m², 11 birds each pen) and four pens were assigned to one of the five experimental treatment groups. The subgroups of 11 chicks (mean weight 39±0.5 g) were kept in a well-ventilated house with wood shavings litter. All the birds were fed the same starter (from day one to day 10 of age), grower (from day 11 to day 24 of age) and finisher (from day 25 to day 42 of age) diets in mash form (Table 1). Vitamin and mineral premixes did not include anticoccidials or antioxidants (except ethoxyquin in the vitamin premix). The birds of different groups received the different water treatments. Control birds consumed the natural (having 5.4 mg/L nitrate) water whereas the birds of other groups received the sodium nitrate (27.4 mg/L) in drinking water alone or along with the different levels of 1.5, 3 and 6 mL/L hydroalcoholic extract of rosemary (Rosmarinus officinalis) from day one to the end of the experiment. The rosemary extract was purchased from Exir Gol Sorkh Company of Mashhad City, Iran. One of the active component of the extract (1, 8-cineol) was determined using a GC chromatograph (PV 4500 Shimaozu GC Chromatograph). For this determination, the extract was dissolved in a polar solvent of dichloromethane to derivate the essential oils. Then the essential oils were injected to GC Chromatograph. The amount of 1.5 mg/kg was obtained for 1, 8-cineol in rosemary extract. Feed and water were provided ad libitum consumption. Birds were exposed to 23 h light and 1 h darkness during the experiment. Birds were exposed to 23 h light and 1 h darkness during the experiment.

Collection of samples and measurements

Body weight gains (BWG), feed intake (FI) and feed conversion ratio (FCR) were determined for the starter, grower, finisher and whole the experimental periods. At the end of the experiment (week 6), four birds per treatment were randomly selected and slaughtered. Before slaughter, one series of venous blood samples were collected in heparinised (20 U/mL) syringes (1 mL, 29 G, 0.33X 12 mm) for determination of blood gas indices. These blood samples were placed on ice and moved to the laboratory in less than 2 h. The blood gas indices of pH, bicarbonate, pCO_2, pO_2 and O_2 saturation then were determined using a pH/blood gas analyzer (Nova biomedical model pHOx plus, USA). At time of slaughter, another series of blood samples was collected in anticoagulant tubes (ethylenediaminetetraacetic acid (EDTA), 7.5%) and transferred immediately to the laboratory, then centrifuged at 2800 × g for 5 min, for plasma separation, and stored at -20 °C for the later analyses. Plasma total antioxidant capacity (TAC) was determined using Randox test kit (Randox Laboratories Ltd., Crumlin, UK).

Table 1 Composition of experimental diets

Ingredients (g/kg)	Starter (0-10 d)	Grower (11-24 d)	Finisher (25-42 d)
Maize	329.9	344.5	392.8
Wheat	200.0	250.0	250.0
Soybean meal	393.3	335.0	282.3
Soybean oil	29.4	29.0	33.0
Dicalcium phosphate	21.0	21.5	21.5
Limestone	11.0	8.60	8.60
L-lysine HCl	2.90	2.20	2.00
DL-methionine	3.80	0.80	1.40
Vitamin and mineral premix[1]	5.00	5.00	5.00
Sodium chloride	3.70	3.40	3.40
Total	1000	1000	1000
Calculated analysis			
Dry matter (g/kg)	859.8	862.1	862.7
Metabolisable energy (kcal/g)	2.86	2.93	3.00
Crude protein (g/kg)	219.9	199.9	179.9
Fat (g/kg)	48.7	49.3	54.7
Fiber (g/kg)	39.6	37.0	34.4
Calcium (g/kg)	10.0	9.0	8.9
Available phosphorus (g/kg)	4.5	4.5	4.4
Calcium/phosphorus	2.22	2.00	2.66
Chloride (g/kg)	3.3	3.0	2.9
Sodium (g/kg)	1.6	1.5	1.5
Methionine (g/kg)	7.0	3.8	4.1
Lysine (g/kg)	14.2	12.4	10.9
Arginine (g/kg)	15.3	13.7	12.2
Methionine + cysteine (g/kg)	10.7	7.3	7.4
Tryptophan (g/kg)	2.9	2.6	2.3
Tyrosine (g/kg)	9.8	8.9	8.1
Threonine (g/kg)	8.5	7.7	6.9

[1] Supplied per kilogram of diet: vitamin A: 9000 mg; vitamin D_3: 2000 mg; vitamin E: 18 mg; vitamin B_{12}: 0.15 mg; Riboflavin: 6.6 mg; Calciumpantothenate: 10 mg; Niacin: 30 mg; Choline: 500 mg; Biotin: 0.1 mg; Thiamine: 1.8 mg; Piridoxin: 3 mg; Folic acid: 1 mg; vitamin K_3: 2 mg; Antioxidant (ethoxyquin): 100 mg; Zinc: 50 mg; Manganese oxide: 100 mg; Copper: 10 mg; Fe: 50 mg; I: 1 mg and Se: 0.2 mg.

Plasma malondialdehyde (MDA) concentration was determined by MDA reaction with thiobarbituric acid followed by extraction with butanol (Kolahi *et al.* 2011). Additionally, plasma creatinine, uric acid and urea were determined spectrophotmetrically (Alcyon 300 USA) using enzymatic kits (Pars Azmon Co., Iran).

Statistical analyses

The data were analyzed based on a completely randomized experimental design using the GLM procedure of SAS (2002).

Duncan's multiple range test was used to separate the means when treatment means were significant (P<0.05).

RESULTS AND DISCUSSION

Performance

The effects of treatments on FI, BWG and FCR during the starter, grower, finisher, and the whole experimental periods are shown in Tables 2, 3 and 4, respectively. No significant difference was observed among the treatments for FI during the starter, grower, finisher and whole the experimental periods (P>0.05). Moreover, there was no significant differences between the treatments for BWG during the starter, grower, and the whole experimental periods (P>0.05), but during finisher period, nitrate diminished the BWG.

Table 2 Feed intake (g) of broiler chickens consumed nitrate (27.4 mg L^{-1} water) alone or along with different levels of rosemary extract (1.5, 3.0 and 6.0 mL/L) in drinking water from day 1 to day 42 of age

Items	Starter	Grower	Finisher	Whole
Control	138.50	1169.55	3018.06	4326.10
Nitrate	135.54	1186.82	2983.86	4306.20
Nitrate and 1.5 mL/L of rosemary extract	124.29	1220.34	3051.92	4396.50
Nitrate and 3 mL/L of rosemary extract	130.46	1074.24	2969.66	4174.40
Nitrate and 6 mL/L of rosemary extract	142.33	1149.57	3026.00	4317.90
Pooled SEM	3.93	23.69	19.08	34.28
P-value	0.67	0.40	0.70	0.37
Rosemary extract *vs.* nitrate	0.76	0.53	0.54	0.91

SEM: standard error of the means.

Table 3 Body weight (g) of broiler chickens consumed nitrate (27.4 mg L^{-1}) alone or along with different levels of rosemary extract (1.5, 3.0 and 6.0 mL/L) in drinking water from day 1 to day 42 of age

Items	Starter	Grower	Finisher	Whole
Control	154.01	699.55	1588.38[a]	2441.94
Nitrate	143.20	693.35	1493.89[b]	2330.43
Nitrate and 1.5 mL/L of rosemary extract	163.19	776.40	1591.63[a]	2531.21
Nitrate and 3 mL/L of rosemary extract	150.96	700.27	1585.67[a]	2436.89
Nitrate and 6 mL/L of rosemary extract	161.94	729.33	1595.01[a]	2486.28
Pooled SEM	3.83	15.37	12.65	24.72
P-value	0.48	0.43	0.03	0.10
Rosemary extract *vs.* nitrate	0.14	0.30	0.002	0.01

The means within the same column with at least one common letter, do not have significant difference (P>0.05).
SEM: standard error of the means.

Table 4 Feed conversion rate of broiler chickens consumed nitrate (27.4 mg L^{-1}) alone or along with different levels of rosemary extract (1.5, 3.0 and 6.0 mL/L) in drinking water from day 1 to day 42 of age

Items	Starter	Grower	Finisher	Whole
Control	0.87[a]	1.68[ab]	1.9[b]	1.77[b]
Nitrate	0.94[a]	1.71[a]	2.0[a]	1.85[a]
Nitrate and 1.5 mL/L of rosemary extract	0.75[b]	1.57[bc]	1.91[b]	1.74[b]
Nitrate and 3 mL/L of rosemary extract	0.86[a]	1.53[c]	1.87[b]	1.71[b]
Nitrate and 6 mL/L of rosemary extract	0.88[a]	1.57[bc]	1.89[b]	1.74[b]
Pooled SEM	0.02	0.02	0.01	0.01
P-value	0.01	0.01	0.003	0.002
Rosemary extract *vs.* nitrate	0.007	0.002	0.0003	0.0001

The means within the same column with at least one common letter, do not have significant difference (P>0.05).
SEM: standard error of the means.

However, supplementation of all the rosemary levels returned BWG to the control level (P<0.05). In orthogonal comparisons, consumption of rosemary improved the BWG as compared to nitrate during both finisher and whole periods (P>0.05).

Meanwhile the FCR was affected by treatments during the all periods (P<0.05). In the starter period, consumption of low rosemary extract level (0.15%) lowered the FCR as compared to the other treatments (P<0.05). During the grower, finisher and whole the experimental periods, all the rosemary levels returned the FCR increment due to nitrate consumption to the normal level (P<0.05).

Blood parameters

There was no significant difference among the treatments for blood urea and MDA concentrations at week 6 of age (P>0.05; Table 5). The blood creatinine was decreased by consumption of 1.5 and 3.0 mL/L rosemary extract of drinking water as compared to other treatments (P<0.05). Uric acid was decreased by nitrate consumption, and consumption of 1.5 and 3 mL/L rosemary extract returned uric acid to a level that was not different from the control. However, the consumption of rosemary extract at 6.0 mL/L caused plasma uric acid to remain at a level similar to consumption of nitrate alone.

Table 5 Blood urea, creatinine, uric acid, total antioxidant capacity (TAC) and malondialdehyde (MDA) of broiler chickens fed different levels of rosemary extract (1.5, 3.0 and 6.0 mL/L) or consuming sodium nitrate (27.4 mg L^{-1}) in drinking water from day 1 to day 42 of age

Treatment	Urea (mg/dL)	Creatinine (mg/dL)	Uric acid (mg/dL)	TAC (nmol/L)	MDA (nmol/L)
Control	1.25	0.23	7.52[a]	1.61[a]	2.77
Nitrate	2.00	0.24	3.52[b]	0.83[b]	2.30
Nitrate and 1.5 mL/L of rosemary extract	1.25	0.21	6.15[ab]	1.3[ab]	2.45
Nitrate and 3 mL/L of rosemary extract	1.00	0.23	6.65[ab]	1.46[a]	2.22
Nitrate and 6 mL/L of rosemary extract	1.00	0.27	4.77[b]	1.08[ab]	3.07
Pooled SEM	0.15	0.02	0.49	0.09	0.25
P-value	0.17	0.32	0.05	0.04	0.83
Rosemary extract vs. nitrate	0.01	0.51	0.04	0.03	0.68

N: 4 for blood indices.
The means within the same column with at least one common letter, do not have significant difference (P>0.05).
SEM: standard error of the means.

Nitrate decreased the blood TAC and consumption of 3 mL/L rosemary returned it to the normal level (P<0.05).

Blood pH and gases indices
Venous blood pH and O_2 saturation and were not affected significantly by the treatments at week 6 of age (P>0.05; Table 6). Bicarbonate was decreased by nitrate in the drinking water, and consumption of all rosemary levels returned bicarbonate to the level of the control or slightly higher (P<0.05). The venous blood pO_2 was decreased by nitrate in drinking water, and consumption of the rosemary extract at 3 and 6 mL/L increased pO_2 to a level similar to the control (P<0.05). In contrast, venous blood pCO_2 was elevated by nitrate in the drinking water, and supplementation of all rosemary extracts decreased pCO_2 to the level of the control (P<0.05).

The results of recent experiment showed that consumption of nitrate in drinking water negatively affected the growth performance, especially BWG and FCR, of broiler chickens. The decreased performance due to nitrate consumption has been reported in many animals especially broiler chickens. A study with commercial broilers showed that nitrate levels greater than 20 mg/L had a negative effect on weight, feed conversion, or performance (Lil, 2009). Reeder (1996) indicated that consumption of sodium nitrate at 2033 mg/kgdiet reduced the growth performance and body weight of broilers. Grizzele et al. (1996) reported that a water nitrate levels of 3.55 and 5.19 mg/L negatively affected broiler growth performance (body weight and feed conversion ratio). Barton et al. (1986) found a negative correlation between nitrate consumption and feed conversion ratio in chickens. The compromised growth performance of nitrate-treated chickens could be related to their lower body antioxidant capacity as suggested by the nitrate-related decreased plasma TAC and uric acid. Safary and Daneshyar (2012) reported the lower plasma TAC content in nitrite-fed laying hens and connected it to the potential for increased peroxide production.

Peroxide is a reactive oxygen specious (ROS), which has the potential to cause oxidative damage, a conclusion supported by the nitrate-associated decrease in plasma total antioxidant activity (TAC). Excessive levels of ROS or peroxynitrite- the most damaging of the reactive nitrogen species (RNS)- results in oxidative stress and causing severe damage to cellular components and lipid peroxidation (Seven et al. 2010). Reeder (1996) observed the decreased uric acid level in chicks consuming 50, 200 or 1000 mg/kg diet of calcium nitrate. The current research showed that addition of rosemary extract to the nitrated drinking water of broiler chickens ameliorated the nitrate-associated negative effects. It has been reported that growth performance parameters of broiler chickens can be enhanced by the addition of aromatic herbs and their extracts to poultry diets (Yesilbag et al. 2011). The current findings agree with those obtained by Radwan (2003) who reported improved body weight and feed conversion ratio of chickens fed rosemary leaves at 5 g/kg diet. Al Kassie (2008) indicated that addition of anise at 10 g/kg and rosemary 10 g/kg diet improved feed conversion ratios in broilers. Yesilbag et al. (2011) reported that addition of rosemary volatile oils at 100, 150 and 200 mg/kgdiet s improved both weight gain and feed conversion ratio in broilers. In Japanese quail, Ciftci et al. (2013) found a decrease in feed conversion ratio by adding rosemary oil at 125 and 250 mg/kg diet when the birds were under a heat stress condition. In broiler chicken fed with poultry fat, adding 0.1 rosemary powder to diet improved the BWG and feed efficiency (Khazaei et al. 2016). It is highly likely that an improved body antioxidant system could be the reason for improved growth performance associated with rosemary treated broilers in this experiment. Both the blood TAC and uric acid were higher in rosemary extract treated broilers, which indicated improved antioxidant capacity of these birds.

Uric acid is the main nitrogenous waste product of birds (Wright, 1995), and it is also a potent antioxidant (Ames et al. 1981).

Table 6 Blood pH, HCO_3, pO_2, pCO_2 and O_2 saturation of broiler chickens fed different levels of rosemary extract (1.5, 3.0 and 6.0 mL/L) or consuming sodium nitrate (27.4 mg/L) in drinking water from day 1 to day 42 of age

Treatment	pH	HCO_3 (nmol/L)	pO_2 (mm Hg)	pCO_2 (mm Hg)	O_2 saturation (m/L)
Control	7.46	31.85[bc]	46.85[a]	48.20[b]	54.60
Nitrate	7.38	30.85[c]	32.30[b]	61.95[a]	47.60
Nitrate and 1.5 mL/L of rosemary extract	7.42	34.16[ab]	37.15[b]	46.75[b]	50.40
Nitrate and 3 mL/L of rosemary extract	7.46	33.53[ab]	48.40[a]	46.10[b]	57.50
Nitrate and 6 mL/L of rosemary extract	7.45	35.33[a]	48.10[a]	47.80[b]	59.45
Pooled SEM	0.01	0.55	2.28	1.88	1.76
P-value	0.33	0.05	0.004	0.0005	0.14
Rosemary extract vs. nitrate	0.097	0.01	0.002	0.0001	0.059

N: 4 for blood gas parameter.
The means within the same column with at least one common letter, do not have significant difference (P>0.05).
SEM: standard error of the means.

Uric acid inactivates the peroxynitrite and consequently decreases the ROS production in the body. It is likely that uric acid is degraded or metabolized when it scavenges peroxynitrite (Santos et al. 1999).

Allantoin, the oxidative product of uric acid, has been proposed as a potential biomarker for in vivo free radical reactions. Since birds do not possess the enzyme urate oxidase, the presence of allantoin in their plasma indicates a direct oxidation of uric acid possibly by peroxynitrite. Strong oxidizers such as hypochlorous acid and hydroxyl radicals rapidly oxidize uric acid into allantoin and other products (Kaur and Halliwell, 1990).

The improved body antioxidant capacity by rosemary extract consumption possibly have caused the lower methemoglobin production and better status of blood gaseous (lower pCO_2 and higher pO2) in current experiment.

The antioxidant ability of rosemary extract or one or more of its components has been reported by some researchers. Krause and Ternes (2000) found the improvement of the oxidative stability of egg yolk when carnosic acid (an antioxidant constituent of rosemary) was used as a dietary supplement in laying hens. Galobart et al. (2001) reported that the dietary supplementation of a rosemary extract to laying hens had no effect on lipid oxidation of eggs. Yasar et al. (2011) detected the highest SOD activity by dietary consumption of 100 mg/kg rosemary oil in broilers. In quails, dietary supplementation of 250 and 500 ppm rosemary essential oils alleviated the heat stress induced testicular lipid peroxidation (Turk et al. 2016). The decreased blood creatinine was the other consequence of rosemary consumption in recent experiment. Creatinine is a chemical waste molecule that is generated from muscle metabolism. The kidneys maintain the blood creatinine in a normal range (Polat et al. 2011). It has been reported that nitrate consumption can impair the kidney functions (Zurovsky and Haber, 1995).

CONCLUSION

Current finding suggests that rosemary extract at levels of 1.5, 3.0 and 6.0 mL/L of drinking water may cause an amendatory effect on antioxidant ability and growth performance of broiler chickens. It can be concluded that the biochemical activities attributed to rosemary extract phenolic compounds and metabolites could be important to chicken and human health. Additional studies are necessary to ascertain plant and plant extracts that have the potential to provide benefits to the animal and reduce the harmful effects of nitrate exposure.

ACKNOWLEDGEMENT

The authors would like to sincerely appreciate the Urmia University for the financial support of this research.

REFERENCES

Al-Kassie G.A.M. (2008). The effect of anise and rosemary on broiler performance. Int. J. Poult. Sci. 7, 243-245.

Ames B.N., Cathcart R., Schwiers E. and Hochstein P. (1981). Uric acid provides an antioxidant defense in humans against oxidant- and radical-caused aging and cancer: a hypothesis. Proc. Natal. Acad. Sci. USA. 78, 6858-6862.

Aruoma O.I., Halliwell B., Aeschbach R. and Loligers J. (1992). Antioxidant and prooxidant properties of active rosemary constituents: carnosol and carnosic acid. Xenobiotica. 22, 257-268.

Atef M., Abo-Norage M.A., Hanafy M.S. and Agag A.E. (1991). Pharmacotoxicological aspects of nitrate and nitrite in domestic fowls. British Poult. Sci. 32, 399-404.

Barton T.L., Hileman L.H. and Nelson T.S. (1986). A survey of water quality on Arkansas broiler farms andits effect on performance. Pp. 36. Proc. 21st Natal. Meet. Poult. Health. Condemnations, University of Arkansas, Fayetteville, Arkansas.

Cheung S. and Tai J. (2007). Anti-proliferative and antioxidant properties of rosemary Rosmarinus officinalis. *Oncol. Rep.* **17**, 1525-1531.

Chow C.K. and Hong C.B. (2002). Dietary vitamin E and selenium and toxicity of nitrite and nitrate. *Toxicology.* **180**, 195-207.

Ciftci M., Şimsek Ü.G., Azman M.A., Çerci İ.H. and Tonbak F. (2013). The effects of dietary rosemary (*Rosmarinus officinalis*) oil supplementation on performance, carcass traits and some blood parameters of Japanese quail under heat stressed condition. *Kafkas Üniv. Vet. Faküll. Derg.* **19**, 595-599.

Galobart J., Barroeta A.C., Baucells M.D., Codony R. and Ternes W. (2001). Effect of dietary supplementation with rosemary extract and α-tocopheryl acetate on lipid oxidation performance, meat yield and feather coverage. *British Poult. Sci.* **45**, 677-683.

Grizzele J.M., Armbrust T.A., Brayan M.A. and Saxton A.M. (1996). The effect of water nitrate and bacterial on broiler growth performance. *J. Appl. Poult. Res.* **6**, 48-55.

Hras A.R., Hadolin M., Knez Z. and Bauman D. (2000). Comparison of antioxidative and synergistic effects of rosemary extract with α-tocopherol, ascorbyl palmitate and citric acid in sunflower oil. *Food Chem.* **71**, 229-233.

Ibanez E., Kubatova A., Senorans F.J., Cavero S., Reglero G. and Hawthorn S.B. (2003). Subcritical water extraction of antioxidant compounds from rosemary plants. *J. Agric. Food Chem.* **50**, 3512-3517.

Jaffe E.R. (1981). Methemoglobin pathophysiology. Pp. 133-151 in Proc. Int. Symp. Erythrocyte Pathobiol. New York.

Kaur H. and Halliwell B. (1990). Action of biologically-relevant oxidizing species upon uric acid. Identification of uric acid oxidation products. *Chem. Biol. Int.* **73**, 235-247.

Khazaei R., Esmailzadeh L., Seidavi A. and Simoes J. (2016). Comparison between rosemary and commercial antioxidant blend on performance, caecal coliform flora and immunity in broiler chickens fed with diets containing different levels of poultry fat. *J. Appl. Anim. Res.* **45**, 263-267.

Kolahi S., Hejazi J., Mohtadinia J., Jalili M. and Farzin H. (2011). The evaluation of concurrent supplementation with vitamin E and omega-3 fatty acids on plasma lipid per oxidation and antioxidant levels in patients with rheumatoid arthritis. *Int. J. Rheumatol.* **7**, 1-7.

Krause E.L. and Ternes W. (2000). Bioavailability of the antioxidative *Rosmarinus officinalis* compound carnosic acid in eggs. *European Food Res. Technol.* **3**, 161-164.

Lil I. (2009). Clean drinking water is crucial in enhancing animal productivity. Pp. 17th in Proc. Ann. Asain Sea Feed Technol. Nutr. Workshop. Imperial Hotel, Hue, Vietnam.

Polat U., Yesilbag D. and Eren M. (2011). Serum biochemical profile of broiler chickens fed diets containing rosemary and rosemary volatile oil. *Poult. Sci.* **5**, 23-30.

Radwan N.L. (2003). Effect of using some medicinal plants on performance and immunity of broiler chicks. Ph D. Thesis. Cairo Univ., Cairo, Egypt.

Reeder J.A. (1996). The Effects on the performance of broilers consuming calcium, potassium and sodium nitrates and nitrites from the drinking water. MS Thesis. Oregon State Univ., Oregon, USA.

Safary H. and Daneshyar M. (2012). Effect of dietary sodium nitrate consumption on egg production, egg quality characteristics and some blood indices in native hens of west Azarbaijan province. *Asian-Australian J. Anim. Sci.* **25**, 1611-1616.

Santos C.X.C., Anjos E.I. and Augusto O. (1999). Uric acid oxidation by peroxynitrite: multiple reactions, free radical formation, and amplification of lipid oxidation. *Arch. Biochem. Biophys.* **372**, 285-294.

SAS Institute. (2002). SAS®/STAT Software, Release 9.1. SAS Institute, Inc., Cary, NC. USA.

Seven I., Aksu T. and Seven P.T. (2010). The effects of propolis on biochemical parameters and activity of antioxidant enzymes in broilers exposed to lead-induced oxidative stress. *Asian-Australian J. Anim. Sci.* **23**, 1482-1489.

Shahidi F. and Wanasundara P.D. (1992). Phenolic antioxidants. *Crit. Rev. Food Sci. Nutr.* **32**, 67-103.

Shuang-Sheng H. and Rong-Liang Z. (2006). Rosmarinic acid inhibits angiogenesis and its mechanism of action *in vitro*. *Cancer Lett.* **239**, 271-280.

Slepchenko V.N. and Rhnelnitskh G.A. (1988). Toxic effects of dietary nitrate on semenproduction in bulls. *Vet. Kiev. USSR.* **63**, 63-67.

Turk G., Çeribaş A.O., Şimsek U.G., Ceribaşı S., Guvenç M., Ozer K.S., Ciftçi M., Sonmez M., Yuce A., Bayrakdar A., Yaman M. and Tonbak F. (2016). Dietary rosemary oil alleviates heat stress-induced structural and functional damage through lipid peroxidation in the testes of growing Japanese quail. *Anim. Reprod. Sci.* **164**, 133-143.

Walker R. (1975). Naturally occurring nitrate/nitrite in foods. *J. Sci. Food Agric.* **26**, 1735-1742.

Williams R.J., Spencer J.P.E. and Rice-Evans C. (2004). Flavonoids: Antioxidants or signalling molecules? *Free Radical. Biol. Med.* **36**, 838-849.

Wright P.A. (1995). Nitrogen excretion: three end products, many physiological roles. *J. Exp. Biol.* **198**, 273-281.

Yasar S., Namik D., Fatih G., Gokcimen A. and Selcuk K. (2011). Effects of inclusion of aeriel dried parts of some herbs in broiler diets. *J. Anim. Plant. Sci.* **21**, 465-476.

Yesilbag D., Eren M., Agel H., Kovanlikaya A. and Balci F. (2011). Effects of dietary rosemary, rosemary volatile oil and vitamin E on broiler performance, meat quality and serum SOD activity. *British Poult. Sci.* **52**, 472-482.

Yu B.P. (1994). Cellular defenses against damage from reactive oxygen species. *Physiol. Rev.* **74**, 139-162.

Zurovsky Y. and Haber C. (1995). Antioxidants attenuate endotoxin- generation induced acute renal failure in rats. *Scandinavian J. Urol. Nephrol.* **29**, 147-154.

20

The Metabolizable Energy Content and Effect of Grape Pomace with or without Tannase Enzyme Treatment in Broiler Chickens

S.K. Ebrahimzadeh[1*], B. Navidshad[1], P. Farhoomand[2] and
F. Mirzaei Aghjehgheshlagh[1]

[1] Department of Animal Science, Faculty of Agricultural Science, University of Mohaghegh Ardabili, Ardabil, Iran
[2] Department of Animal Science, Faculty of Agriculture, Urmia University, Urmia, Iran

*Correspondence E-mail: sk.ebrahimzadeh@uma.ac.ir

ABSTRACT

Two experiments were conducted to determine the chemical composition, metabolizable energy content and effect of grape pomace (GP) with or without tannase enzyme treatment on growth performance of broiler chickens. In experiment 1, the apparent metabolizable energy (AME) and apparent metabolizable energy corrected for nitrogen (AMEn) of GP were determined as 2642.19 and 2641.08 kcal/kg, respectively. Supplementation of tannase enzyme at 0.05 and 0.1 percent of experimental diets significantly ($P \leq 0.01$) affected the metabolizable energy content of the diets and the highest improvement was observed with diets containing 0.01 percent tannase enzyme ($P \leq 0.01$). True metabolizable energy (TME) and true metabolizable energy corrected for nitrogen (TMEn) were determined based on Sibbald's procedure. The TME and TMEn of grape pomace were 1844.14 and 1839.83 kcal/kg, respectively. In experiment 2, four dietary treatments (50 birds/treatment) were conducted to study the effect of inclusion of 10% GP with or without tannase enzyme supplementation on growth performance in broiler chicks (0 to 42 days of age). At 10 d of age, the body weight (BW) of the control groups was higher ($P < 0.05$) compared with other experimental groups. Furthermore, broilers fed the control diet had higher ($P < 0.05$) average daily gain (ADG) in the starter (days 0-10) experimental periods. Addition of GP in the chicken diets did not impair growth performance (BW, ADG, average daily feed intake (ADFI) and feed-to-gain ratio (F:G)) of birds fed the grower (days 11-24), finisher (25-42) and the overall (days 0-42) experimental periods compared with other treatments. Among diets supplemented with GP, feed intake and body weight of broilers feeding diet containing 0.1% tannase were higher and the addition of tannase enzyme, improve the metabolizable energy content of GP.

KEY WORDS grape pomace, growth performance, metabolizable energy, tannase enzyme.

INTRODUCTION

Population growth, economic and social development caused higher demand for livestock products in many developing countries. A large portion of agricultural by-products has no direct human consumption, but it can be used indirectly to produce human food. Effective usage of agricultural by-products as animal feed depends on some factors such as nutrient composition compared to animal needs (McDonald et al. 1995). In order to enhance energy efficiency and optimize poultry performance it is suggested that the proximate analysis and metabolizable energy value of feedstuffs must be measured before feed formulation (McDonald et al. 1995). Grapes (Vitis vinifera) are one of the world largest fruit crops with annual production of 77 million metric tons (FAO, 2013). Esteeming, crushing, and pressing of grapes during processing for ethanol, fruit juice and wine production results in huge quantities of grape pomace (GP) including stems, skins, seeds, and peels and accounts for about 20% of the weight of the grape processed into wine (Llobera and Canellas, 2007). The GP are rich in a wide range of polyphenols like flavonoids, monomeric phenolic compounds, catechins, and epicatechins (Dorri *et al.* 2012). The available GP as a by-product of food processing industry in Iran exceeds 50000 ton/year, causing problems in both economical and ecological terms. Thus, any useful application for these by-products could represent an interesting advance in the maintenance of the environmental equilibrium and also an economic revaluation of the raw material (Abarghuei *et al.* 2010). The GP contains some antinutritional factors such as tannins that may decrease the GP feeding value (Pirmohammadi *et al.* 2007). A variety of protocols has been suggested by some scientists to improve the feeding value of tanniniferous feeds such as storage, drying, ensiling and adding exogenous enzymes (Makkar, 2003). Enzyme supplementation is a technique with increasing applicability for improving the nutritional characteristic of by-products and it is widely used in animal nutrition. However, results of GP utilization on growth performance in livestock were inconsistent.

For example, Hughes *et al.* (2005) found the addition of grape seed extract (90.2% total phenolics) at the level of 30 g/kg decreased the growth performance in chickens. However, Brenes *et al.* (2008) and Goni *et al.* (2007) found GP supplementation enhanced antioxidant capacity without any negative effect on growth performance in broilers (4.86% total phenolics; added at 30 g/kg). The aim of this study was to determine the chemical composition, metabolizable energy, and effect of GP with or without tannase enzyme addition on growth performance of broiler chickens.

MATERIALS AND METHODS

Experiment 1
Grape pomace preparation and analysis

The red GP was obtained from fruit juice factory located in the Urmia city (TATAO factory, Urmia city, Iran) and were completely dried under sunlight. Three samples were taken and ground to pass through a 1 mm screen for chemical analyses. Proximate composition (Table 1) of GP samples were analyzed according to procedures described by the Association of Official Analytical Chemists (AOAC, 2000). Gross energy (GE) content was determined by a Parr adiabatic calorimetric bomb. Extraction of the phenolic compounds from GP was performed by ultrasonication, using methanol/HCl 99/1 (v/v) as extraction solvent. Total phenolic content of extracts was determined by using Folin-Ciocalteu colorimetric method (Singleton and Rossi, 1965).

Gallic acid was employed as a calibration standard and results were expressed as gallic acid equivalents (mg gallic acid/g of dried samples).

Apparent metabolizable energy assay

In this experiment, 16 male broilers (Ross 308) from 37 to 45 d of age were used. Birds were distributed randomly to 4 treatments with 4 replicates in battery cages. Basal diets were formulated to meet or exceed NRC (1994) nutrient recommendations for broilers (Table 2). The amount of 10% GP without enzyme or supplemented with 0.05 or 0.1% tannase enzyme (supplied by Kikkoman Foods Products Company, Edogawa Plant, Japan, containing tannin acylhydrolase, 500 U/g, EC 3.1.20) was substituted by the corn and soybean in the experimental diets. Then the experimental diets were as follows: 1) GP free diet (100% basal corn, soybean diet), 2) 90% basal diet + 10% GP, 3) 90% basal diet + 10% GP + 0.05% tannase, 4) 90% basal diet + 10% GP + 0.1% tannase. Birds were fed *ad libitum*. The following procedures were common to all experiments. Body weight was determined when broilers were allocated to battery cages and also at the end of experimentation to ensure that dietary treatments did not limit growth. A 72-h total excreta collection period was conducted to evaluate AME and AMEn of GP. After a 3-d acclimation period, feed refusal and feed allocation were weighed daily throughout the 72-h collection period. The total amount of excreta voided at the end of the collection was weighed (wet basis). Multiple subsamples were collected from the total amount of excreta and homogenized, and then a 250-g representative sample was placed in a plastic bag for analysis. Representative samples of feed and excreta were frozen and subsequently dried at 70 °C. Dry matter and nitrogen content of feed and excreta were determined according to AOAC (2000). The GE was determined by a Parr adiabatic calorimetric bomb. Feed consumption and excreta weights during the 72-h collection period were used to calculate energy and nitrogen intake and excretion. AME and AMEn experimental diets were calculated using the following equations:

$$AME/g \ of \ feed = [(Fi \times GEf) - (E \times GEe)] \ / \ Fi$$
$$AMEn/g \ of \ feed = [(Fi \times GEf) - (E \times GEe) - (NR \times K)] \ / \ Fi$$
$$NR = (Fi \times Nf) - (E \times Ne)$$

Where:

AME: apparent metabolizable energy (kcal/g).

AMEn: apparent metabolizable energy corrected for nitrogen (kcal/g).

F_i: feed intake (g).

E: excreta (g).

GE_f: gross energy of feed sample (kcal/g).

GE_e: gross energy of excreta (kcal/g).

NR: nitrogen retention (g).

K: nitrogen retention correction coefficient (8.22 kcal/g for each g N).

NF: feed nitrogen (%).

N_e: fecal nitrogen (%).

The AME and AMEn content of GP calculated using the following modified equations (Anison *et al.* 1996):

AME_{GP}= $AME_{testdiet}$ -   [$AME_{basaldiet}$ ×
(corn+soybean$_{inclusionrate}$)] / grape pomace $_{inclusion\ rate}$

$AMEn_{GP}$= $AMEn_{testdiet}$ -   [$AMEn_{basaldiet}$ ×
(corn+soybean$_{inclusionrate}$)] / grape pomace $_{inclusion\ rate}$

Where:

AME_{GP}: apparent metabolizable energy of grape pomace (kcal/g).

$AMEn_{GP}$: apparent metabolizable energy corrected for nitrogen for grape pomace (kcal/g).

Corn + soybean$_{inclusion\ rate}$= 82. 74%.

Grape pomace$_{inclusion\ rate}$= 10%.

True metabolizable energy assay

The precision-fed cockerel assay of Sibbald *et al.* (1986) was used for determining the true metabolizable energy (TME and TMEn) of the GP. Seven adults Leghorn roosters were housed in individual metabolism cages that were 0.40-m wide, 0.40-m long, and 0.50-m high. Following a period of 24 h without feed, 25 g of the GP samples were fed by intubation to 6 adult Leghorn roosters. Another rooster was fasted to estimate endogenous losses. Total excreta voided over the following 48-h period was dried and ground for subsequent analyses. The GP and excreta samples were freeze-dried before analysis and the DM and Nitrogen content of feed and excreta were determined according to AOAC (2000). The GE was determined by a Parr adiabatic calorimetric bomb. The TME and TMEn content of GP samples were calculated using the following equations:

TME/feed= {[(Fi×GEf) - (E×GEe)] + (FEm+UEe)} / Fi
TMEn/feed= {[(Fi×GEf) - (E×GEe) - (NR×K)] + [(FEm+UEe) + (NR×K)]} / Fi
NR= (Fi×Nf) - (E×Ne)

Where:

TME: true metabolizable energy (kcal/g).

TMEn: true metabolizable energy corrected for nitrogen (kcal/g).

F_i: feed intake (g).

E: excreta (g).

GE_f: gross energy of feed sample (kcal/g).

GE_e: gross energy of excreta (kcal/g).

FE_m: metabolic fecal energy (kcal/g).

UE_e: indigenous urinary energy (kcal/g).

NR: nitrogen retention (g).

K: nitrogen retention correction coefficient (8.22 kcal/g for each g N).

NF: feed nitrogen (%).

N_e: fecal nitrogen (%).

Experiment 2
Dietary treatments and feeding schedules

A total of 200 newly hatched male broiler chicks (Ross 308) were purchased from a local hatchery. On arrival, all birds were weighed and randomly assigned to 1 of 4 treatments, with 5 replicate pens/treatment and 10 chickens/pen.Dietary treatments (Table 6) were formulated to meet or exceed the nutrient requirements of broilers provided by Ross Broiler Manual (Aviagen, 2014). The MEn value for GP was assumed 1839.83 kcal/kg based on experimen1. Experimental diets were as follows: 1) control corn, soybean diet (C); 2) C + 10% of GP; 3) C + 10% GP + T1 (500 mg/kg tannase enzyme) and 4) C + 10% GP + T2 (1000 mg/kg tannase enzyme).

The straw was substituted by GP in the experimental diets.The feeding regimen consisted of a starter (1 to 10 d), grower (11 to 24 d), and finisher diet (25 to 42 d). Chickens were raised in floor pens (100×120 cm) and had free access to feed and water for the entire experimental period (days 0-42). The room temperature gradually decreased from 33 to 22 °C on day 28 and then remained constant thereafter. The lighting program was a period of 20 h of light and 4 h of darkness. All experimental procedures were evaluated and approved by the Institutional Animal Care and Ethics Committee of the Mohaghegh Ardabili University. An enzyme with tannase activity (T) supplied by Kikkoman Foods Products Company (Edogawa Plant, Japan) containing tannin acylhydrolase (500 U/g, EC 3.1.20) was used.

Performance data

Birds and feed were weighed at 10, 24 and 42 d of age. The values of AFDI and ADG were recorded in different periods, and the feed-to-gain ratio (F:G) was calculated. Mortality was also recorded as it occurred. However, AFDI and F:G was corrected for the mortality of related groups.

Statistical analysis

Data were analyzed in a completely randomized design using the General Linear Models procedures of SAS (2001). When the differences were significant (P<0.05), mean values between treatments were compared using the Duncan test.

Table 1 Chemical composition (%) , gross energy (kcal/kg) and total polyphenol (mg gallic acid equivalent/g) of grape pomace samples

Item	Grape pomace samples			Mean	SD	CV
	1	2	3			
Dry matter	91.82	91.07	91.46	91.45	0.37	0.41
Crude protein	9.13	8.87	8.83	8.94	0.16	1.82
Ether extract	6.71	6.28	8.03	7	0.91	13.01
Ash	3.75	3.09	2.94	3.26	0.43	13.21
Crude fiber	31.08	27.94	31.58	30.20	1.97	6.53
Calcium	0.59	0.56	0.42	0.52	0.09	17.23
Phosphorus	0.39	0.26	0.24	0.29	0.08	27.45
Gross Energy	4346.72	4264.88	4581.28	4397.63	164.22	3.73
Total polyphenol	34.17	34.36	33.22	33.92	0.48	0.56

SD: standard deviation and CV: coefficient of variation.

Table 2 Ingredients of basal and experimental diets (experiment 1)

Ingredients (%)	Dietary treatment[1]			
	Diet 1	Diet 2	Diet 3	Diet 4
Corn	61.38	55.24	55.24	55.24
Soybean meal (44% CP)	30.55	27.50	27.50	27.50
Soybean oil	4.28	4.28	4.28	4.28
Dicalcium phosphate	1.4	1.4	1.4	1.4
Limestone	1.17	1.17	1.17	1.17
Sodium chloride	0.25	0.25	0.25	0.25
Vitamin and mineral premix[2]	0.5	0.5	0.5	0.5
DL-methionin	0.251	0.251	0.251	0.251
L-lysin, HCL	0.144	0.144	0.144	0.144
Grape pomace (GP)[3]	-	10	10	10
Enzyme	-	-	+	+

[1] The experimental diets were as follows: 1) GP free diet (100% basal corn, soybean diet); 2) 90% basal diet + 10% GP; 3) 90% basal diet + 10% GP + 0.05% tannase; 4) 90% basal diet + 10% GP + 0.1% tannase.

[2] Each kg of vitamin and trace mineral premix provided: vitamin A: 900 IU; vitamin D: 2000 IU; vitamin E: 18 IU; vitamin K: 2 mg; vitamin B_1: 1.8 mg; vitamin B_2: 6.6 mg; vitamin B_6: 3 mg; vitamin B_{12}: 15 µg; Niacin: 30 mg; Pantothenic acid: 10 mg; Biotin: 0.1 mg; Folic acid:1.25 mg; Choline chloride: 200 mg; Fe: 50 mg; Cu:10 mg; Mn: 100 mg; Zn: 85 mg; I: 0.8 mg and Se: 0.2 mg.

[3] 10% GP without enzyme or supplemented with 0.05 or 0.1% tannase enzyme was substituted by the corn and soybean in the experimental diets.

RESULTS AND DISCUSSION

Experiment 1

The chemical composition of GP samples is presented in Table 1. The mean value of dry matter (DM), crude protein (CP), crude fiber (CF), ash, ether extract (EE), calcium (Ca), phosphorus (P), gross energy and total polyphenols content were 91.45, 8.94, 30.2, 3.26, 7, 0.52, 0.29%, 4397.63 kcal/kg and 33.92 mg gallic acid/g ,respectively. The obtained data from this study was similar to values reported by Pirmohammadi et al. (2007). Unfortunately, there are no data for a GP in the NRC (1994) to compare with the results of this study.

Mean CP, CF, EE and total polyphenols content, obtained in the present study was less (8.94, 30.2, 7 and 33.92 vs. 13.79, 32.5, 10.26 and 48.70, respectively) than that reported by Goni et al. (2007). Brenes et al. (2016) reported the Ca and P content of GP samples to be 0.5 and 0.3%, respectively. In the current study, mean Ca and P content was 0.52 and 0.29%, respectively. Ash content averaged 3.26% and was higher (3.26% vs. 2.41%) than that reported by Goni et al. (2007).

However, any differences in chemical composition and mineral content between plant material were due to the agronomic and climatological conditions of the area that the plants are grown. Feed intake, nitrogen retention, AME and AMEn of experimental diets are presented in Table 3. The feed intake of the birds that fed the diet containing 10 % GP was less than those fed other diets (P<0.01). The enzyme supplementation increased feed intake (P<0.01) and birds fed on a diet containing 0.1% tannase had an increased feed intake compared to birds fed diet containing 0.05% tannase enzyme (P<0.01). Similarly, AME and AMEn were significantly reduced (P<0.01) on the diet containing 10% GP but, increased by diets with enzyme supplementation (P<0.01). Nitrogen retention was significantly affected by experimental diets (P<0.01). The nitrogen retention of birds fed the diet containing 10% GP, was significantly less than those fed other diets (P<0.01). The GP contains high levels of fiber and polymeric polyphenols as tannins with the capacity to bind and precipitate both dietary and endogenous proteins, and therefore the incorporation of GP at high doses in chicken diets might impair feed intake and nitrogen retention (Chamorro et al. 2015).

Table 3 Feed intake, nitrogen retention and apparent metabolizable energy of experimental diets

Experimental diets[1]	Feed intake(g)	AME (kcal/kg)	AMEn (kcal/kg)	Nitrogen retention (g)
Diet 1	541.162[a]	3215.95[a]	3203.77[a]	0.802[ab]
Diet 2	410.659[c]	2859.68[d]	2852.52[d]	0.359[c]
Diet 3	492.098[b]	2901.66[c]	2890.01[c]	0.697[b]
Diet 4	551.452[a]	3012.07[b]	2998.31[b]	0.924[a]
SEM	14.655	35.880	35.530	0.057
P-value	< 0.01	< 0.01	< 0.01	< 0.01

[1] The experimental diets were as follows: 1) GP free diet (100% basal corn, soybean diet); 2) 90% basal diet + 10% GP; 3) 90% basal diet + 10% GP + 0.05% tannase; 4) 90% basal diet + 10% GP + 0.1% tannase.
AME: apparent metabolizable energy and AMEn: apparent metabolizable energy corrected for nitrogen.
The means within the same row with at least one common letter, do not have significant difference (P>0.05).
SEM: standard error of the means.

Although the information on enzyme application to GP is scarce, the results of the present study showed that utilization of enzymes with the capacity to hydrolyse polymeric polyphenols present in GP, might improve the feed intake and AME content of the diets.

Apparent metabolizable energy content(AME) and metabolizability(AMEn/GE) of GP samples are presented in Table 4. The AME and AMEn content of GP samples were significantly affected by experimental diets (P<0.01). The AME and AMEn of GP in diet with 10% GP, was significantly less than GP samples in diets containing 90% basal diet + 10% GP + 0.05% tannase and 90% basal diet + 10% GP + 0.1% tannase (P<0.01). The mean value of AME and AMEn content was 2642.19 and 2641.08 kcal/kg, respectively.

The metabolizability in diet containing 90% basal diet + 10% GP + 0.1% tannase was higher (79.34%) than other diets (P<0.01). The addition of tannase enzyme, improved the AME content and metabolizability (AMEn/GE) of GP samples. Moreover, AME determination can be a highly variable measurement (Dozier et al. 2001; Batal and Dale, 2006). In addition, variability associated with feed intake and excreta measurements in balance experiments can make differences due to treatments with low inclusion levels of a test ingredient.

True metabolizable energy content and metabolizability of GP samples are presented in Table 5. The mean value of TME and TMEn content of GP was 1844.14 and 1839.83 kcal/kg, respectively. The metabolizability (TMEn/GE) of GP was 56.41%. Unfortunately, there are no data for AME and TME of GP in the NRC (1994) to be compared with the results of this study. It seems that high level of fiber and tannins are the main cause of low TMEn and metabolizability of GP. Fiber may interfere with protein and mineral digestion, whereas tannin binds and precipitate both dietary and endogenous proteins (Pirmohammadi et al. 2007).

Experiment 2

The effect of feeding diets containing GP and tannase enzyme on broiler BW, ADG, AFDI, and F:G is shown in Table 7.

At 10 d of age, the BW of the control groups was higher (P<0.05) compared with other experimental groups. Furthermore, broilers fed the control diet had higher (P<0.05) ADG in the starter (days 0-10) experimental periods compared with other treatments, but the addition of GP in the chicken diets did not impair growth performance (BW, ADG, ADFI and F:G) of birds fed the grower (days 11-24), finisher (25-42) and the overall (days 0-42) experimental periods compared with other treatments.

There are few references in the literature in relation to the use of grape byproducts in chicken feed. Hughes et al. (2005) and Lau and King (2003) reported growth depression in chickens fed diets containing grape seed extract. The effect of polyphenols has also been studied in chickens using ingredients like sorghum and faba bean. In general, relatively high dietary concentrations of polyphenols by the addition of these ingredients reduced performance in chickens as well as other livestock (Gualtieri and Rapaccini, 1990; Jansman et al. 1989; Nyachotti et al. 1997). Polyphenolic compounds are also known for their ability to interact with different molecules such as proteins. Binding of polyphenolic compounds to both dietary and endogenous proteins, such as digestive enzymes and proteins located at the luminal side of the intestinal tract, have been used to explain the reduced apparent digestibility of protein in the polyphenol-containing diets. GP contains a high level of fiber and polymeric polyphenols as procyanidins with the capacity to bind and precipitate both dietary and endogenous proteins, and therefore the incorporation of GP at high doses in chicken diets might impair nutrient digestion and growth. Thus, in the present study, we used high doses of GP and because all diets were formulated to contain the same fiber content, any difference should be attributed to the polyphenol content. The effect of feeding diets containing GP supplemented with the tannase enzyme on growth performance of chickens is shown in Table 7. No effect (P>0.05) of dietary treatments were observed on body weight and feed consumption. There are several methods which can be used to reduce or neutralize the negative effects of tannins in poultry feeds so as to increase the efficiency of feed utilization.

Table 4 Apparent metabolizable energy content and metabolizability of grape pomace samples*

Grape pomace samples	AME (kcal/kg)	AMEn (kcal/kg)	Metabolizability (%)[1]
Diet 2	2030.1[c]	2001[c]	75.47[c]
Diet 3	2420.8[b]	2405.1[b]	76.47[b]
Diet 4	3504.8[a]	3488[a]	79.34[a]
SEM	198.69	193.47	0.512
P-value	< 0.01	< 0.01	< 0.01
Mean	2642.19	2641.08	77.09

[1] AMEn/GE.
* The experimental diets were as follows: 1) GP free diet (100% basal corn, soybean diet); 2) 90% basal diet + 10% GP; 3) 90% basal diet + 10% GP + 0.05% tannase; 4) 90% basal diet + 10% GP + 0.1% tannase.
AME: apparent metabolizable energy and AMEn: apparent metabolizable energy corrected for nitrogen.
The means within the same row with at least one common letter, do not have significant difference (P>0.05).
SEM: standard error of the means.

Table 5 True metabolizable energy (TME) content and metabolizability of grape pomace samples*

Energy content (kcal/kg)	Grape pomace samples						Mean	SD	CV
	1	2	3	4	5	6			
TME	1831.71	1827.14	1799.16	1939.88	1891.63	1775.33	1844.14	61.029	3.30
TMEn	1827.60	1822.83	1794.90	1935.52	1887.23	1770.92	1839.83	61.006	3.31
Metabolizability (%)[1]	56.04	55.89	55.04	59.35	57.87	54.30	56.41	1.871	3.31

[1] TMEn/GE.
* The experimental diets were as follows: 1) GP free diet (100% basal corn, soybean diet); 2) 90% basal diet + 10% GP; 3) 90% basal diet + 10% GP + 0.05% tannase; 4) 90% basal diet + 10% GP + 0.1% tannase.
TME: True metabolizable energy and TMEn: True metabolizable energy corrected for nitrogen.
SD: standard deviation and CV: coefficient of variation.

Table 6 Ingredients and nutrient composition of basal diets (experiment 2)

Ingredients	Starter (days 0-10)	Grower (days 11-24)	Finisher (days 25-42)
Corn	42.53	44.52	48.75
Soybean meal (44% CP)	41.18	37.87	32.76
Soybean oil	5.99	7.27	7.77
Straw[1]	5.93	6.40	7.06
Grape pomace[2]	0	0	0
Dicalcium phosphate	1.75	1.57	1.40
Limestone	1.10	0.99	0.94
Sodium chloride	0.35	0.35	0.35
Vitamin and mineral premix[3]	0.5	0.5	0.5
DL-methionine	0.36	0.32	0.29
L-lysine, HCl	0.22	0.15	0.14
L-threonine	0.10	0.07	0.05
Calculated analysis			
Metabolizable energy	2900	3000	3100
Crude protein	22.22	20.81	18.89
Methionin + cystine	1.04	0.96	0.88
Lysine	1.39	1.25	1.12
Ether extract	7.87	9.19	9.80
Crude fiber	6.20	6.20	6.20
Ca	0.93	0.84	0.77
Available P	0.46	0.42	0.38

[1] The straw was substituted by grape pomace in the experimental diets.
[2] Experimental diets were as follows: 1) Control corn, soybean diet (C); 2) C + 10% of GP; 3) C + 10% GP + T1 (500 mg/kg tannase enzyme); 4) C + 10% GP + T2 (1000 mg/kg tannase enzyme).
[3] Each kg of vitamin and trace mineral premix provided: vitamin A: 900 IU; vitamin D: 2000 IU; vitamin E: 18 IU; vitamin K: 2 mg; vitamin B_1: 1.8 mg; vitamin B_2: 6.6 mg; vitamin B_6: 3 mg; vitamin B_{12}: 15 µg; Niacin: 30 mg; Pantothenic acid: 10 mg; Biotin: 0.1 mg; Folic acid:1.25 mg; Choline chloride: 200 mg; Fe: 50 mg; Cu:10 mg; Mn: 100 mg; Zn: 85 mg; I: 0.8 mg and Se: 0.2 mg.

The effectiveness of enzymes to enhance nutrient digestibility has been reported (Choct, 2006). However, Torki and Farahmand-Pour (2007) observed no significant effect of enzyme supplementation on performance of broiler chickens fed sorghum based diets.

The utilization of enzymes with the capacity to hydrolyse complex polyphenols present in GP, might allow the use of higher doses of GP in chicken diets.

Table 7 The effects of different dietary treatments on broiler body weight (BW), average daily gain (ADG), average daily feed intake (ADFI) and feed-to-gain ratio (F:G)

Item	Dietary treatment[1]				SEM	P-value
	Control	C + 10 GP	C + 10 GP+T1	C + 10 GP + T2		
BW, g						
10 d	248.32[a]	214.32[b]	202.64[b]	217.04[b]	5.836	0.022
24 d	979.68	902.61	910.87	912.96	15.754	0.291
42 d	2416.20	2272.46	2275.92	2358.74	40.280	0.560
ADG, g						
Starter (0-10 d)	20.52[a]	17.15[b]	15.96[b]	17.17[b]	0.580	0.019
Grower (11-24 d)	52.24	49.16	50.58	49.70	1.089	0.796
Finisher (25-42 d)	79.80	76.09	75.83	80.31	2.273	0.867
Overall (0-42 d)	56.50	53.08	53.16	55.13	0.959	0.560
ADFI, g						
Starter (0-10 d)	32.80	24.60	23.37	21.88	1.587	0.052
Grower (11-24 d)	101.68	93.96	98.17	97.34	1.544	0.392
Finisher (25-42 d)	161.95	150.30	151.45	165.28	5.435	0.732
Overall (0-42 d)	113.46	101.90	104.26	109.24	2.087	0.207
F:G, g:g						
Starter (0-10 d)	1.60	1.42	1.46	1.26	0.058	0.235
Grower (11-24 d)	1.96	1.91	1.94	1.96	0.026	0.933
Finisher (25-42 d)	2.03	1.98	1.98	2.05	0.038	0.909
Overall (0-42 d)	2.00	1.92	1.96	1.98	0.015	0.241

[1] Experimental diets were as follows: 1) Control corn, soybean diet (C); 2) C + 10% of GP; 3) C + 10% GP + T1 (500 mg/kg tannase enzyme); 4) C + 10% GP + T2 (1000 mg/kg tannase enzyme).
The means within the same row with at least one common letter, do not have significant difference (P>0.05).
SEM: standard error of the means.

Thus, in the present study, we used high doses (10%) of GP and hypothesized that the use of enzymes with tannase activities might hydrolyze the complex polyphenol structure of grape pomace into more simple phenols reduce the negative effects of complex polyphenols in feeds. In summary, the degradation and absorption of polyphenols within the gastrointestinal tract depend on the nature not only of the phenolic compound, but also of the intestinal microflora, which fermentative effect on other dietary components will be affected, conversely, by the type of polyphenolic compound (Bravo, 1998). Among diets supplemented with GP, feed intake and body weight of broilers feeding diet containing 0.1% tannase were higher. It seems that the tannase supplementation increased feed intake compared to birds fed diet containing 0.05% or unsupplemented tannase enzyme. Higher BW of broiler feed diet supplemented with 0.1% tannase can related to improving in ME content of the diet that we detected in experiment 1.

CONCLUSION

The results of this research showed that GP has 2030.1 and 1839.83 kcal/kg AMEn and TMEn, respectively. The addition of 10 percent GP in the chicken diets did not impair growth performance and the supplementation of tannase enzyme, improve the metabolizable energy content of GP and increase the nutritive value of grape pomace in poultry diets.

ACKNOWLEDGEMENT

The authors are grateful to the Mohaghegh Ardabili University of Ardabil, Iran, for providing the experimental facilities and financial support for this experiment.

REFERENCES

Abarghuei M.J., Rouzbehan Y. and Alipour D. (2010). The influence of the grape pomace on the ruminal parameters of sheep. *Livest. Sci.* **132,** 73-79.

Anison G., Hughes R.J. and Choct M. (1996). Effect of enzyme supplementation on the nutritive value of dehulled lupins. *British Poult. Sci.* **37,** 157-172.

AOAC. (2000). Official Methods of Analysis. Vol. I. 17th Ed. Association of Official Analytical Chemists, Arlington, VA, USA.

Aviagen. (2014). Ross 308: Broiler Nutrition Specification.. Aviagen Ltd., Newbridge, UK.

Batal A.B. and Dale N.M. (2006). True metabolizable energy and amino acid digestibility of distillers dried grains with solubles. *J. Appl. Poult. Res.* **15,** 89-93.

Bravo L. (1998). Polyphenols: chemistry, dietary sources, metabolism, and nutritional significance. *Nutr. Rev.* **56,** 317-333.

Brenes A., Viveros A., Goni I., Centeno C., Sayago-Ayerdi S.G., Arija I. and Saura-Calixto F. (2008). Effect of grape pomace concentrate and vitamin E on digestibility of polyphenols and antioxidant activity in chickens. *Poult. Sci.* **87,** 307-316.

Brenes A., Viveros A., Saura C. and Arija I. (2016). Use of polyphenol-rich grape by-products in monogastric nutrition. *Anim.*

Feed Sci. Technol. **1**, 1-17.

Chamorro S., Viveros A., Rebolé A., Rica B., Arija I. and Brenes A. (2015). Influence of dietary enzyme addition on polyphenol utilization and meat lipid oxidation of chicks fed grape pomace. *Food Res. Int.* **4**, 215-224.

Choct M. (2006). Enzymes for the feed industry: past, present and future. *World Poult. Sci. J.* **62**, 5-15.

Dorri S., Tabeidian A.S., Toghyani M., Jaha-nian R. and Behnam-nejad F. (2012). Effect of different levels of grape pomace on blood serum and biochemical parameters of broiler chicks at 29 and 49 days of age. Pp. 68-72 in Proc. 11th Int. Natl. Congr. Recy. Organic Waste. Agric. Isfahan, Iran.

Dozier W.A., Moran E.T. and Kidd M.T. (2001). Male and female responses to low and adequate dietary threonine on nitrogen and energy balance. *Poult. Sci.* **80**, 926-930.

FAO (2013). Food and Agriculture Organization of the United Nations (FAO), Rome, Italy.

Goni I., Brenes A., Centeno C., Viveros A., Saura-Calixto A., Rebole´I. and Arija R. (2007). Effect of dietary grape pomace and vitamin E on growth performance, nutrient digestibility, and susceptibility to meat lipid oxidation in chickens. *Poult. Sci.* **86**, 508-516.

Gualtieri M. and Rapaccini S. (1990). Sorghum grain in poultry feeding. *World's Poult. Sci. J.* **46**, 246-254.

Hughes R.J., Brooker J.D. and Smyl C. (2005). Growth rate of broiler chickens given condensed tannins extracted from grape seed. Pp. 56-68 in Proc. Poult. Rese. Found. University of Sidney, Sidney, Australia.

Jansman A.J.M., Huisman J. and van der Poel A.F.B. (1989). Recent Advances in Research of Antinutritional Factors in Legume Seeds. Pudoc Publication, Wageningen, The Netherlands.

Lau D.W. and King A.J. (2003). Pre- and post-mortem use of grape seed extract in dark poultry meat to inhibit development of thiobarbituric acid reactive substances. *J. Agric. Food Chem.* **51**, 1602-1607.

Llobera A. and Canellas J. (2007). Dietary fiber content and antiox-idant activity of Manto Negro red grape (*Vitis vinifera*): pomace and stem. *Food Chem.* **101**, 659-666.

Makkar H.P.S. (2003). Effects and fate of tannins in ruminant animals, adaptation to tannins, and strategies to overcome detrimental effects of feeding tannin-rich feeds. *Small Rumin. Res.* **49**, 241-256.

McDonald P., Edwards R.A., Greenhalgh J.F.D. and Morgan C.A. (1995). Animal Nutrition. Published by Prentice Hall, New York.

NRC. (1994). Nutrient Requirements of Poultry, 9th Rev. Ed. National Academy Press, Washington, DC., USA.

Nyachotti C.M., Atkinson J.L. and Leeson S. (1997). Sorghum tannins: a review. *World's Poult. Sci. J.* **53**, 5-21.

Pirmohammadi R., Golgasemgarebagh A. and Mohsenpur Azari A. (2007). Effects of ensiling and drying of white grape pomace on chemical composition, degradability and digestibility for ruminants. *J. Anim. Vet. Adv.* **6**, 1079-1082.

SAS Institute. (2001). SAS®/STAT Software, Release 9.1. SAS Institute, Inc., Cary, NC. USA.

Sibbald I.R. (1986). The TME system of feeding evaluation. Research branch contribution. Pp. in 43-86 Proc. Anim. Res. Center. Agriculture Canada, Ottawa, Ontario, Canada.

Singleton V.L. and Rossi J.A. (1965). Colorimetry of total phenolics with phosphomolybdicphosphotungstic acid reagents. *American J. Enol. Viticult.* **16**, 144-158.

Torki M. and Farahmand-Pour M. (2007). Use of dietary enzyme inclusion and seed germination to improve feeding value of sorghum for broiler chicks. Pp. 643-646 in Proc. 16th European Symp. Poult. Nutr. Strasbourg, France.

21

The Effect of Prebiotic and Types of Feed Formulation on Performance, Intestinal Microflora and Cecum Gas Production of Laying Hens

H. Jahanian Najafabadi[1*], A.A. Saki[1], Z. Bahrami[1], A. Ahmadi[1], D. Alipour[1] and M. Abdolmaleki[1]

[1] Department of Animal Science, Faculty of Agriculture, Bu Ali Sina University, Hamedan, Iran

*Correspondence E-mail: hjahanian@yahoo.com

ABSTRACT

In this experiment, the effect of prebiotic Bio-Mos® (MOS) and types of feed formulation (total and digestible amino acids) were evaluated on performance and the small intestine microflora in laying hens for a duration of 10 wks. A total of 168 Hy line W-36 laying hens, with an initial age of 73 wks, were randomly allocated to 4 dietary treatments, 7 replicates and 6 birds in each replicate. The study was conducted in a completely randomized design with 2×2 factorial arrangement of treatments including 2 preboitic levels (0, 0.5 kg/ton of diet) and 2 types of AA feed formulation (total and digestible). Egg production (EP), egg weight (EW), egg mass (EM), average daily feed intake (ADFI) and feed conversion ratio (FCR) were not affected significantly by MOS, types of AA feed formulation and their interaction. Egg quality parameters including, specific gravity (SG), shell thickness (ST), shell weight (SW), haugh unit (HU) and yolk color (YC) did not affected by P-BM and types of AA feed formulation (P>0.05). No significant effects were observed on ileum *Lactobacillus* count by MOS, types of AA feed formulation and their interaction (P>0.05). There were no significant differences in the rate and gas production volume of experimental treatments (P>0.05). In conclusion, feed additive used in this study did not significantly affect the laying hens performance; intestinal *Lactobacillus* count, cecum gas production and microbial population.

KEY WORDS amino acid (AA), *Lactobacillus*, laying hens, microflora, prebiotic.

INTRODUCTION

In recent years, more attention has been focused on antibiotics which have been used as therapeutic level in poultry production due to development of multiple drug resistance bacteria and their residuals in animal products. Many of the antibiotics in the poultry industry have been used in human medicine as well. Therefore, producers compelled to use alternatives of antibiotics in poultry industry as well as human. Alternatives to use of antibiotics such as probiotics, prebiotics, immunostimulants and medicinal plants are used as growth promoters. A prebiotic component is defined as non-digestible food ingredients which can be limited number of the intestinal bacteria by beneficially affect the host (Gibson and Roberfroid, 1995). It has been hypothesized that supplementation with a prebiotic, could improve the detrimental effects of low or high protein diets. Prebiotics are components non-digestible carbohydrates which more of these are short chains of monosaccharide (MOS), called oligosaccharides. Some oligosaccharides are seemed to improve the intestinal and its function as competition for sites of attachment to pathogenic bacteria. It has been concluded that MOS exert the beneficially effects by modification of the intestinal microflora, reduction in turnover rate of the intestinal mucosa and modulation of the immune system. These effects have the potential to enhance growth rate, feed efficiency and live ability in commercial broiler and turkeys, and egg production in layers (Shane, 2001).

Attempts have been made to improve performance and egg quality by supplementation diet with common performance enhancers, especially prebiotics (Zarei et al. 2011; Kim et al. 2011; Hajati and rezaie, 2012; Shahir et al. 2014).

Based on some reports (Berry and Lui, 2000; Shashidhara and Devegowda, 2003), improvement in egg shell quality observed by supplementing MOS to older breeder females diets. A positive effect of the prebiotics on some egg shell quality parameters in laying hens had been reported by Swiatkiewicz et al. (2010).

Amino acids are important components of poultry diets; recently there has been much interest in formulating diets based on a digestible amino acid in this respect. Formulating diets in this fashion can result in a decrease of excess nutrients being excreted into the environment. Feed safety margins (ei nutrient requirements) are commonly used in commercial feed formulations and reducing these safety margins can help reduce nutrient excretion into the environment. Reducing these feed safety margins can also decrease feed costs, which is an integral input in poultry production. However, there is utilization a lack of information regarding the amino acids content and digestibility in poultry industry (Garcia et al. 2007). On the other hand prebiotic MOS is a commercial product containing yeast cells of *Saccharomyces cerevisiae*.

This experiment was designed because of the lack of enough information to evaluate the effects of prebiotic MOS in diets based on total or digestible amino acids on performance, eggs qualities, small intestine microflora and the cecum gas production of laying hens.

MATERIALS AND METHODS

Birds, diets and experimental design

All procedures were used during this study approved by Animal Care Committee of Bu-Ali Sina University, Hamedan, Iran. In total, 168 Single Comb White Leghorn (SCWL) hens Hy Line W-36, with an initial age of 73 weeks allocated at random into 4 treatments, 7 replicates and 6 birds in each. This study was conducted in a 2×2 factorial arrangement including 2 levels of prebiotic (0, 0.5 kg/ton of diet) and types of AA feed formulation, (total and digestible). Experimental basal diet was formulated according to Hy-Line International (2011) recommendation (Table 1). Each experimental diet was offered for 10 weeks. The light was provided 16 h in daily and the temperature was maintained at 21-26 °C. All birds were maintained under similar management conditions throughout the experimental period in 73-83 weeks of age.

Egg production (EP) and EW were recorded daily and FI was recorded weekly.

This information was used to calculate ADFI, EP, EM and FCR. Egg mass were calculated by multiplying percentage EP and EW for each replicate. Egg quality traits were determined on a biweekly basis. These eggs were individually weighed and their external and internal qualities were tested.

To measure shell weight (SW), the shell was separated from the yolk and albumen weighed after drying overnight at 60 °C as indicated by Grobas et al. (2001).

Shell thickness (ST) was measured using a digital micrometer (Echometer 1061, Robotmation Company, Tokyo, Japan). Shape index (SI), egg surface area (ESA) and specific gravity (SG) were determined based on methods from Yannakopoulos and TserveniGousi (1986) and Paganelli et al. (1974). Haugh unit (HU) was calculated from egg weight and albumen height as indicated by Haugh (1973). It is shown by the following equation:

$$HU= 100 \, Log \, (AH+7.57-1.7 \, EW^{0.37})$$

Where:
HU: haugh unit.
AH: albumen height (mm).
EW: egg weight (g).

Determination of dietary total and digestible amino acid total amino acid profile ingredients (corn and soybean meal) were measured in Tehran Evonic Degosa Company by Infrared Spectrometer, then standardized ileal digestibility coefficients were calculated by Opapeju et al. (2006) and MacLeod et al. (2008) methods. It is shown by following equation:

Concentration of digestible amino acid (g/kg)= [(concentration of total amino acid of ingredients (g/kg)) × (standardized ileal digestibility coefficients of ingredients (%)] × 100

Lactobacillus population measurement was performed according to Mathlouthi et al. (2002) methods. Briefly, fresh samples of ileum were collected immediately from healthy laying hens at 83 wk after slaughter (2 birds per replicate). Samples were separately in normal saline tubes, kept on ice, and transferred to laboratory. Samples were serially diluted in normal saline (from 10^{-1} to 10^{-10}) plated onto Man Rogosa and Sharp (MRS) (Merck, Germany) medium and incubated anaerobically at 37 °C for 36 hour.

Gas production test

At the end of 83 weeks, two birds were randomly selected per replicate.

Table 1 Amino acids profile total and digestible of corn and soybean meal (%)

Amino acids profile	Corn (total)	Corn (digestible)	Soybean meal (total)	Soybean meal (digestible)
Methionine	0.156	0.147	0.592	0.539
Cysteine	0.175	0.152	0.661	0.542
Methionine + cysteine	0.333	0.299	1.250	1.075
Lysine	0.237	0.218	2.636	2.372
Threonine	0.283	0.241	1.710	1.453
Tryptophan	0.060	0.049	0.592	0.527
Arginine	0.357	0.332	3.096	2.879
Isoleucine	0.269	0.256	1.948	1.734
Leucine	0.938	0.882	3.267	2.908
Valin	0.369	0.340	2.064	1.816
Histidine	0.223	0.211	1.156	1.063
Phenylalanine	0.381	0.358	2.168	1.929

Table 2 Diet formulation and composition (%)

Ingredients (g/kg)	Diet 1	Diet 2	Diet 3	Diet 4
Corn	61.82	62.83	61.81	61.82
Soybean Meal	22.77	21.93	22.77	22.77
Soybean oil	2.64	2.45	2.64	2.64
Dicalcium Phosphate	1.37	1.38	1.37	1.37
Oyster shells	10.43	10.43	10.43	10.43
Common salt	0.34	0.34	0.34	0.34
Vitamin premix[1]	0.25	0.25	0.25	0.25
Mineral premix[2]	0.25	0.25	0.25	0.25
DL-methionine	0.13	0.13	0.12	0.13
L-lysine HCl	-	0.02	-	-
Prebiotic Bio-Mos®	0.05	0.05	-	-
Cost diets (IRR)	8930	8899	8740	8750
Total	100	100	100	100
Nutrient composition (%)				
Analyzed				
Metabolizable energy	2800	2800	2800	2800
Crude protein (N×6.25)	14.56	14.3	14.56	14.56
Calcium	4.37	4.37	4.37	4.37
Available phosphor	0.39	0.39	0.39	0.39
Sodium	0.17	0.17	0.17	0.17
Chlorine	0.24	0.24	0.24	0.24
Dietary cation anion balance (DCAB)	173.52	168.92	173.50	173.52
Arginine	0.92	0.83	0.86	0.92
Isoleucine	0.60	0.54	0.55	0.60
Methionine	0.36	0.33	0.33	0.36
Lysine	0.74	0.67	0.67	0.74
Methionine + cysteine	0.62	0.55	0.55	0.62
Threonine	0.56	0.47	0.48	0.56
Tryptophan	0.17	0.14	0.15	0.17
Valin	0.69	0.61	0.62	0.69

[1] Vitamin premix supplied per kg of diet: vitamin A: 8.8 IU; vitamin D_3: 2.5 IU; vitamin E: 6.6 g; vitamin B_1: 1.5 g; vitamin B_2: 4.4 g; vitamin B_3 (calcium panthotenate) 8 g; vitamin B_5 (niacin): 20 g; vitamin B_6: 2.5 mg; vitamin B_{12}: 0.08 g; Biotin: 0.15 g; Cholin chloride: 400 mg and vitamin K_3: 2.5 g.
[2] Mineral premix supplied per kg of diet: Phosphorus: 18.7%; Calcium: 22%; Sodium: 39%; Manganese (oxide): 64 g; Iron (sulfate): 100 g; Zinc (oxide): 44 g; Copper (sulfate): 16 g; Iodine (iodate calcium): 0.64 g and Selenium (1%): 8 g.
Prebiotic MOS is a mannan oligosaccharide (processed from cell wall of saccharomyces yeast) and product of ALtech American.
Diet 1: diet based on total amino acid with prebiotic MOS (0.5 kg/ton of diet).
Diet 2: diet based on digestible amino acid with prebiotic MOS (0.5 kg/ton of diet).
Diet 3: diet based on digestible amino acid without prebiotic Bio-Mos.
Diet 4: diet based on total amino acid without prebiotic Bio-Mos.

The contents of the cecum was removed under aseptic condition and placed in falcon tubes (50 mL) after slaughter. Samples were collected immediately, kept on ice, and transferred to laboratory.

Gas production test was carried out according to Menke and Steingass (1988) method.

Statistical analysis

All data were analyzed with prebiotic levels and amino acid (total and digestible) levels, as factorial 2×2 using a completely randomized design by the GLM procedure of SAS (SAS, 2004). Treatments means were compared with Duncan's multiple range tests (Duncan, 1955). All differences were considered significant at $P \leq 0.05$. There was evaluated the normal distribution of data using Shapiro-Wilk test.

RESULTS AND DISCUSSION

Performance

The effects of dietary treatments on laying performance are shown in Table 3. Over the entire period, EW, EP, EM, ADFI and FCR were not affected (P>0.05) by MOS and types of feed formulation (total and digestible AA). The interaction between week with MOS and type of feed formulation on performance were not significant (P>0.05).

The possible reason for this, it may be age of hens, which with increasing of age, physiological conditions of the digestive tract were developed and, morphological conditions and gastrointestinal microbial were stable. In addition, it could be offered that prebiotics more effective in specific condition such as diseases, stress, density and poor environmental management which may occur in poultry industry. Furthermore, various responses to these different additives, may be due to the age and feed formulation, gut microflora, types of prebiotic dietary or others environmental conditions (Patterson and Burkholder, 2003; Hajati and rezaie, 2010).

This agrees with results of current study (Zarei et al. 2011) where found no significant effects on EP, FCR, FI and EM supplemented with two probiotics (Thepax and Yeasturer) and two prebiotics (Fermacto A-Max) and one synbiotic (Biomin) for six weeks.

Bozkurt et al. (2012) reports that production performance of laying hens were not affected by adding mannanoligosaccharides (MOS) and essential oils mixture (EOM) in the diet.

However, Chen et al. (2005) report that commercial prebiotics improved laying hens' performance. Berrin (2011), has reported that additives of probiotic and prebiotic to the quail breeder diets improved egg production and egg shell thickness and positively affected hatchability in quail breeders. Mostafa et al. (2015), indicated that MOS supplemented and forms of its inclusion in starter and grower diets were significantly affected chick performance.

It is reported that the body weight, body weight gain, feed consumption, FCR, mortality and percentage of carcass yield did not affected by the dietary inclusion of prebiotic, probiotic, and synbiotic compared with unsupplemented control in commercial broiler chicken (Sarangi et al. 2016).

Egg quality

The effects of dietary treatments on egg quality are shown in Table 4. Numerical differences in HU, ESA, USSR, SW, ST, SG and YC were apparent although not statistically significant (P>0.05). It seems that ST more influenced by the environment temperature, diet and age of birds. Based on the results in this study, the utilization of prebiotics can be a way to increase egg shell quality. The beneficial microorganisms increased absorption of vitamins and minerals especially calcium (Ca) and magnesium (Mg). This effect resulted to increased body weight and eggshell thickness (Roberfroid, 2000).

Also, Berrin (2011) suggested that the improvement in egg shell quality can be resulted from the increased mineral absorption. This agrees with earlier works (Nahashon et al. 1994; Mohan et al. 1995), which were reports a little improvement ST in hens supplemented with prebiotics for 10 weeks. Sharifi et al. (2011) have concluded that shell thickness significantly increased, it was due to the high Ca absorption and deposition and pH reduction of gastrointestinal tract by prebiotics which could have effect on egg shell, which this results in accordance with Swiatkiewicz et al. (2010) studies. In addition, it is demonstrated that some of the microbial species such as Lactobacillus sporogenes more increased absorption and concentration of Ca in the blood, therefore increasing of egg shell thickness (Panda et al. 2008).

This also agrees with Shahir et al. (2014) works; that found no significant effects on egg quality supplemented with commercial prebiotics. Zarei et al. (2011) reports that feed additives did have beneficial effects on egg quality characteristics in terms of egg shell weight and shell thickness.

However, Bozkurt et al. (2012) reports that egg quality except strength thickness were significantly affected by feed additives which additive, it is important or environmental temperature or both.

Intestinal Lactobacillus count

The effects of prebiotic MOS and types of feed formulation on ileum Lactobacillus count are shown in Table 5. In the present study, MOS, types of feed formulation and their interaction have no significant effects on population of lactic acid bacteria in the ileum (P>0.05). This is probably due to bird age, digestive tract evolution and absence of heat stress. Intestinal beneficial bacteria, especially lactic acid bacteria, reduced intestinal pH by the production organic acids. They weakened conditions for Salmonella and Colibacillus by alkaline pH for optimum their activity. They reduced proliferations of Salmonella and Colibacillus and their survival in the gastrointestinal tract (Nurmi et al. 1992).

Table 3 Effects of prebiotic Bio-Mos® and types of feed formulation on laying hens performance

Effect / index	ADFI (g/d)	EP (%)	EW (g)	EM (g/d)	FCR
Prebiotic					
0	100.26	81.47	52.56	43.72	2.42
0.5 (kg/ton)	101.37	78.92	51.25	41.28	2.62
SEM	0.30	0.84	0.58	0.84	0.083
Type formulation diet					
Total	100.72	78.57	50.72	40.79	2.68
Digestible	100.91	81.83	53.09	44.21	2.37
SEM	0.30	0.84	0.58	0.84	0/083
Prebiotic					
0.5 × total amino acid	101.42	78.12	50.24	40.05	2.74
0.5 × digestible amino acid	101.31	79.73	52.25	42.50	2.50
0 × total amino acid	100.50	83.92	53.92	45.92	2.23
0 × digestible amino acid	100.02	70.01	51.21	41.53	2.62
SEM	0.43	1.19	0.82	1.19	0.11
P-value					
Prebiotic	0.30	0.26	0.38	0.25	0.36
Types of feed formulation	0.85	0.16	0.13	0.11	0.16
Prebiotic × types of feed formulation	0.77	0.46	0.81	0.63	0.72
Week	< 0.0001	< 0.0001	0.0002	< 0.0001	< 0.0001
Prebiotic × week	0.28	0.34	0.25	0.35	0.16
Week × types of feed formulation	0.89	0.62	0.80	0.77	0.58
Prebiotic × types of feed formulation × week	0.76	0.42	0.55	0.47	0.53

ADFI: average daily feed intake; EP: egg production; EW: egg weight; EM: egg mass and FCR: feed conversion ratio.
The means within the same column with at least one common letter, do not have significant difference (P>0.05).
SEM: standard error of the means.

Table 4 Effects of prebiotic MOS and types of feed formulation on egg quality

Effect / index	SG	ST (mm)	SW (g)	HU	YC
Prebiotic					
0	1.07	0.31	5.40	79.47	5.08
0.5 (kg/ton)	1.07	0.75	5.56	79.17	5.20
SEM	0.001	0.31	0.06	0.66	0.003
Types of feed formulation					
Total	1.07	0.76	5.49	78.47	5.12
Digestible	1.07	0.30	5.47	80.18	5.16
SEM	0.001	0.31	0.06	0.66	0.003
Prebiotic					
0.5 × total amino acid	1.08	0.30	5.48	81.42[a]	5.21
0.5 × digestible amino acid	1.07	1.21	5.63	76.93[b]	5.18
0 × total amino acid	1.07	0.31	5.50	80.01[a]	5.13
0 × digestible amino acid	1.07	0.31	5.31	78.93[ab]	5.03
SEM	0/001	0/44	0.09	0.93	0.006
P-value					
Prebiotic	0.41	0.33	0.16	0.81	0.29
Types of feed formulation	0.47	0.32	0.87	0.19	0.72
Prebiotic × types of feed formulation	0.27	0.32	0.13	0.04	0.55
Week	0.005	0.43	< 0.0001	< 0.0001	< 0.0001
Prebiotic × week	0.31	0.43	0.002	0.18	0.0001
Week × type of feed formulation	0.32	0.44	0.28	0.13	0.01
Prebiotic × types of feed formulation × week	0.28	0.44	0.14	0.0001	0.28

SG: specific gravity; ST: shell thickness; SW: shell weight; HU: haugh unit and YC: yolk color.
The means within the same column with at least one common letter, do not have significant difference (P>0.05).
SEM: standard error of the means.

This agrees with Yang *et al.* (2007) reports that MOS was not significant affected on intestinal microbial composition.

Also, Ceylan *et al.* (2003) concluded that cecal microflora population did not affect by feed additives such as probiotics, organic acids, MOS and antibiotics.

Table 5 The effects of prebiotic MOS and types of feed formulation on ileum *Lactobacillus* count

Effect / index	Number of *lactobacillus* bacterial (Log10 cfu¹/mL)
Prebiotic	
0	7.77±1.37
0.5 (kg/ton)	7.64±3.01
SEM	0.87
Types of feed formulation	
Total	7.86±2.19
Digestible	7.56±2.48
SEM	0.87
Prebiotic	
0.5 × total amino acid	7.36±2.72
0.5 × digestible amino acid	7.92±3.69
0 × total amino acid	8.35±1.77
0 × digestible amino acid	7.19±0.62
SEM	1.23
P-values	
Prebiotic	0.92
Types of feed formulation	0.81
Prebiotic × types of feed formulation	0.50

The means within the same column with at least one common letter, do not have significant difference (P>0.05).
SEM: standard error of the means.

Table 6 Effects of prebiotic MOS and types of feed formulation on cecum microbial population of laying hens

Effect / index	Gas production rate (A)	Gas production volume (C)
Prebiotic		
0	0.046	281.93
0.5 (kg/ton)	0.043	255.83
SEM	0.003	10.02
Types of feed formulation		
Total	0.03	273.23
Digestible	0.05	264.53
SEM	0.003	10.02
Prebiotic		
0.5 × total amino acid	0.052	280.97
0.5 × digestible amino acid	0.040	282.90
0 × total amino acid	0.038	263.57
0 × digestible amino acid	0.048	248.10
SEM	0.005	14.18
P-value		
Prebiotic	0.58	0.10
Types of feed formulation	0.08	0.55
Prebiotic × types of feed formulation	0.93	0.64

A: potential ofgas production (mL/g) and C: rate of gas production (mL/h).
SEM: standard error of the means.

In addition, Fernandez *et al.* (2002) concluded that the bird intestinal microflora (*Bifidobacterium* spp. and *Lactobacillus* spp.) enhanced by supplementing diets with mannoseoligosaccharide (MOS) or palm kernel meal (PKM).

However, Donalson *et al.* (2008) report that laying hen cecal lactic acid bacteria enhanced with combination of fructooligosaccharide prebiotics with alfalfa or a layer ration. It is reported that additives of mannanoligosaccharides (Bio-Mos or PKE) in diets did not affected population of the ileal *Lactobacilli*, *Enterococcus* or *Enterobacteriaceae* family (Bahman *et al.* 2015).

Gas production

There were no significant differences in the rate and gas production volume by experimental treatments (P>0.05). This is probably due to the number and low activity of cecum microbial population than the rumen of ruminants. In general, in total experimental period, the rate and gas production volume were not affected by MOS, types of feed formulation and their interaction (P>0.05). Although, Guo *et al.* (2003) shown that extract of polysaccharide increased gas production in cecum of chickens. Short chain fatty acid (SCFA) (such as acetate, propionate, butyrate and lactic acid) and gasses (CO_2, CH_4 and H_2) produced by prebiotic fermentation in hindgut. Acid production causes release of toxic NH_3 (and amines) to produce NH_4^+. This NH_4^+ is no-permeation and result to decreases the blood NH_3 level. *Lactoacillus* and *Bifidobacterium* are capable of using NH_3 as their N source and reduce its concentration both in the gut and in the blood.

Fecal pH and NH_3 concentration did not affected by supplementation of MOS in dog's diets (Pawar *et al.* 2008; Kore *et al.* 2009) but faecal lactate, propionate and butyrate concentrations tended to increase (Kore *et al.* 2009). Also, there was a significant increase ($P<0.05$) in the faecal lactate and SCFA contents with a reduction in the faecal NH_3 content in dietary of dogs (Samal *et al.* 2012). However, Zentek *et al.* (2002) found an increased faecal NH_3 concentration after lactulose supplementation.

Previous studies have demonstrated that additives of non starch polysaccharide (NSP) in poultry diets could decline detrimental gas production (Wang, 2009).

CONCLUSION

Prebiotic MOS and types of feed formulation did not significant effect on laying hen's performance, intestinal *Lactobacillus* number and gas production of cecum microbial population. Egg quality did not affected by P-BM and types of feed formulation based on total AA and ileal digestibility.

ACKNOWLEDGEMENT

Our special thanks to Bu-Ali Sina University for providing facilities and financial support for this study. We also wish to thank to the staff of the Department of Animal Science for their excellent scientific collaboration.

REFERENCES

Bahman N., Juan B.L, Mohammad F.J, Amir A. and Norhani A. (2015). A comparison between a yeast cell wall extract (Bio-Mos®) and palm kernel expeller as mannan-oligosac-charides sources on the performance and ileal microbial population of broiler chickens. *Italian J. Anim. Sci.* **14(1)**, 34-52.

Berrin K.G. (2011). Effects of probiotic and prebiotic (mannanoligosaccharide) supplementation on performance, egg quality and hatchability in quail breeders. Ankara Üniv. Vet. Fak. Derg. **58**, 27-32.

Berry W.D. and Lui P. (2000). Egg production, egg shell quality and bone parameters in broiler breeder hens receiving Bio-Mos and Eggshell 49. *Poult. Sci. J.* **79(1)**, 124-131.

Bozkurt M., Kucukyilmaz K., Catli A.U., Cinar M., Bintas E. and Covent F. (2012). Performance, egg quality and immune response of laying hens fed diets supplemented with mannan-oligosaccharide or essential oil mixture under moderate and hot environmental conditions. *Poult. Sci. J.* **91**, 1379-1386.

Ceylan N., Çiftçi Y'.C. and lhan Z.Y'. (2003). The effects of some alternative feed additives for antibiotic growth promoters on the performance and gut micro flora of broiler chicks. *Turkian J. Vet. Anim. Sci.* **27**, 727-733.

Chen Y.C., Nakthong C. and Chen T.C. (2005). Improvement of laying hen performance by dietary prebiotic chicory oligofructose and inulin. *Int. Poult. Sci. J.* **4**, 103-108.

Duncan D.B. (1955). The multiple range and F-tests. *Biometrics.* **11**, 1-24.

Donalson L.M., Kim W.K., Chalova V.I., Herrera P., McReynolds J.L., Gotcheva V.G., Vidanovic D., Woodward C.L., Kubena L.F., Nisbet D.J. and Ricke S.C. (2008). *In vitro* fermentation response of laying hen cecal bacteria to combinations of fructooligosaccharide prebiotics with alfalfa or layer ration. *Poult. Sci. J.* **87**, 1263-1275.

Fernandez F., Hinton M. and van Gils B. (2002). Dietary mannan-oligosaccharides and their effect on chicken caecal microflora in relation to *Salmonella enteritidis* colonization. *Avian Pathol.* **31**, 49-58.

Garcia A.R., Batal A.B. and Dale N.M. (2007). A comparison of methods to determine amino acid digestibility of feed ingredients for chickens. *Poult. Sci.* **86**, 94-101.

Gibson G.R. and Roberfroid M.B. (1995). Dietary modulation of the human colonic microbiota: introducing the concept of probiotic. *J. Nutr.* **125**, 1401-1412.

Grobas S., Mednez J., Lazaro T.R., DeBlase C. and Mateos G.G. (2001). Influence of source and percentage of fat added to diet on performance and fatty acid composition of egg yolks two strains of laying hens. *Poult. Sci. J.* **80**, 1171-1179.

Guo F.C., Williams B.A., Kwakkel R.P. and Verstegen M.W.A. (2003). *In vitro* fermentation characteristics of two mushroom species, an herb and their polysaccharide fractions, using chicken cecal contents as inoculum. *Int. J. Poult. Sci.* **82**, 1608-1615.

Hajati H. and Rezaei M. (2010). The application of prebiotic in poultry production. *Int. J. Poult. Sci.* **9**, 298-304.

Haugh R.R. (1937). The haugh unit for measuring egg quality. *US Egg. Poult. Mag.* **43**, 552-555.

Hy-line International Publication. (2011). Hy-Line Variety W-36 Commercial Management Guide. West Des Moines, Iowa. USA.

Kim G.B., Seo Y.M., Kim C.H. and Paik I.K. (2011). Effects of dietary prebiotic supplemention on the performance, intestinal microflora, and immune response of broilers. *Poult. Sci. J.* **90**, 75-82.

Kore K.B., Pattanaik A.K., Das A. and Sharma K. (2009). Evaluation of alternative cereal sources in dog diets: effect on nutrient utilisation and hindgut fermentation characteristics. *J. Sci. Food. Agric.* **89**, 2174-2180.

MacLeod M.G., Valentine J., Cowan A., Wade A., McNeill L. and Bernard K. (2008). Naked oats: Metabolisable energy yield froma range of varieties in broilers, cockerels and turkeys. *Br. Poult. Sci.* **49**, 368-377.

Mathlouthi N., Saulnier L., Quemene B. and Larbier M. (2002). Xylanase a-glucanase and other side enzymatic activities have greater effects on the viscosity of sevsral feed stuff than xylanase and a-glucanase used alone or in combination. *J. Agric. Food Chem.* **50**, 5121-5127.

Menke K.H. and Steingass H. (1988). Estimation of the energetic feed value obtained from chemical analyses and *In vitro* gas production using rumen fluid. *Anim. Res. Dev.* **28**, 47-55.

Mohan R., Kocabagle N., Alp M., Acar N., Eren M. and Gezen S.S. (1995). Effect of probiotic supplementation on serum/yolk cholesterol and on egg shell thickness in layers. *Br. Poult. Sci.* **36**, 799-803.

Mostafa M.M.E., Thabet H.A. and Abdelaziz M.A.M. (2015). Effect of Bio-Mos utilization in broiler chick diets on performance, microbial and Hhistological alteration of small intestine and economic efficiency. *Asian J. Anim.* Vet. *Adv.* **10**, 323-334.

Nahashon S.N., Nakaue H.S. and Mirosh I.W. (1996). Performance of single comb white leghorn fed a diet supplemented with a live microbial during the growth and egg laying phases. *Anim. Feed Sci. Thechnol.* **57**, 25-38.

Nurmi E., Nuotio L. and Schncitz C. (1992). The competitive exclusion concept: Development and future. *Int. J. Food Microbiol.* **15**, 237-240.

Opapeju F.O., Golian A., Nyachoti C.M. and Campbell L.D. (2006). Amino acid digestibility in dry extruded-expelled soybean meal fed to pigs and poultry. *J. Anim. Sci.* **84**, 1130-1137.

Paganelli Charles V., Olszowka A. and Amos A. (1974). The avian egg surface area, volume, and density. *Condor.* **76**, 319-325.

Panda A.K., Rao M.V.L.N. and Sharma S.S. (2008). Effect of probiotic (*Lactobacillus sporogenes*) feeding on egg production and quality, yolk cholesterol and humoral immune response of white Leghorn layer breeders. *J. Sci. Food Agric.* **88**, 43-47.

Patterson J.A. and Burkholder K.M. (2003). Application of prebiotic and probiotics in poultry production. *Poult. Sci. J.* **82**, 627-631.

Pawar M.M., Pattanaik A.K., Kore K.B., Sharma K. and Sinha D.K. (2008). Evaluation of mannan-oligosachharides as a functional food in Spitz pups fed on homemade diet: influence on nutrient metabolism, gut health and immune response. Pp. 102-111 in Proc. 5th SAARC Congr. Canine Practice and Conven. Indian Soc. Adv. Canine Pract. Chennai, India.

Roberfroid M.B. (2001). Preferential substrates for specific germs. *Am. J. Clin. Nut.* **73**, 406-409.

Samal L., Chaturvedi V.B., Baliyan S., Saxena M. and Pattanaik A.K. (2012). Jerusalem artichoke as a potential prebiotic: in-

fluence on nutrient utilization, hindgut fermentation and immune response of Labrador dogs. *Anim. Nutr. Feed Technol.* **12**, 343-352.

Sarangi N.R., Babu L.K., Kumar A., Pradhan C.R., Pati P.K. and Mishra J.P. (2016). Effect of dietary supplementation of prebiotic, probiotic and synbiotic on growth performance and carcass characteristics of broiler chickens. *Vet. World.* **9**, 313-319.

SAS Institute. (2004). SAS®/STAT Software, Release 9.1. SAS Institute, Inc., Cary, NC. USA.

Shahir M.H., Sharifi M., Afsarian O. and Mousavi S.S. (2014). A comparison of the effects of commercial prebiotic (safmannan®, biomos ® and fermacto®) on performance, egg quality and antibody titer of Avian Influenza and Newcastle disease in laying hens. *J. Vet. Res.* **69**, 79-84.

Shane S.M. (2001). Mannan oligosaccharides in poultry nutrition: mechanism and benefits. Pp. 65-77 in Science and Tecnology in the Feed Industry. T.P. Lyons and K.A. Jacques, Eds. Nottingham University Press, Nottingham, United Kingdom.

Sharifi M., Shahir M.H., Safamehr A.R. and Abdi Shishegaran K. (2011). The effects of commercial prebiotics on egg qualitative characteristics. Pp. 55-58 in Proc. Nat. Conf. Modern. Agric. Sci. Technol. Zanjan, Iran.

Shashidhara R.G. and Devegowda G. (2003). Effect of dietary mannan oligosaccharide on broiler breeder production traits and immunity. *Poult. Sci. J.* **82**, 1319-1325.

Swiatkiewicz S., Koreleski J. and Arczewska A. (2010). Layikg performance and eggshell quality in laying hens fed diets supplemented with prebiotics and organic acids. *Czech J Anim. Sci.* **55**, 294-306.

Wang Y. (2009). Prebiotics: present and future in food science and technology. *Food Res. Int.* **42**, 8-12.

Yang Y., Iji P.A., Kocher A., Mikkelsen L.L. and Choct M. (2007). Effects of mannanoligosaccharide on growth performance, development of gut microflora and gut function of broiler chickens Raised on new litter. *J. Appl. Poult. Res.* **16**, 280-288,

Yannakopoulos A. and Tserveni-Gousi A. (1986). Quality characteristics of quail eggs. *Br. Poult. Sci.* **27(2)**, 171-176.

Zarei M., Ehsani M. and Torki M. (2011). Effects of adding various feed additives to diets of laying hens onproductive performance and egg quality traits. *J. Anim. Prod.* **13(2)**, 61-71.

Zentek J., Marquart B. and Pietrzak T. (2002). Intestinal effects of mannanoligosaccharides, transgalactooligosaccharides, lactose and lactulose in dogs. *J. Nutr.* **132**, 1682-1684.

22

Effects of Black Plum (*Vitex doniana*) Leaf Meal Inclusion on Performance, Haematology and Serum Biochemical Indices of Cockerels

A.O. Adeyina[1*], K.M. Okukpe[1], A.S. Akanbi[1], M.D. Ajibade[1], T.T. Tiamiyu[1] and O.A. Salami[1]

[1] Department of Animal Production, Faculty of Agriculture, University of Ilorin, Ilorin, Nigeria

*Correspondence E-mail: aadeyina@unilorin.edu.ng

ABSTRACT

A total of 105, 14 weeks old cockerels weighing 1.49 ± 0.023 kg were used to investigate the effects of *Vitex doniana* leaf meal (VDLM) on performance, haematological and serum biochemical parameters in cockerels over a seven-week experimental trials. The cockerels were divided into three (3) treatment groups of 0%, 5% and 10% inclusion levels of VDLM in a completely randomized design layout. The result showed significant increase (P<0.05) in final body weight, average daily weight gain and feed to gain ratio. There was no significant difference (P>0.05) in all haematological parameters except for the absolute content of granulocyte which, significantly increased (P<0.05) with increase in VDLM inclusion. Serum total protein significantly increased (P<0.05) with increase in VDLM inclusion while serum potassium in the control was significantly (P<0.05) higher compared to 5% and 10% VDLM inclusions It was concluded that VDLM could be included up to 10% in cockerels diet with an improvement in performance and without detrimental alteration in haematological and serum biochemical parameters.

KEY WORDS cockerels, haematology, performance, *Vitex doniana*.

INTRODUCTION

The inclusion of leaves in diet of poultry is becoming adaptable due to its availability and phytochemical constituents responsible for medicinal or organoleptic properties of the plant (Ugwu *et al.* 2013). Profitable livestock enterprise depends on availability and affordability of feedstuff (Adeyina *et al.* 2014). Surprisingly, the cost of producing conventional feed that supports improved performance, haematology and serum biochemistry of animal has been on the increase in Nigeria over the last three decades. This is attributable to inadequate production of grains coupled with competition between man, industry and livestock over the available feed materials. With increasing interest in

foliage plant as feed ingredient, several plants have been assessed with respect to their effects on performance and blood parameters in poultry. Some of which are *Napoleon imperialis*, *Ipomea asorfolia* and *Ipomea purpurea* (Adeyina *et al.* 2014). With the current emphasis on improvement of livestock production in Nigeria, foliage plants have found an application without compromising nutritional standard. In the list of possible alternatives are *Leucaena leucocephala*, *Lablab purpureus*, *Tithonia diversifolia*, to mention but a few (Ekenyem *et al.* 2003). *Vitex doniana* is among plant leaves with potential for improving animal productive performance, haematology and serum parameters. It is an indigenous tropical plant distributed across tropical sub-saharan, Africa's coastal savannas and savanna woodland. The tree is none domesticated, but it is often found at the centre of West African villages. There is little scientific information on the tree but African horticulturist and livestock stand to benefit from output of research and commercial development in Nigeria. *Vitex doniana* is commonly known as Black plum (English), 'Dinya' (Hausa), 'Oriri' (Yoruba) and 'Uchakoro' (Igbo) where the bark, leaves and roots of the plant are used in ethnomedicine for the management and treatment of numerous disorders such as microbial infection, cancer, rheumatism, hypertension and inflammatory diseases (Atawodi, 2005). The bark of the stem is aromatic and serves as blood tonic (Sofowora, 1993). An extract of *Vitex doniana* plant lowered blood pressure in rat (Olusola *et al.* 1997). These attributes make *Vitex doniana* one of the potential plants that may depict significant influence on animal. In view of this, performance haematological and serum biochemical parameters of cockerels fed *Vitex doniana* leaf meal (VDLM) were examined.

MATERIALS AND METHODS

The experiment was conducted at the Teaching and Reasearch Farm, University of Ilorin, Kwara State, Nigeria with an average mean temperature of 32.5 °C and annual rainfall of about 900 mm.

Fresh matured leaves were harvested from *Vitex doniana* trees located within the study location premises in the month of February, 2015. The leaves were identified at the herbarium, Department of Plant Biology, University of Ilorin. The leaves were sun-dried in the open field for three days. The leaves were later milled using a burr mill machine to get desired particle size for inclusion in experimental diet. Samples of sun-dried leaf meal were analyzed for proximate composition (Table 1). The leaf meal was then used in formulating the experimental diets.

Three diets with comparable energy and protein content were formulated with *Vitex doniana* leaf meal at 0%, 5% and 10% levels of inclusion (Table 3) in a completely randomized design. One hundred and five, 14-weeks-old cockerels from Teaching and Research Farm, University of Ilorin were used for the experiment. The birds were allocated at random to the three dietary treatments in battery cages. Each treatment group consisted 5 replicates with 7 birds in a replicate. Experimental diets and water were offered to the birds *ad libitum* under the natural day length of 14 hours over a period of seven weeks.

At the end of the experiment, blood samples were collected from the vein under the wingweb of two birds per replicate into bottles containing ethylenediaminetetraacetic acid (EDTA) for haematological parameters and into bottles without EDTA for some serum biochemical indices.

Chemical analysis

Analysis of the nutrients composition of the leaf meal was carried out by the methods of AOAC (2000). Metabolizable energy of VDLM was calculated according to the method of Pauzenga (1985) and nutrient composition of the experimental diets was calculated using nutrients table (Aduku, 1993).

Haematological indices were determined using haematology analyzer (HA) model 6000. Analysis of the biochemical indices was conducted using the clinical chemistry semi-auto-analyser and a commercial biochemical assay kit.

Enzyme activities of aspartate amino transferase (AST), alanine aminotransferase (ALT) and alkaline phosphatase (ALP) were analyzed by the spectrophotometric linked reaction methods (Reiching and Kaplan, 1988). Serum electrolytes were determined by the method of Young (2001).

Qualitative and quantitative phytochemical screening of the leaf meal was carried out by the methods of Trease and Evans (1989).

Statistical analysis

All data obtained were subjected to statistical analysis procedure using analysis of variance (ANOVA) following a completely randomized design (SAS, 1999) and level of significance ($P<0.05$) were seperated using Duncan multiple-range test (Duncan, 1955).

RESULTS AND DISCUSSION

The proximate composition of dried *Vitex doniana* leaf meal (VDLM) and the phytochemical analysis are shown in Tables 1 and 2. VDLM contained crude protein (11.66%) which is similar to 10% reported by Nnamani *et al.* (2007) and crude fibre (7.41%). These nutrients in VDLM are comparable to wheat offal and maize bran. This suggests that VDLM has potential as good feed ingredient for cockerels. The phytochemical analysis revealed that VDLM is rich in phenol as well as saponin, tannin and flavenoids which corroborated Adejumo *et al.* (2013) who had also reported that VDLM contains tannin, saponin and flavonoids.

The performance parameters of cockerels on experimental diets are shown in Table 4. The average feed intake of cockerels on experimental and control diets are comparable indicating that VDLM based diets are acceptable and palatable to the cockerels.

According to Cheeke and Skull (1985), reduction in feed intake in poultry occurred due to unpalatability and poor acceptability of meal which contained saponin that had bitter taste.

Table 1 Proximate composition of sun-dried *Vitex doniana* leaf meal (VDLM)

Parameters	Sun-dried VDLM (%)
Dry matter	88.99±0.05
Crude protein	11.10±0.08
Crude fibre	7.20±0.01
Ether extract	2.52±0.02
Ash	9.51±0.05
Nitrogen free extract	58.42±0.06
Calculated nutrient*	
Metabolizable energy (kcal/kg)	2691

* Calculated value (37×crude protein) + (81.8×ether extract) + (35.5×nitrogen free extract).

Table 2 Phytochemical composition of *Vitex doniana* leaf meal (VDLM)

Phytochemicals	Qualitative	Quantitative (mg/100 g)
Alkaloids	Present	2.61
Saponins	Moderately present	6.48
Tannins	Present	1.45
Flavonoids	Moderately present	20.82
Terponids	Present	0.21
Phenols	Highly present	96.14
Steroids	Present	2.02
Anthraquinones	Present	0.04

Table 3 Dietary composition of experimental diets

Ingredients/VDLM inclusion(%)	0%	5%	10%
Maize	38.25	38.25	38.25
Groundnut cake	11	11	11
Palmkernel cake	16	16	16
Wheat offal	20	20	20
Maize milling waste	10	5	0
VDLM	0	5	10
Methionine	0.25	0.25	0.25
Lysine	0.25	0.25	0.25
Bonemeal	3	3	3
Vitamin and mineral premix*	0.5	0.5	0.5
Salt	0.5	0.5	0.5
Palm oil	0.25	0.25	0.25
Total	100	100	100
Calculated nutrient			
Crude protein (%)	16.16	16.17	16.17
Metabolizable energy (kcal/kg)	2600	2601	2606

* Premix supplied: vitamin A: 30789 IU; vitamin: D 36 IU; vitamin E: 115 IU; vitamin K: 77 mg; Thiamine: 39 mg; Pyridoxine: 39 mg; Riboflavin: 115 mg; Calcium panthothenate: 173 mg; Nicotinic acid: 346 mg; vitamin B12: 0.31 mg; Folic acid: 19 mg; Manganese: 3 g; Zinc: 2 g; Iron: 1 g; Copper: 115 g; Iodine: 38 mg; Cobalt: 8 mg; Selenium: 4 mg; Antioxidant: 4 g and Chloride: 8 g.

The saponin and other phytochemicals in VDLM based diets did not affect the feed intake of the cockerels. There was a significant (P<0.05) increase in final weight with VDLM inclusion. Cockerels fed VDLM had higher final weight compared to cockerels on control diet. The improvement in weight of birds on VDLM based diets could be as a result of better utilization of the feed and availability of growth improvement factors such as vitamins and microminerals contained in VDLM as reported by Nnamani *et al.* (2007) as well as phytochemicals. There was a significant (P<0.05) decrease in Feed:Gain ratio (F:G) with the control diet having the highest F:G compared to VDLM based diets.

The decreased values of F:G for cockerels fed VDLM based diet suggests better feed utilization as activated by the presence of tolerable levels of phytochemicals. According to Oleszek *et al.* (2001), constituents of phytochemical may contribute to the beneficial properties of VDLM in feed improvement.

Saponin and polyphenols are anti-inflammatory agents that reduce formation of cytokines and diversion of nutrients from growth to immune response. The saponin and phenols in VDLM could have shifted nutrient from the synthesis of cytokine to body mass production in the cockerels to effects growth performance and feed conversion efficiency.

Table 4 Effect of *Vitex doniana* (VDLM) on performance of cockerels

Parameters/VDLM inclusion (%)	0%
Initial weight (g)	1491
Final weight (g)	1814[b]
Weight gained (g)	323[c]
Average weight gained (g/bird/day)	6[c]
Average feed intake (g//bird/day)	122
Feed:gain	18[a]

The means within the same row with at least one common letter do not have si nificant difference

Table 5 Effect of *Vitex doniana* (VDLM) on haematological and serum biochemical parameters of cockerels

Parameters/VDLM inclusion (%)	0%	5%	10%	SEM
WBC ($\times10^9$/L)	149.20	147.10	153.80	4.09
GRAN* (cell/μL)	29.16[a]	37.26[b]	37.68[b]	1.91
RBC ($\times10^{12}$/L)	2.81	3.23	3.51	0.29
Hb (g/L)	173.80	167.20	180.60	12.82
MCV (f/L)	119.20	116.00	115.40	1.93
PCV (%)	38.60	36.40	43.00	7.75
Total protein (g/dL)	46.75[ab]	42.00[a]	47.75[b]	1.520
Albumin (g/dL)	24.50	22.00	25.25	1.510
ALT (IU/L)	11.50	9.75	10.00	3.290
AST (IU/L)	10.75	17.25	15.75	2.21
Potassium (mmol/L)	5.63[b]	5.00[a]	5.30[ab]	0.181

* GRAN: absolute content of granulocyte.
WBC: white blood cell; RBC: red blood cell; Hb: haemoglobin; MCV: mean corpuscular volume; PCV: packed cell volume; ALT: alanine aminotransferase and AST: aspartate aminotransferase.
The means within the same row with at least one common letter, do not have significant difference (P>0.05).
SEM: standard error of the means.

The influence of dietary treatment of VDLM based diet on the haematological and serum biochemical parameters of cockerels is shown in Table 5.

Dietary treatment did not significantly (P>0.05) alter the parameters observed except for the value of absolute content of granulocyte where there was significant (P<0.05) increase with increase in VDLM.

Lack of dietary effect of VDLM based diets and control on the values of haemoglobin (Hb) across the treatments is an indication that inclusion of VDLM up to 10% did not inhibit oxygen availabilty in the blood. Haemoglobin is known to function as oxygen and carbon dioxide carrier within the body of the animal and low level of hemoglobin indicates anaemic condition. The values of Hb observed in this study implies that the cockerels are not anaemic since the values are within the normal range (70-186 g/L) reported for chickens (Mitruka and Rawnsley, 1977).

The hemoglobin and RBC values obtained in this study indicated that the birds were healthy and not anemic. This is in line with Akan and Sodipo (2012) who reported that albino rats administered with aqueous root-back extract of *Vitex doniana* were not anaemic.

Packed cell volume (PCV) which is also an indicator of anaemia was similar in the experimental birds across the treatments. This indicates that oxygen circulation was not inhibited in birds fed 5% and 10% VDLM. These values are within the range(38.4%-40.67%) for healthy chickens reported by Sebastian *et al.* (2013).

The values Mean Corpuscular Volume (MCV) observed were also in agreement with the recommended range of 90.00-140.00 fL (Bounous and Stedman, 2000).

White blood cells (WBC) are cells in the blood concerned with the recognition and subsequent removal or deactivation of foreign bodies and dead cells in the body.

The values of WBC in experimental birds remained within the normal range as stated by Bounous and Stedman, (2000). This implies that VDLM is not toxic to cockerels. This increase in absolute content of granulocyte could be as a result of possible stimulation of immune system (Kashinath, 1990).

Lack of dietary effect on the values of platelets with increase in inclusion level of VDLM is an indication that the plant did not affect the blood clotting ability of the birds. There was no significant difference (P>0.05) in serum biochemical parameters examined except for total protein and potassium. The change in serum protein could have been caused by a change in globulin. This observation signifies balanced nutritional status of birds fed VDLM based diets. Blood serum examination plays an important role in the physiology, nutritional and pathological status of an animal (Aderemi, 2004). The ALT and AST values of birds fed 5% and 10% VDLM inclusion levels are comparable to that of the control. These enzymes usually respond to the presence of toxic substances in diet (Iyayi, 1994). The observed values indicate the ability of the birds to tolerate the phytochemicals in VDLM. This suggests that the liver and bone of the birds are not affected. The mean values for most of heamatological and serum biochemical indices are within the normal range for chicken.

CONCLUSION

The result of this study established that *Vitex doniana* leaf meal (VDLM) produced improvement in growth, feed:gain and other parameters compared with the control. Based on these findings, it is apparent that VDLM is an available novel plant ingredient with potentials that can be utilized in cockerels' diet without negative effect on haematological and serum biochemical parameters.

ACKNOWLEDGEMENT

The authors are acknowledges the support and assistance of Mr Famolu and Mr Adebiyi of the Teaching and Research Farm, University of Ilorin especially during the feeding trials.

REFERENCES

Adejumo A.A., Alaye S.A., Ajagbe R.O., Abi E.A. and Adedokun F.T. (2013). Nutritional and anti-nutritional composition of black plum (*Vitex doniana*). *J. Nutr. Sci.* **13**, 144-148.

Aderemi F.A. (2004). Effects of replacement of wheat bran with cassava root sieviate supplemented or unsupplemented with enzyme on the haematology and serum biochemistry of pullet chicks. *Trop. J. Anim. Sci.* **7**, 147-153.

Adeyina A.O., Akanbi A.S., Sanusi S.B., Olaniyi B.T., Adegoke A.G., Hassan K.T., Olaoye T.S., Salako A.O. and Adeyina O.A. (2014). Reproductive response to inclusion of graded levels of *Ipomea purpurea* leaf meal (*Morning glory*) in diets of laying chickens. *J. Agric. Sci. Environ.* **14**, 36-41.

Aduku A.O. (1993). Tropical Feedstuff Analysis Table. Ahamadu Bello University, Zaria, Nigeria.

Akan J.C. and Sodipo O.A. (2012). Effect of aqueous root-bark extract of *Vitex doniana* sweet on haematological parameters in rats. *Int. J. Chem.* **1**, 13-20.

AOAC. (2000). Official Methods of Analysis. Vol. I. 18th Ed. Association of Official Analytical Chemists, Arlington, VA, USA.

Atawodi S.E. (2005). Comparative *in vitro* trypanocidal activities of petroleum ether, chloroform, methanol and aqueous extracts of some Nigerian savannah plants. *African J. Biotechnol.* **4**, 177-182.

Bounous D.I. and Stedman N.L. (2000). Normal avian hematology: chicken and turkey. Pp. 1147-1154 in Schalm'S Veterinary Hematology. B.F. Feldman, J.G. Zinkl and N.C. Jain, Eds. Williams And Wilkisn, Lippincott, Philadelphia.

Cheeke P.R. and Skull L.R. (1985). Natural Toxicant in Feed and Poisonous Plants. AUI Publishing Company, Westport, USA.

Duncan D.B. (1955). Multiple range and multiple F-tests. *Biometrics.* **11**, 1-42.

Ekenyem B.U., Iheukwumere F.C., Iwuji T.C., Akanmu N. and Nwugo O.H. (2003). Evaluation of *Microdermis puberula* leaf meal as feed ingredients in broiler chicks production. *Pakistan J. Nutr.* **5(1)**, 46-50.

Iyayi E.A. (1994). Supplemented effects of low and high cyanide cassava on performance, nutrient digestibility and serum metabolites of growing pigs. *J. Agric. Trop. Subtrop.* **95**, 199-205.

Kashinath R.T. (1990). Hypolipimedic effect of disulphide in rat fed with high lipid and / or ethanol. Ph D. Thesis. University of Bangalore, Bangalore, India.

Mitruka B.M. and Rawnsley H.M. (1977). Clinical, Biochemical and Haematological Reference Values in Normal Experimental Animal and Normal Humans. Masson Publishing, USA.

Nnamani C.V, Oselebe H.O. and Okporie E.O. (2007). Ethnobotany of indigenous leafy vegitables of Izzi clan in Ebonyi State, Nigeria. Pp. 111-114 in Proc. 20th Ann. Natal Conf. Biotechnol. Soc. Abakaliki, Nigeria.

Oleszek W., Sitek M., Stochmal A., Placente S., Pizza C. and Cheeke P. (2001). Relationship between saponin content in alfalfa and other browse folders. *J. Agric. Food Chem.* **49**, 747-752.

Olusola L., Zebulon S.C. and Okoye F.U. (1997). Effects of *Vitex doniana* stem bark on blood pressure. *Nigerian J. Nat. Prod. Med.* **1**, 19-20.

Pauzenga U. (1985). Feeding parent stock. *Zootec. Int.* **11**, 22-24.

SAS Institute. (1996). SAS®/STAT Software, Release 6.11. SAS Institute, Inc., Cary, NC. USA.

Sebastian K., Detro-Dassen S., Rinis N., Fahrenkamp D., Muller-Newen G., Merk H.F., Schmalzing G., Zwadlo-Klarwasser G. and Baron J.M. (2013). Characterization of SLCO5A1/OATP5A1 a solute carrier transport protein with non classical function. *PLoS One.* **8(12)**, e83257.

Sofowora A.E. (1993). Medicinal plants and traditional medicine in Africa. Spectrum Books Ltd., Nigeria.

Trease G.E. and Evans W.C. (1989). Trease and Evans Pharmacology. Bailliere Press, London, United Kingdom.

Ugwu Okechukwu P.C., Nwodo Okwesili F.C., Joshua, Parker E., Bawa Abubakar, Ossai Emmanuel C. and Odo Christain E. (2013). Pyto chemical and anti toxicity study of Moringa Olifera Ethanol leaf extract. *Int. J. Life Sci. Biotechnol. Pherma Res.* **2(2)**, 66-70.

Young D.S. (2001). Effect of Diseases on clinical laboratory test. American Association of Clinical Chemistry Press, Washington, D.C., USA.

23

Investigation on Biochemically Processed Castor Seed Meal in Nutrition and Physiology of Japanese Quails

A.A. Annongu[1], J.O. Atteh[1], J.K. Joseph[2], M.A. Belewu[1], A.O. Adeyina[1*],
A.S. Akanbi[1], A.T. Yusuff[1], F.E. Sola-Ojo[1], S.O. Ajide[3], V.O. Chimezie[3] and
J.H. Edoh[1,4]

[1] Department of Animal Production, Division of Nutritional Biochemistry and Toxicology, P.M.B. 1515 University of Ilorin,
   Ilorin, Nigeria
[2] Department of Home Economics and Food Science, University of Ilorin, Ilorin, Nigeria
[3] Landmark University, Km 4 Ipetu, Omu Aran, PMB 1001, Omu Aran, Kwara State of Nigeria
[4] Laboratoire de Recherche Avicole et de Zoo-Economie (LaRAZE), Département de Production Animale, Faculté des
   Science Agronomiques, Université d'Abomey-Calavi, 01 BP 526 Cotonou, Benin

*Correspondence E-mail: aadeyina@unilorin.edu.ng

ABSTRACT

Native de-oiled and treated castor seed meal was subjected to proximate analysis and quantification of anti-nutrients (phytochemicals). Seed cake was treated by biochemical technique of solid state fermentation with *Aspergillus niger* and addition of calcium oxide (CaO) to give treated castor seed meal (TCSM). One hundred and twenty Japanese quails (*Coturnix coturnix japonica*) were fed four (4) iso-nitrogemous and iso-caloric diets containing 0, 2.5, 5.0 and 7.5% TCSM corresponding to the diet 1, 2, 3 and 4, respectively. While the feeding trial lasted for 56 days, feed and water were supplied *ad libitum*. Data on proximate composition showed that raw seed, defatted residue (cake) and the processed castor seed meal contained valuable nutrients like dry matter, crude protein, fat, fibre, mineral matter and soluble carbohydrate (NFE). Phytochemical quantification gave high levels of the anti-nutrients such as ricin, allergens, ricinine in the raw seed. However, levels of these phytochemicals were reduced by defatting and treatments of the cake meal by solid state fermentation, *A. niger* and CaO. Performance traits indicated decreases in feed intake, weight gain, growth and increases in mortality rates especially on the diet with the highest (7.5%) inclusion of TCSM compared to the control diet (P<0.05). Nutrients retentions on the test feedstuff were not comparable with values on the reference diet on soluble carbohydrate values which decreased with increasing CSM (P<0.05) relative to the control diet. In haematological parameters packed cell volume (PCV) and mean corpuscular hemoglobin (MCH) values on diets with TCSM were exceptionally high relative to the control diet (P<0.05). However, biochemical indices (serum protein, albumin, globulin, albumin:globulin ratio and alkaline phosphatase (ALP) activities were not influenced by dietary CSM (P>0.05). Enzyme activity of aspartate aminotransferase (AST) showed decreasing trend with increasing level of CSM in diets (P<0.05). Profiling electrolytes in the fed quails showed significant variations in concentrations of Ca^{++} and HCO_3^- on TCSM based diets (P<0.05) comparable with the control diet values. Conclusively, despite treating CSM by solid state fermentation with *A. niger* and CaO addition in this trial, TCSM addition still appears to induce toxic and deleterious effects on the quails. Subsequent works to enable inclusions at acceptable and higher levels after treatments are on-going.

KEY WORDS   blood-composition, castor seed meal, Japanese quail, performance, solid-state-fermentation.

INTRODUCTION

Meeting the demand for animal protein requirement in de veloping countries of the world is becoming very difficult due to daily increment in human population. For instance, in Nigeria, an average person consumes in diet only eight grams of animal protein per day which is less than the minimum requirement of 65g per adult per day recommended by the National Research Council (FAO, 2014). To overcome the problem, animal producing industries are making efforts by researches to increase population of livestock especially in the poultry sector with short generation interval in production. In addition, poultry are highly prolific, good feed converter (Smith, 2001) and the production involves the least hazardous and arduous process relative to other farmed animals (Oluyemi and Roberts, 2007). However, achieving maximum performance and cost return benefits in poultry production depend on quality and quantity of feeds. The quantity of feed requirement for intensive livestock production accounts for more than 75% of the total cost of production especially the production of monogastric animals like poultry that share the same staple foodstuffs with humans. To reduce production cost, increase novel micro livestock and animal protein in diets of the common Nigerian, alternative sourcing to the scare conventional foodstuff necessitated research on Japanese quails using castor bean residue or cake. Japanese quail (*Corturnix corturnix japonica*) is a small-sized bird with early maturing and prolific rate in addition to its unique traits and merits such as early attainment of sexual maturity, short generation interval, attainment of market weight of 150-180 g within 5-6 weeks of age and a high rate of egg production between 180-250 eggs per year (Shwarts and Allen, 1981; Garwood and Diehl, 1987). Furthermore, quails require less floor space, 8 to 10 adult quails can be housed in a space meant for just one adult chicken (Haruna *et al.* 1997). Quails require 20-25 g less feed per day, their meat and egg are high in protein quality with low cholesterol content and their meat is tender, tasty and highly acceptable making the meat a choice protein for high blood pressure persons as well as their eggs (Haruna *et al.* 1997). Japanese quails also can be produced with small capital and short day length period.

On the other hand, inadequacy and escalating cast of traditional feeding stuffs in Nigeria are what prompted research on alternative and use of castor seed residue after oil extraction.

The defatted seed cake has high protein content and other nutrients warranting it a potential rich protein source for food animals. Despite the relative abundance of *Ricinus communis* and its high nutrients content, use of the residue in diet for domestic animals is hindered by toxic phytochemicals namely ricin (1.5 w/w defatted), the most lethal of toxins (Audi *et al.* 2005). The other phytochemicals are ricinine-an agglutinin, allergen CB-IA, cyanide, phytic acid, and lipase. Attempts to detoxify castor residues for use in animal nutrition are usually aimed at removing of ricin, ricinine and allergens in the raw beans (Audi *et al.*

2005; Ani and Okorie, 2005). This investigation therefore attempted the use of solid state fermentation with *Aspergillus niger* as a bio-degrader in addition of calcium oxide to potentiate detoxification of castor seed meal in nutrition of *Corturnix corturnix japonica*.

MATERIALS AND METHODS

Castor seeds obtained from castor plants grown on Ilorin soil, Ilorin, Kwara State of Nigeria, were de-hulled using a de-hulling machine and ground in a hammer mill into paste. The paste was de-fatted by mechanical hydraulic oil press and residue (cake) subjected to solid state fermentation with the spores of *Aspergillus niger*. *Aspergillus niger* spores were generated by suspension in yeast extract broth at a concentration of 9×10^6 spores/mL (Pandey and Larroche, 2008). After the inoculation, 20 kg of the cake meal mixture was placed in a 100-litre bowl and covered with a muslin cloth and left to incubate for 7-days at room temperature (27 °C). The one week period was for the *A. niger* to effect degradation of the castor bean cake anti-nutritional factors. Incubation was terminated on the seventh day by ovendrying the solid state fermented stuff at 100 °C. 40g/kg CaO was added to the fermented cake meal for use in the diet mixtures. Removal or detoxification of castor bean antinutrients is easier when carried out in an alkaline environment (Pandey and Larroche, 2008).

One hundred and twenty Japanese quail chicks averaging 7.40g initial weight/chick at day old hatched from eggs incubated in the Departmental hatcher (incubator) were used for the experiment. The day old birds were first fed a commercial broiler starter mash for one week for adaptation. After the 7-days acclimatization, they were weighed, wing branded and randomly allotted to the four dietary treatments in triplicate lots of 10 chicks per replicate in a single factor design experiment. The diets contained 0.00, 2.50, 5.00 and 7.50% of the TCSM at the expense of the conventional protein source, soybeans, corresponding to diets 1, 2, 3 and 4, respectively. The control diet contained maize and soybean meals as basic ingredients. Birds were fed *ad libitum* during a feeding trial that lasted 56 days (8 weeks). Use of the quail in this trial followed the Ethical Protocols of the Committee of Animals Use of the University of Ilorin and that of Global Standard Practices. The composition of the experimental diets on as fed basis and its analysed nutrient contents of the diets is presented in Table 2.

Data collection / response criteria
Data on the nutritional composition of native milled seed, defatted and treated castor seed meals (TCSM) collected were subjected to descriptive statistics.

Table 1 Nutritional and phytochemical composition of milled raw seed, defatted and treated seed residue, cake

Parameters (%)	Raw seed	Defatted CSM	Treated CSM
Dry matter	94.70	99.24	89.14
Crude protein	36.40	35.66	35.09
Ether extract	51.19	36.87	22.09
Total Ash	4.28	7.04	11.67
Crude fibre	0.89	5.67	3.54
Soluble carbohydrate	8.44	14.76	47.61
Gross energy (kcal/g)	7.26	6.36	6.24
Phytochemicals (%)			
(Anti-nutrients)	Raw	Defatted cake	Treated
Ricin	0.10	0.070	0.065
Allergenic protein	0.30	0.240	0.230
Ricinine	0.10	0.048	0.046

CSM: castor seed meal.

Table 2 Composition of the experimental diets with analyzed nutrient contents (kg/100 kg diet)

Diets (%)	1	2	3	4
Treated castor seed meal	0.00	2.50	5.00	7.50
Maize	52.60	48.60	48.00	47.60
Soybean meal	25.00	22.50	21.00	18.00
Fish meal	4.00	5.00	6.00	6.50
Wheat offal	15.00	18.00	17.00	17.00
Dicalcium phosphate	2.00	2.00	2.00	2.00
Oyster shell	0.80	0.80	0.40	0.80
Salt	0.25	0.25	0.25	0.25
Vitamin and mineral premix[1]	0.20	0.20	0.20	0.20
Methionine	0.15	0.15	0.15	0.15
Total (%)	100	100	100	100
Analysed diet nutrient content (%)				
Dry matter	90.33	89.31	90.93	89.50
Crude protein	20.17	21.94	22.94	21.25
Ether extract	4.73	4.97	5.23	3.12
Ash	5.55	12.94	12.97	13.10
Crude fibre	4.29	6.81	10.50	11.61
Nitrogen free extract (NFE)	65.26	53.34	48.36	50.92
Gross energy (GE) (kcal/g)	4.53	4.26	4.25	4.15

[1] Premix supplied: vitamin A: 200000 IU; vitamin D_3: 400000 IU; vitamin E: 8.00 g; vitamin K: 0.40 g; vitamin B_{12}: 0.32 g; vitamin B_2: 0.96 g; vitamin B_6: 0.56 g; vitamin C: 2400 mg; Folic acid: 0.16; Biotin: 8.00 mg; Choline: 48.00 g; Calcium panthothiolate: 1.6 g; Manganese: 16.00 mg; Fe: 8.00 mg; Zinc: 7.20 g; Copper: 0.32 mg; Iodine: 0.25 mg; Cobalt: 36.00 mg; Selenium: 16.00 mg and butylated hydroxy toluene (BHT): 32.00 g.

In the course of the feeding trial, data were collected on daily feed consumption, weekly body weight gain while growth was calculated. Mortality rates (number of dead birds per replicate/diet) and efficiency of feed utilization (F/G) were also computed. A week before termination of the experiment, nutrients digestibility trials were conducted for a 72 h period and fecal samples collected from each replicate of a treatment containing feces as well as urinary paste were properly air-dried and analyzed for the determination of apparent nutrients retention. Percent of nutrients retention was calculated using the formula (Van Soest, 1982):

((Nutrient in feed-Nutrient in faeces)/(Nutrient in feed)) × 100

Data for haematological and serum biochemistry
Blood collection made from the external jugular vein of the quail for the determination of haematological and serum

biochemistry was conducted in the morning of the following day (between 8-9 a.m.). Blood samples for haematological indices were collected in heparinized (EDTA) sample bottles while samples for biochemical parameters were taken in bottles without the anticoagulant, allowed to clot at room temperature before centrifuging at 3000 rpm to obtain clear sera. The sera were stored at -20 °C for the analysis of biochemical determinants.

Chemical analyses
Proximate and quantification of CSM anti-nutrients in the raw, defatted and treated cake meal, and proximate composition of diets and fecal samples were carried out by the methods of AOAC (2000). Haematological parameters were determined with the auto-haemocytometer while mean corpuscular haemoglobin (MCH), mean corpuscular volume (MCV) and mean corpuscular haemoglobin concentration (MCHC) values were calculated using the formula descried by Schalm et al. (1975).

Table 3 Dietary effect of treated CSM on performance of quail

Diets	1	2	3	4	SEM
Treated castor seed meal inclusion (%)	0.00	2.50	5.00	7.50	
Feed intake (g/bird/day)	16.74^c	11.23^b	9.38^a	7.10^a	0.29^*
Weight gain (g/bird/day)	18.39^c	12.16^{ab}	9.95^a	8.94^a	0.31^*
Growth rate	84.66^c	77.27^b	74.35^b	68.10^a	0.82^*
Feed efficiency (F/G)	0.91	0.92	0.94	0.78	0.01^{ns}
Mortality (birds/diet)	0	5	7	10	

The means within the same row with at least one common letter, do not have significant difference (P>0.05).
* (P<0.05).
SEM: standard error of the means.
NS: non significant.

Table 4 Nutrients retention in quail fed processed dietary castor seed meal (CSM)

Diets	1	2	3	4	SEM
Treated castor seed meal inclusion (%)	0.00	2.50	5.00	7.50	
Dry matter intake	67.82	73.32	69.00	71.21	0.10^{ns}
Crude protein	48.10	60.30	55.41	62.17	0.69^{ns}
Crude fat	78.59	89.35	89.78	82.02	0.51^{ns}
Crude fibre	71.33	74.47	94.10	72.68	1.98^{ns}
Soluble carbohydrate	279.60^a	163.73^b	128.03^b	98.87^b	0.002^*

The means within the same row with at least one common letter, do not have significant difference (P>0.05).
* (P<0.05).
SEM: standard error of the means.
NS: non significant.

Analysis of the biochemical indices was conducted using the clinical chemistry semi-auto-analyzer and a commercial biochemical assay kit. Enzyme activities of aspartate amino transferase (AST), alanine aminotransferase (ALT) and alkaline phosphatase (ALP) were analyzed by the spectrophotometric linked reaction methods (Reichling and Kaplan, 1988). Serum electrolytes were determined by the method of Young (2001).

Statistical analysis

Data on proximate composition and quantification of CSM anti-nutrients in the Raw, de-oiled and treated cake meal was subjected to descriptive statistics while data on performance, haematological and serum biochemistry were analysed by analysis of variance (ANOVA) following the design of a one-way classification. Significant differences between treatments means were separated by Duncan multiple range test as described by Steel and Torrie (1980).

RESULTS AND DISCUSSION

Result on proximate analysis of castor seed (Table 1) shows that the seed contains valuable nutrients such as dry matter, crude protein, crude fat, crude fibre, soluble carbohydrate (NFE). These nutrients could be made available to the fed animals for utilization when the anti-nutrients present in the unprocessed seed are removed or reduced to minimum threshold. Treating the milled raw seed by solid state fermentation with *Aspergillus niger* microbes as bio-degrader and addition of calcium oxide (CaO) to potentiate detoxification of the castor seed cake meal (Table 1) resulted in reductions of the castor phytotoxins, ricin, ricinine and the

allergens phytochemicals determined to give low levels of 0.065, 0.046 and 0.23, respectively, relative to the values on the native or raw milled seed. Detoxification of castor seed anti-nutrients is better made in alkaline medium. Dietary effects of treated CSM on performance of quail are shown in Table 3. Castor seed meal treatments reduced feed consumption, body weight gain, growth and increased casualties (mortality rate) in the course of the four weeks feeding experiment. Decreases in feed intake, weight gain, growth rate and survival rate responded correspondingly as the inclusion level of the seed meal increased in diets (P<0.05), suggesting, but not proving the residual toxic effects of the castor seed meal subjected to fermentation/CaO treatment since processing of the castor seed was done only in one batch. Corrections for birds being removed (by death) within the interval of the feeding trial were made by calculations. Retention of nutrients in quails (Table 4) however, was not influenced by castor seed meal inclusions in diets hence results on the test feedstuff diets and that of the control diet was comparable (P>0.05). An exception to this finding was retention of soluble carbohydrate (NFE) which decreased concomitantly with increasing dietary levels of castor seed meal in comparison with the reference diet (P<0.05). Data on haematological parameters (Table 5) indicated that treating CSM by the methods adopted caused variations in values obtained on haematopoiesis. Most of the indices on blood composition (red blood cell (RBC), hemoglobin (Hb), platelet, white blood cell (WBC), with its differential counts as well as MCV, MCHC) appeared comparable with those on the control diet (P>0.05) but PCV and MCH values presented increases with increasing dietary levels of TCSM relative to the reference diet (P<0.05).

Table 5 Influence of processed castor seed meal (CSM) on blood composition in quail

Diets	1	2	3	4	SEM
Treated castor seed meal inclusion (%)	0.00	2.50	5.00	7.50	
Red blood cell (RBC) ($\times 10^{12}$/L)	2.00	2.55	2.53	2.10	0.14^{ns}
Hemoglobin (Hb) (g/dL)	6.93	7.70	10.70	9.90	5.58^{ns}
Packed cell volume (PCV) (%)	0.73^a	1.00^b	1.10^b	1.10^b	0.34*
Mean corpuscular volume (MCV) (Pg)	131.67	161.67	177.67	154.67	1.00^{ns}
Mean corpuscular haemoglobin (MCH) (Pg)	33.77^a	44.00^b	42.63^b	43.70^b	3.75*
Mean corpuscular haemoglobin concentration (MCHC) (%)	26.27	27.63	24.20	28.00	5.19^{ns}
Platelet ($\times 10^9$/L)	353.67	447.00	308.33	222.67	0.70^{ns}
White blood cell (WBC) ($\times 10^9$/L)	10.76	9.74	9.10	7.787.78	2.69^{ns}
White blood cell (WBC) differential counts					
Monocytes (%)	1.00	0.67	2.00	2.00	0.80^{ns}
Eosinophils (%)	0.33	0.00	0.33	0.00	0.37^{ns}

The means within the same row with at least one common letter, do not have significant difference (P>0.05).
* (P<0.05).
SEM: standard error of the means.
NS: non significant.

Table 6 Biochemical Indices in quail fed dietary treated castor seed meal (CSM)

Diets	1	2	3	4	SEM
Treated castor seed meal inclusion (%)	0.00	2.50	5.00	7.50	
Blood glucose (mmol/L)	9.70^a	9.50^a	12.96^b	19.00^c	3.64*
Serum protein (g/L)	62.33	48.33	59.00	56.67	0.62^{ns}
Serum albumin (g/L)	29.33	30.33	27.33	30.67	3.65^{ns}
Serum globulin (g/L)	16.00	15.00	23.00	16.00	0.24^{ns}
Albumin/globulin ratio	1.98	2.11	1.37	1.97	0.18^{ns}
Aspartate amino transferase (AST) (IU/L)	180^d	175^c	94^b	79.66^a	4.82*
Alanine aminotransferase (ALT) (IU/L)	89.37^b	102.26^c	109.10^d	65.56^a	6.34*
Alkaline phosphatase (ALP) (IU/L)	1424	1716	1749	2271	4.27^{ns}

The means within the same row with at least one common letter, do not have significant difference (P>0.05).
* (P<0.05).
SEM: standard error of the means.
NS: non significant.

Table 7 Dietary impact on blood electrolytes in quail given processed castor seed meal (CSM)

Diets	1	2	3	4	SEM
Treated castor seed meal inclusion (%)	0.00	2.50	5.00	7.50	
Ca^{2+} (mmol/L)	2.70^c	2.00^c	1.54^b	0.99^a	0.92*
Na^+ (mmol/L)	139	132	141	134	1.12^{ns}
K^+ (mmol/L)	5.63^b	6.33^b	4.17^a	3.90^a	2.37*
Cl- (mmol/L)	53.00	71.00	72.00	45.00	3.02^{ns}
Mg^{2+} (mmol/L)	0.73	1.00	1.07	1.10	0.15^{ns}
HCO_3^- (mmol/L)	27.67^c	26.33^c	21.67^b	17.67^a	0.35*

The means within the same row with at least one common letter, do not have significant difference (P>0.05).
* (P<0.05).
SEM: standard error of the means.
NS: non significant.

Deviation in these values from the control value may be inimical to the wellbeing of the birds.

Biochemical determinants in quails (Table 6) receiving dietary treated CSM indicated increment in blood glucose level (P<0.05) compared to the conventional diet and values on serum total protein, albumin, globulin, albumin/globulin ration were not different from those on the control diet (P>0.05). Feeding the treated seed meal caused decreases in the activities of the transaminases (AST and ALT) especially at the highest inclusion of 7.50% CSM (P<0.05). Reduction in activity of the enzymes may reduce the functions they are supposed to perform in the animal body.

Profiling electrolytes in the quails given dietary treated CSM (Table 7) revealed that treatment resulted in changes in mineral constituent in the birds. Treatment failed to improve availability and utilization of calcium, potassium and bicarbonate electrolytes (Ca^{2+}, K^+ and HCO_3^-) as values on these electrolytes decreased in response to increasing level of CSM in diets (P<0.05).

CONCLUSION

This experiment established that in spite of treatment by solid state fermentation with addition of CaO, residual toxic effect of castor seed meal prevailed to elicit adverse effects on the fed animals. Further works to detoxify and determine acceptable levels of inclusion of the castor seed in diets of food animals are in progress.

ACKNOWLEDGEMENT

The authors are earnestly grateful to the Department of Animal production, Faculty of Agriculture, University of Ilorin, Nigeria for providing equipment and appliances needed for hatching the day old quails and their brooding.

REFERENCES

Ani A.O. and Okorie A.U. (2005). The effect of graded levels of dehulled and cooked Castor oil bean (*Ricinus communis*) meal on performance of broiler starters. *Nigerian J. Anim. Prod.* **32(1),** 54-60.

AOAC. (2000). Official Methods of Analysis. Vol. I. 18th Ed. Association of Official Analytical Chemists, Arlington, VA, USA.

Audi J., Belson M., Patel M., Schier J. and Osterloh J. (2005). Ricin poisoning : a comprehensive review. *J. Am. Med. Assoc.* **294(18),** 2342-2351.

FAO. (2014). Food and Agriculture Organization of the United Nations (FAO), Rome, Italy.

Garwood A.A. and Diehl R.C.J. (1987). Body volume and density of live *Coturnix* quail and associated genetic relationships. *Poult. Sci.* **66 (8),** 1269-1272

Haruna E.S., Musa U., Okewole P.A., Shemaki D., lombin L.H., Molokwu J.U., edache J.A. and Kaarsin P.D. (1997). Protein requirement of quail chicks in plateau state, Nigeria. *Nigerian Vet. J.* **18,** 108-113.

Oluyemi A.J. and Roberts F.A. (2007). Poultry Production in Warm West Climates. Spectrum Books Ltd., Ibadan. Nigeria.

Pandey A. and Larroche C. (2008). Current Development in Solid State Fermentation. Published by Springer, Germany.

Reichling J.J. and Kaplan M.M. (1988). Clinical use of serum enzymes in liver diseases. *Digest. Dis. Sci.* **33,** 1601-1614.

Schalm O.W., Jain N.C. and Carroll E.J. (1975). Haematology. Publisher: Lea and Febiger, Philadelphia, US.

Smith A.J. (2001). Poultry. Macmillan Publishers Ltd., London, United Kingdom.

Steel R.G.D. and Torrie J.H. (1980). Principles and Procedures of statistics, a Biometrical Approach. Publisher: McGraw-Hill Companies, New York.

Van Soest P.J. (1982). Nutritional Ecology of the Ruminant. Cornell University Press, Ithaca, New York.

Young D.S. (2001). Effect of Diseases on clinical Laboratory Tests. American Association for Clinical Chemistry Press, Washington, D.C.

24

Evaluation of Environmental Impacts in Turkey Production System in Iran

K. Kheiralipour[1*], Z. Payandeh[1] and B. Khoshnevisan[2]

[1] Department of Mechanical Engineering Biosystems, Ilam University, Ilam, Iran
[2] Department of Agricultural Machinery, University of Tehran, Karaj, Iran

*Correspondence E-mail: k.kheiralipour@ilam.ac.ir

ABSTRACT

Poultry industry is an important production system due to providing remarkable portion of the human food and protein needs. Considering the necessity of environmental protection, the amount of environmental impacts of a turkey production unit in Iran was determined using life cycle assessment method. The required information were collected through questionnaires and interviews with farm owners. In this research, the system boundary was poultry farm gate and functional unit was considered as onetonne of turkey meat. The amount of indicators including abiotic depletion, abiotic depletion (fossil fuels), global warming, ozone layer depletion, human toxicity, fresh water aquatic ecotoxicity, marine aquatic ecotoxicity, terrestrial ecotoxicity, photochemical oxidation, acidification potential and eutrophication potential were found to be 1.61 kg Sb eq, 20.19 MJ, 3.63 kg CO_2 eq, 1.90 kg CFC -11 eq, 67.60 kg 1.4 DB eq, 4.55 kg 1.4 DB eq, 1.04 kg 1.4 DB eq, 1.17 kg 1.4 DB eq, 0.0005 kg C_2H_4 eq, 0.024 kg SO_2 eq and 0.0094 kg PO4 eq, respectively. The results showed that the feed input has the highest emissions in comparison with other inputs.

KEY WORDS emission, environment, life cycle assessment, turkey production.

INTRODUCTION

Awareness about environment protection and demand for environmentally friendly products in recent decades has led agricultural researchers to pay more attention to clean production (Khoshnevisan et al. 2015). Now, the environment is one of the main elements in the global macro policies. For this reason, the most important factor and prerequisite for many activities at the macro level are compatibility with environment (OECD, 1999). In this regard, some appropriate indexes have been introduced in order to assess the sustainability of agricultural production methods from the point of environmental aspects (Brentrup et al. 2004). In the recent decade, life cycle assessment (LCA) is one of the main tools for assessing environmental impacts. In fact, LCA is an environmental management tool via environmental performance evaluation (Guinée, 2002). In addition, the method is suitable for comparison of different agricultural production or processing systems (Bojacá et al. 2014; Khoshnevisan et al. 2014). Livestock and poultry industries have high importance in terms of providing human required food and protein. Due to the appropriate growing characteristics such as weight gain and high growth rate, low feed conversion ratio and high nutritional value, industrial production of turkey is expanding around the world (Anonymous, 2008). In 2007, the United States, Europe, Brazil and Canada were ranked from first to fourth turkey meat production contries in the world, respedtively. In breeding poultry in Iran, turkey production has the highest economic aspect after chicken production.

Industrial turkey production in Iran was began since 1976. Turkey meat production in 1996 was 25 tonne and in 1997 was increased to 350 tonne per year. In 2014, it was increased to 1700 tonnes per year (Anonymous, 2012; Anonymous, 2013).

Some studies have been conducted in the context of the application of LCA method in agriculture for crop and food production (Rebolledo-Leiva et al. 2017; Benis and Ferrão, 2017; Llorach-Massana et al. 2017). In livestock and animal production, LCA methodwas used to study the milk production (Cederberg and Mattsson, 2000; Thomassen et al. 2008), pig production (Basset-Mens and Vanderwrf, 2005; Nielsen et al. 2013) and egg production (Sefeedpari et al. 2012; Leinonen et al. 2014). Some researches were reported in different countries using LCA method in chicken production, such as: Pelletier et al. (2008) in the US, Nielsen et al. (2011) in Denmark, Bengtsson and Seddon (2013) in Australia, Ewemoje et al. (2013) in Niger, Da Silva et al. (2013) in Brazil and French and Gonzalez-Garcia et al. (2014) in Portogate. However, there wasn't reported a study about application of life cycle assessment method in turkey production system.

According to importance of preserving environment natural resources in livestock production, the aim of this study was to investigate environmental impacts in turkey production in terms of resource use and environmental impact loads using LCA method.

MATERIALS AND METHODS

Data collection
This research was conducted in 2015. The required data was collected from a turkey production unit with capacity of 16000 chicks and 120 day production period in Najafabad Township, Isfahan, Iran. The township is 26 km far from Isfahan city at 32 degrees and 38 minutes north latitude along the equator and 51 degrees and 21 minutes east of Greenwich meridian. Najafabad is one of the main poultry production townships in Iran.

Life cycle assessment
International Organization for Standardization (ISO) has introduced the life cycle assessment (LCA) as a method for collection and evaluation of inputs, process, outputs and potential environmental impacts of a system over its life cycle. According to ISO, each LCA project includes four stages: 1) goal and scope definition, 2) inventory analysis, 3) impact assessment and 4) interpretation (ISO, 2006a). In this study, LCA method was used to analyze the environmental impacts of turkey. The LCA was carried out based on ISO standard (ISO, 2006b).

Emissions related to inputs and outputs of the turkey production unit were considered to be determined. The emissions were those emitted into water, soil and air be determined were. In this study, inputs were feed, fuel and electricity and outputs were chicken meat and manure. To determine the environmental impacts of the turkey production system, SimaPro Software was used. The collected data were interred to the software and analyzed by CML-IA baseline V3.01 / EU25 model. LCA method was conducted in four stages as follows. The first stage in LCA study is defining the purpose and scope. Defining the goal and scope should be clear and in compatible with the purpose of the study. This stage describes the studied product, goal and scope through its boundaries.

Selecting functional unit in this stage is an important phase. Functional unit is a reference unit that connects the input and output of a system (Sonesson et al. 2010). In this study, the functional unit was considered as one tonne of turkey meat in the production unit, i.e. all emissions due to consumed inputs for production of one tonne turkey meat was calculated. Also, the boundary of the studied system was gate of the turkey production unit.

Inventory analysis stage is most laborious and sensitive stage of the LCA study and should be conducted very carefully because further stages are highly dependent on the results of this stage (Leap, 2014). Inventory is actually a set of data that includes the creation of methods to calculate the inputs and outputs resources or materials in the process. Inventory data should be in compatible with the functional unit that it was created in the previous stage. At this stage, basically a collection of data is gathered to be obtained a quantitative analysis of the environmental impacts. To achieve this goal in the presented research, the calculations were done based on the data related to electricity, fuel and feed consumption and producing the manure which the information was provided by the farm owners.

Electricity was used for providing water, lighting, ventilation and powering the feeding system in the production unit. Electricity is supplied from the Isfahan power plant. The environmental impacts of electricity supply depend on used fuel in the power plant to generate electricity. In the power plant 99% of the electricity supply was generated from natural gas and 1% from rresidual fuel (mazut).

The studied turkey production unit consumes diesel fuel and natural gas for heating and transportation. Therefore, the environmental impacts of diesel fuel and natural gas were calculated in this study.

To calculate the environmental impacts of feed materials, amounts of consumed feeds by birds were calculated based on the information provided by the farm owners.

Turkey manure is used as a fertilizer for growing plants. Because of the low rate of nitrogen in turkey manure, it is an alternative to artificial fertilizers. Greenhouse gas emissions of livestock systems are due to enteric fermentation and manure and that of poultry is from manure only. Poultry manure produces direct and indirect nitrous oxide (N_2O) and methane emissions (Nielsen *et al.* 2011). Methane is emitted through manure storing. The nnitrous oxidedirectly is released from manure surface in unit floor whereas indirect emission of nitrous oxide is result of nitrogen leaching and evaporation (IPCC, 2006).

The environmental impacts of input and output materials in previous stage were determined in impact assessment stage. Impact categories should be established according to objective criteria. There are different methods; some methods state the impacts on human health. For example, Eco Indicator 99 is a method that focuses on global effects such as ecosystems (acidification, eutrophication, land use and toxicity), resources (minerals and fossil fuels) and health (carcinogenic, climate change and ozone layer). Another method evaluates the environmental strategies in terms of life expectancy, morbidity, potential growth of crops, meat or fish production potential, the growth potential of tree and so on.

This stage is automatically done by LCA software. In this study 11 environmental indicators were evaluated including: abiotic depletion, abiotic depletion (fossil fuels), global warming, ozone layer depletion, human toxicity, fresh water aquatic eco toxicity, marine aquatic ecotoxicity, terrestrial ecotoxicity, photochemical oxidation, acidification potential and eutrophication potential. Environmental impacts were assessed based on CML-IA baseline V3.02 / EU25 / Characterization.

In interpretation stage LCA results are scrutinized according to the goals of study, for example, analyzing the results and codificating some conclusions and recommendations in order to minimize the environmental impacts of the studied system (Weiler, 2013). In this section, the results were discussed in order to draw conclusions and provide solution.

RESULTS AND DISCUSSION

The amount of each input and output for the studied turkey production unit has been listed in Table 1. These values were considered to calculate the environmental impacts of the turkey production in the farm. According to Table 1, 457000 kg feed, 20000 l diesel fuel and 41322 kWh electricity was used in the studied turkey production unit. On the other hand, production of turkey meat and manure were 201600 and 180000 kg, respectively. Payandeh (2016) reported the average feed consumption, diesel fuel, gas and electricity to produce chicken in poultry production as 5104 kg, 602l, 1084 m³ and 1433 kWh, respectively.

Output of the poultry was chicken meat and manure with amount of 2400 and 1691 kg, respectively. Based on the above data, the ratio of meat production to feed consumption in turkey production was 0.44 kg meat/kg feed that was lower than that of chicken production (0.47 kg meat/kg feed). These results show that by consuming 1 kg feed, 0.44 kg turkey meat is produced but in poultry production 0.47 kg meet is obtained. Although the difference between the ratio of meat production to feed consumption in turkey and poultry units is low (0.03), but to produce more turkey meat form the consumed feed, the management level of input consumption in the turkey production unit must be improved.

The ratio of meat production to fuel and electricity consumption in turkey production unit were 10.08 kg meet/L diesel and 4.88 kg meet/kWh electricity, respectively, whereas these amounts in poultry production were 3.98 kg meet/L diesel, 2.21 kg meet/m³ gas and 1.66 kg meet/kWh electricity, respectively. These results show that the consumption of fuel and electricity in turkey production was lower than that of poultry production unit.

In this research, the amount of abiotic depletion, abiotic depletion (fossil fuels), global warming, ozone layer depletion, human toxicity, fresh water aquatic ecotoxicity, marine aquatic ecotoxicity, terrestrial ecotoxicity, photochemical oxidation, acidification potential and eutrophication potential indicators were calculated. The obtained values of the environmental indicators for producing one tonne of turkey meat have been listed in Table 2. As can be seen in Table 2, the amounts of above indicators were 1.61 kg Sbeq, 20.19 MJ, 3.63 kg CO2 eq, 1.90 kg CFC -11 eq, 67.60 kg 1.4 DB eq, 4.55 kg 1.4 DB eq, 1.04 kg 1.4 DB eq, 1.17 kg 1.4 DB eq, 0.0005 kg C_2H_4 eq, 0.024 kg SO_2 eq and 0.0094 kg PO_4 eq, respectively. Payandeh (2016) investigated the environmental impacts of poultry production systems in Isfahan, Iran. The amounts of the mentioned indicators in producing 1000 kg meat were0.0022 kg Sbeq, 40924.976 MJ, 5782.380 kg CO2 eq, 4.225 kg CFC -11 eq, 41447.050 kg 1.4 DB eq, 5866.113 kg 1.4 DB eq, 32057072.3 kg 1.4 DB eq, 1952.126 kg 1.4 DB eq, 1.237 kg C_2H_4 eq, 35.755 kg SO2 eq and 9.881 kg PO4 eq, respectively. Although the amounts of the environmental impacts of turkey were lower than those of the poultry production system, but some inputs were not studied in turkey production such as one-day chick and feeding equipment. This result is due to lower consumption of electricity and fuel in turkey compare to poultry production unit. The role of effective factors in environmental indicators was determined and has been shown in Figure 1. The contribution of feed input in all environmental indicators is higher than those other factors.

Table 1 Amount of input/outputmaterial for onetonne turkey production

Input/output	Unit	Amount
Input		
Food	kg	457000
Fuel	l	20000
Electricity	kWh	41322
Output		
Meat	kg	201600
Manure	kg	180000

Table 2 Amount of emissions in turkey production unit

Impact category	Unit	Amount
Abiotic depletion	kg Sbeq	1.61
Abiotic depletion (fossil fuels)	MJ	20.19
Global warming (GWP 100a)	kg CO_2 eq	3.63
Ozone layer depletion (ODP)	kg CFC-11 eq	1.90
Human toxicity	kg 1,4-DB eq	67.60
Fresh water aquatic ecotoxicity	kg 1,4-DB eq	4.55
Marine aquatic ecotoxicity	kg 1,4-DB eq	1.04
Terrestrial ecotoxicity	kg 1,4-DB eq	1.17
Photochemical oxidation	kg C_2H_4 eq	0.0005
Acidification potential	kg SO_2eq	0.024
Eutrophication	kg PO_4 eq	0.0094

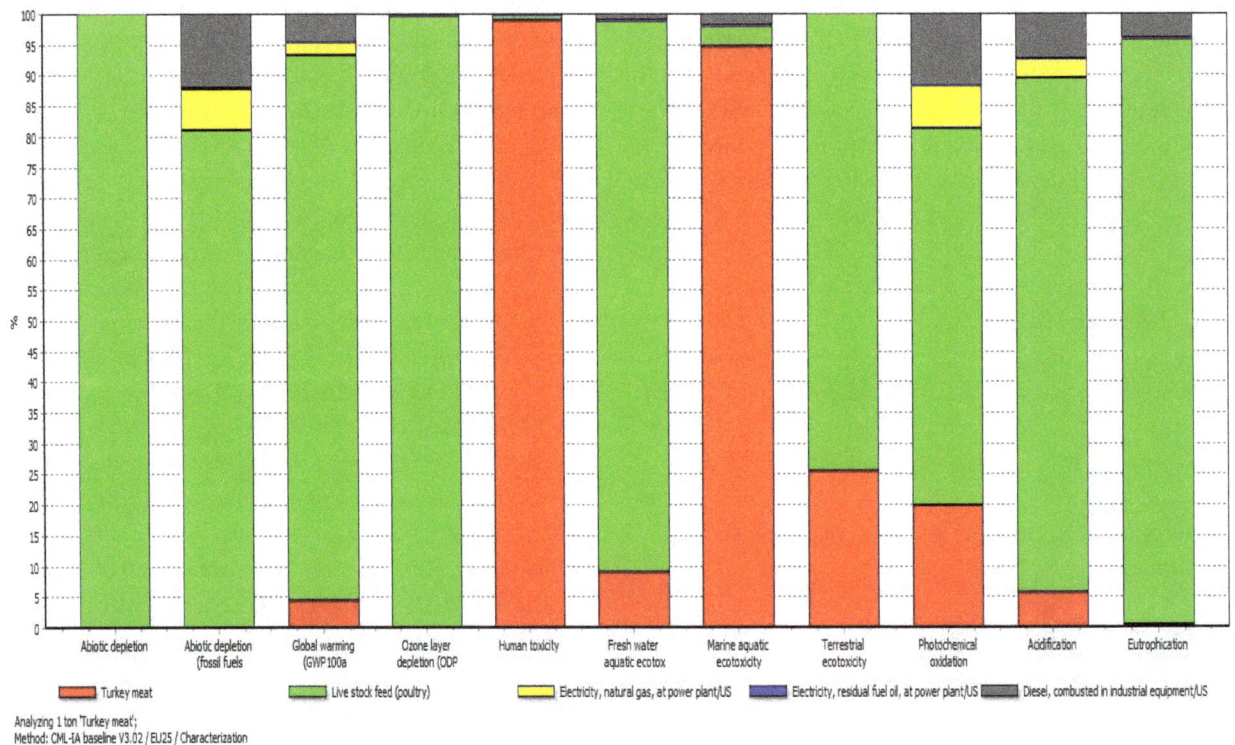

Analyzing 1 ton 'Turkey meat';
Method: CML-IA baseline V3.02 / EU25 / Characterization

Figure 1 Contribution of resources to environmental impact categories for one tonne turkey produced

This result is in agreement with the results of previous studies by Payandeh (2016), Da Silva *et al*. (2013), Nielsen *et al*. (2011) and Bengtsson and Seddon (2013). The reason of this result is due to use fossil fuels in production of agricultural products that causes increasing of greenhouse gases in feed production process.

As can be seen in Figure 1 the portion of feed contribution to most of indicators was higher that other inputs and outputs. The portion of feed contribution to abiotic depletion, abiotic depletion (fossil fuels), global warming, ozone layer depletion, marine aquatic ecotoxicity, photochemical oxidation, acidification potential and eutrophication potential indicators were 100, 81, 89, 99.8, 89, 61, 84 and 95% respectively. This result shows that the management of input consumption in feed production step (i.e. in crop production farms) must be improved to reduce input materials and environmental impacts.

CONCLUSION

In this study the amounts of inputs and outputs and environmental impacts of turkey production system in Iran were determined. Also influenced of fuel, electricity, enteric fermentation and feed on environmental indicators were evaluated. There was concluded that feed input had the highest effect on the determined environmental indicators. The farmers can apply better input use management, improve the feed productivity and use efficient equipment to increase efficiency, decrease losses and finally decrease environmental impacts. As the emissions related to feed input were higher than others, applying appropriate methods of feed consumption is recommended. The farmers are recommended to use renewable energy such as solar, wind and biomass to reduce environmental emissions in turkey production units and feed production farms. Using intelligent systems to control temperature, humidity and ventilation can reduce fuel consumption.Relevant agencies can help turkey production owners to conduct practical programs to promote emission reduction skills of farmers.

ACKNOWLEDGEMENT

The authors acknowledge the farmers, Najafabad Township, Isfahan, Iran, who provided the useful information for this research. Also Ilam University is appreciated for financial supports of the research.

REFERENCES

Anonymous. (2008). Energy balance sheet. Iranian Energy Productivity Organization. Available at: http://www.saba.org.ir.

Anonymous. (2012). Agricultural statistics. Center of Statistics and Information Technology, Ministry of Agriculture.

Anonymous. (2014). Agricultural statistics. Poultry Information Institute. Available at: http://www.infopoultry.net.

Asheri E. and Karimzadeh Y. (2010). Calculation of production factors productivity in broiler units of West Azerbaijan. *Anim. Sci. J. (Pajouhesh and Sazandegi).* **86**, 2-7.

Basset-Mens C. and Vanderwrf H.M.G. (2005). Scenario-based environmental assessment of farming systems: the case of pig production in France. *Agric. Ecosyst. Environ.* **105**, 127-144.

Bengtsson J. and Seddon J. (2013). Cradle to retailer or quick service restaurant gate life cycle assessment of chicken products in Australia. *J. Clean. Prod.* **41**, 291-300.

Benis K. and Ferrão P. (2017). Potential mitigation of the environmental impacts of food systems through urban andperi-urban agriculture (UPA)-a life cycle assessment approach. *J. Clean. Prod.* **140**, 784-795.

Bojacá C.R., Wyckhuys K.A.G. and Schrevens E. (2014). Life cycle assessment of Colombian greenhouse tomato production based on farmer-level survey data. *J. Clean. Prod.* **69**, 26-33.

Brentrup F., Küsters J., Kuhlmann H. andLammel J. (2004). Environmental impact assessment of agricultural production systems using the life cycle assessment methodology: I. Theoretical concept of a LCA method tailored to crop production. *European J. Agron.* **20(3)**, 247-264.

Cederberg C. and Mattsson B. (2000). Life cycle assessment of milk production-a comparison of conventional and organic farming. *J. Clean. Prod.* **8**, 49-60.

Da Silva V., Wander Werf H., Soareso S. and Corson M. (2013). Environmental impacts of French and Brazilian broiler chicken production scenarios: a LCA approach. *J. Environ. Manage.* **133**, 222-231.

Ewemoje T.A., Omotosho O. and Abimbola O.P. (2013).Cradle-to-gate LCA of poultry production system in a developing country-the case of Nigeria. *Int. J. Agric. Sci.* **3**, 323-332.

Gonzalez–Garcia S., Gomez–Fernandez Z., Dias A., Feijoo G., Moreira T. and Arroja l. (2014). Life cycle assessment of broiler chicken production: a Portouguese case study. *J. Clean. Prod.* **71**, 125-134.

Guinée J.B. (2002). Handbook on Life Cycle Assessment Operational Guide to the ISO Standards. Kluwer Academic, New York.

IPCC. (2006). Intergovernmental Panel on Climate Change (IPCC) Guidelines for National Greenhouse Gas Inventories, Prepared by the National Greenhouse Gas Inventories Programme. Published by the Institute for Global Environmental Strategies (IGES), Hayama, Japan.

ISO 14040. (2006a). Environmental Management – Life Cycle Assessment – Principles and Framework. International Standards Organization (ISO), Geneva, Switzerland.

ISO 14044. (2006b). Environmental Management – Life Cycle Assessment – Requirements and Guidelines. International Standards Organization (ISO), Geneva, Switzerland.

Khoshnevisan B., Bolandnazar E., Barak S., Shamshirband S., Maghsoudlou H., Altameem T.A. and Gani A. (2014). A clustering model based on an evolutionary algorithm for better energy use in crop production. *Stoch. Environ. Res. Risk. Assess.* **29**, 1921-1935.

Khoshnevisan B., Bolandnazar E., Shamshirband S., Motamed H., Badrul N., Mat L. and Kiah M.L.M. (2015). Decreasing environmental impacts of cropping systems using life cycle assessment (LCA) and multi-objective genetic algorithm. *J. Clean. Prod.* **86**, 67-77.

LEAP. (2014). Greenhouse gas emissions and fossil energy use from poultry supply chains: guidelines for quantification, Livestock Environmental Assessment and Performance Partnership. FAO, Rome, Italy. Available from: http://www.fao.org/3/a-mj752e.pdf.

Leinonen I., Williams A., Wiseman J., Guy J. and Kyriazakis I. (2014). Predicting the environmental impacts of chicken systems in the United Kingdom through a life cycle assessment: egg production systems. *Poult. Sci.* **91(1),** 26-40.

Llorach-Massana P., Muñoz P., Riera M.R., Gabarrell X., Rieradevall J., Ignacio Montero J. and Villalba G. (2017). N₂O emissions from protected soilless crops for more precise food and urban agriculture life cycle assessments. *J. Clean. Prod.* **149,** 1118-1126.

Nielsen N.I., Jqrgensen M. and Bahrndorff S. (2011). Greenhouse gas Emission from the Danish Broiler Production Estimated via LCA Methodology. Knowledge Center for Agriculture, Denmark.

OECD. (1999). Environmental Indicators for Agriculture. Organisation for Economic Co-operation and Development (OECD) Publications Service, Paris, France.

Payandeh Z. (2016). Life cycle assessment of poultry production in Isfahan Provience. MS Thesis. Ilam Univ., Ilam, Iran.

Pelletier N. (2008). Enviromental performance in the US broiler poultry sector: life cycle energy use and greenhouse gas, ozone depleting, acidifying and eutrophying emission. *Agric. Syst.* **98,** 67-73.

Rebolledo-Leiva R., Angulo-Meza L., Iriarte A. and González-Araya M.C. (2017). Joint carbon footprint assessment and data

envelopment analysis for the reduction of greenhouse gas emissions in agriculture production. *Sci. Total Environ.* **593(1),** 36-46.

Sefeedpari P., Rafiee S.H. and Akram A. (2012). Comparison of energy consumption and greenhouse gas emissions in dairy-cows and egg laying hen farms in Tehran province. Pp. 65-67 in 1ˢᵗ Nat. Conf. Polic. Toward Sustain. Dev. Tehran, Iran.

Sonesson U., Berlin J. and Ziegler F. (2010). Environmental Assessment and Management in the Food Industry. Woodhead Publishing Ltd., Cambridge, United Kingdom.

Thomassen M.A., Van Calker K.J., Smits M.C.J., Iepema G.L. and de Boer I.J.M. (2008).Life cycle assessment of conventional and organic milk production in the Netherlands. *Agric. Syst.* **96,** 95-107.

Weiler V. (2013). Carbon footprint (LCA) of milk production considering multifunctionality in dairy systems: a study on smallholder dairy production in Kaptumo, Kenya. MS Thesis. Wageningen Univ., Wageningen, The Netherlands.

25

Performance of Japanese Quails (*Coturnix coturnix japonica*) on Floor and Cage Rearing System in Sylhet, Bangladesh

A. Razee[1], A.S.M. Mahbub[1], M.Y. Miah[1], M.R. Hasnath[1], M.K. Hasan[1], M.N. Uddin[2,3] and S.A. Belal[1,4*]

[1] Department of Poultry Science, Sylhet Agricultural University, Sylhet, Bangladesh
[2] Department of Livestock Production and Management, Sylhet Agricultural University, Sylhet, Bangladesh
[3] Department of Animal Science, Chonbuk National University, Jeonju 561-756, Republic of Korea
[4] Department of Animal Biotechnology, Chonbuk National University, Jeonju 561-756, Republic of Korea

*Correspondence E-mail: belalsa.dps@sau.ac.bd

ABSTRACT

A total number of 66 day old Japanese quail chicks divided into 2 treatment groups (33 in each treatment) with 3 replications in each having 11 birds (male, 5 and female, 6) were reared on floor and in cage system for a period of 5 weeks to know the effect of rearing system on growth performance and carcass characteristics. At the age of 35 days, average body weight and feed intake were 102.15 and 320.7 g/quail for cage and 78.41 and 146.02 g/quail for floor system, respectively. Feed conversion ratio (FCR) was 3.89 and 4.10 for cage and floor system, respectively ($P<0.01$), at the end of study period. Body weight, feed intake and FCR were significantly ($P<0.01$) different between cage and floor rearing system. At the age of 21 and 35 days survivability were 72.72 and 72.72% for cage and 63.63 and 60.60% for floor, respectively. There was higher survivability in cage system. In case of meat yield characteristics, average weight of breast, thigh, wing, drum stick were 20.92, 7.37, 5.42 and 5.72 g for cage and 20.84, 7.35, 5.39 and 5.63 g for floor, respectively. There were no significant difference among average weight of breast, thigh and drum stick between two rearing system. In case of sex average, wing weight differed among sexes. It was concluded that cage reared quails showed better performance compared to littered floor rearing system.

KEY WORDS comparative study, floor and cage system, performance, quail.

INTRODUCTION

Quail farming is very easy and entertaining because it is one of the smallest species of poultry birds and is very profitable similar to chicken and duck farming. Almost all types of weather condition are suitable for quail farming. Quail eggs are more nutritious than other poultry egg because it contains comparatively much more protein, lipid and carbohydrates (Anca *et al.* 2012). Quail farming, needs small capital and labor can play a vital role to meet up the demand of nutrition. Japanese quail growth experienced great development in recent decades due to biological character-

istics, which determines the high level of production and economic efficiency as well as the market requirement for quail eggs and meat, with recognized quality (highly nutritional and biological value, particular taste) (Ayasan, 2013). Among the main productive characteristics of the quail is fast growth rate (reach adult weight to 5-6 weeks after hatching), early sexual maturity, short interval between generations, high laying rate and low feed and low spaces accommodation. Quail is recently getting popular as a profitable enterprise. People are now becoming interested in commercial quail production in a densely populated country such as Bangladesh due to its short generation interval, quick business return, low investment and small living space. The immense potentiality of quail for meat and egg production is a new dimension in poultry farming providing gainful employment, supplementary income and a valuable source of animal protein for human diet. However enough scientific information about the management practices of quail under Bangladesh conditions yet to be disseminated among the people who are showing interest to rear quails as a commercial venture. Management practices specially housing are most important for optimum production of quail. In addition, proper housing of quails makes it possible to control flying and to manage them more carefully and efficiently. However rearing methods and housing during the growth periods are vitally related to the cost of producing quails. Karousa et al. (2015) demonstrated that the entire farmer in Bangladesh rear quails in cages. This is probably due to comparative availability of information on cage rearing rather than on other methods of rearing. Research comparing the growth performances of quails in different rearing systems is rather limited. Some investigators, Padmakumar et al. (2000a), Ayorinde (1994), Sharma and Panda (1978) and Huque et al. (1992) reported a nonsignificant effect on housing system (Cage vs. littered floor) on body weight gain and survivability, but reported variable results on feed consumption and feed conversion. On the other hand some researcher observed significantly high growth rate in floor system than cage system (Ojedapo and Amao, 2014; Dogan and Tulin, 2012).

Some experimental works have been done comparing cage and littered floor. However, effects on the production performances of quails on these two systems under Bangladesh condition are not known. Therefore, it is necessary to compare these two systems of rearing of growing quails at the present time when more emphasis is given on cage systems. The construction of floor for rearing quails is easier and involves less cost in Bangladesh in comparison to cages due to locally available cheaper materials like bamboos, wood frame packing box, wooden saw dust, rice husk and etc. which may be comfortable housing system for quails for economic productivity. Considering the above fact and circumstances, the present study was conducted to investigate the effect of two different management systems (cage and floor) on the performance and survivability of quails and also to recommend which one is the suitable management system out of these two systems in Sylhet.

MATERIALS AND METHODS

Statement of the experiment
The present experiment was conducted with growing Japanese quails (Coturnix coturnix japonica) at Abul Kashum quail farm, Khadimnagor, Sylhet to study the effect of two

different management systems (cage and floor) on their growth, feed intake, feed conversion efficiency, survivability and interaction of sex and two rearing system on meat yield characteristics of Japanese quail. Before starting experiment, we have conducted academic committee, Department of Poultry Science, Sylhet Agricultural University for approval of the experimental protocol according to the animal ethics guidelines and received protocol approval.

Preparation of experiment house
The experimental house was cleaned, washed and disinfected and kept empty for two weeks. The cages were also washed and disinfected.

Experimental birds and diets
A total of 66 growing Japanese quail were used for conducting the experiment. Sixty six (66) quails were divided into 2 treatment groups (33 in each treatment) with 3 replications having 11 birds (male, 5 and female, 6) in each. Initial body weight was 6.90 ± 0.50 g and birds were randomly assigned to either in cage or floor. Diet was formulated using locally available feed ingredients on dry mater basis. The nutrient requirements (meta bolizable energy (ME): 2758.6 kcal/kg; crude protein (CP): 27.04%; Ca: 1.56%; P: 1.12%; lysine: 1.29% and methionine: 0.75%) were satisfied according to the recommendations of Singh and Panda (1988). The diet was prepared weekly and proper mixing of ingredients was ensured (Table 1).

Table 1 Composition of the diets fed to the experiment quails

Feed ingredients	Amount in 100 kg mixed feed
Crushed wheat	45.50
Rice polish	08.00
Til oil cake	16.00
Soybean meal	16.00
Meat and bone meal	14.00
Common salt	0.25
Rhodivit G.S (vita. mineral premix)	0.25
Total	100

Management practices
Before commencement of the study, the birds were brooded on floor for first week in a previously cleaned and disinfected brooder house under necessary care and management. Then the chicks were brooded under 23 hour lighting and one hour darkness period per day and this lighting schedule was maintained until the end of the experiment. After successful brooding on floor at one week of age, the birds were divided into 2 treatment groups: first group in cage and second group on floor. Floor space in cage and on floor provided according to the recommendation of Panda et al. (1987). Clean dry and fine sawdust at a depth of 5 cm used as litter material on floor.

The feeders and drinkers were set in such way that the experimental bird could eat and drink easily. All mash dry feed was supplied *ad libitum* to the birds throughout the experimental period. Fresh drinking water was made available at all times. Feeders were cleaned every week end and the drinkers were cleaned every morning. Strict hygienic and sanitation procedure were also followed. The trial was conducted with identical care and management for all the treatments.

Recording of temperature and relative humidity

Room temperature was recorded three times daily in the morning, at noon and in the evening. Relative humidity was calculated from dry and wet bulb thermostat reading.

Processing of quails

At the end of the experiment, 2 quails from each replication group were randomly selected and slaughtered for processing. To facilitate slaughtering, all birds had their feed and water removed 12 hours period to killing. The selected birds were slaughtered using halal method. After slaughter, the birds were immersed in hot water (51 °C) for 2 minutes in order to loosen the feather. Final processing was performed by removing of head, shanks, viscera, oil gland, kidney, lungs and heart of the carcass. Heart and liver were cut loose from the viscera. The gizzard was removed by cutting it loose in front of the proventriculus and then cutting both in coming and going tracts. Then the carcass was opened with knife, emptied, washed and the lining was removed by hand.

Data collection and record keeping

During 5 weeks rearing period following records were kept:
a) Body weight: initially and weekly for each replication.
b) Feed intake: weekly for each replication.
c) Mortality: recorded when death occurred for each replication.
d) Dressed weight: blood weight, feather weight, heart weight, liver weight, gizzard, shank weight, head weight, abdominal fat and skin weight and shank length.

Based on above records, feed intake, feed conversion ratio, survivability and dressing percentage were calculated. All dressing yield related parameters were converted into the percentage of live weight of respective quail slaughtered.

Statistical analysis

All recorded and calculated data were analyzed using completely randomized design (CRD). An analysis of variance compared the all the parameters among treatments. The significant differences were identified by Duncan's new multiple range test (DMRT). Data on body weight, feed intake and feed conversion were analyzed by paired t-test with the help of MINITAB (USA). All recorded data and calculated data were analyzed using M-Stat C method and the design was CRD ANOVA.

RESULTS AND DISCUSSION

Performance of quails
Body weight

The result on weekly body weight of quail chicks are shown in Table 2. The data indicate that the body weight of quails reared in cage and littered floor were 102.15 and 78.41 g at 35 days of age, respectively. Significantly (P<0.01) higher body weight gain were found in cage birds compared to the littered floor birds. It is also evident that females were always heavier than males. However, during the experiment, cage birds gained weight faster and reached higher body weight than birds on littered floor. This result are in agreement with the previous findings of Jatoi *et al.* (2013), Roshdy *et al.* (2010), Padankumar *et al.* (2000a) and Ahuja *et al.* (1998). They all found significantly higher body weight in cage housed quails both male and female birds than in littered floor on different density floor housed quail birds.

On the other hand the present results contradict the observation of Ojedapo and Amao (2014) and Dogan and Tulin (2012) who reported that floor housed quail birds had significantly heavier weight gain that cage housed quails.

Feed intake

It is revealed from Table 2 that quails on floor had lower feed intake than cage housed quails at all stage. The feed intake of both cage and floor housed quails increased gradually with age. Feed intake (g/bird) was affected (P>0.01) by rearing system in all stage except 14 days of age.

However, feed intake of quails were significantly (P>0.01) higher at 21, 28, 35 days of age on cage rearing that floor rearing. The feed intake results are supported by Fouzder *et al.* (1999), who found higher feed intake in cage quails than floor quails. On the other hand, this study contradicts the finding of Ayorinde (1994) and Huque *et al.* (1992) who found higher feed intake by floor reared quails compared to those reared in cages.

However, Padmakumar *et al.* (2000a) and Ahuja *et al.* (1998) showed that feed intake were not influenced by two rearing system (cage and floor).

Feed conversion

It is evident from Table 2 that feed conversion ratio (FCR) was higher in floor quails than cage reared quails at all ages.

Table 2 Effect of rearing system on body weight, feed intake and feed conversion ratio (FCR)

Variable	Age	Rearing		LSD and level of significant*
		Cage	Floor	
Body weight (g/quail)	Day old	6.90	6.90	NS
	7	19.75^a	19.42^b	$(0.337)^*$
	14	27.21^a	24.06^b	$(3.288)^*$
	21	42.36^a	31.45^b	$(2.738)^{**}$
	28	76.48^a	62.32^b	$(3.659)^{**}$
	35	102.15^a	78.41^b	$(1.737)^{**}$
Feed intake (g/quail)	Up to 14	33.72	33.43	$(0.766)^{ns}$
	Up to 21	103.5^b	76.12^a	$(2.549)^{**}$
	Up to 28	190.46^b	146.02^a	$(2.829)^{**}$
	Up to 35	320.71^b	241.89^a	$(1.592)^{**}$
FCR	Up to 14	4.5^b	7.20^a	$(2.360)^{**}$
	Up to 21	4.6^b	6.30^a	$(1.453)^{**}$
	Up to 28	3.36	3.40	$(0.337)^{ns}$
	Up to 35	3.89^b	4.10^a	$(0.101)^{**}$

The means within the same row with at least one common letter, do not have significant difference (P>0.05) and (P>0.01).
* (P<0.05) and ** (P<0.01).
NS: non significant.
LSD: least significant difference.

FCR value of quails in two rearing system (cage and floor) were statistically (P<0.01) higher in floor rearing system at 14, 21 and 35 days. Cage housed quails showed superior FCR compared to litter or floor house quails. It may be due to that quails housed on cage utilized feed more efficiently than floor housed quails.

The result of present study demonstrated that quails reared on cage housed showed superior efficiently of feed than quails reared on floor similarly to the observation of Alam et al. (2008) and Narahari et al. (1986). In contrast, Padmakumar et al. (2000b) found non-significant variation in FCR in cage and littered reared birds.

Survivability
The survivability of quails is shown in Table 3. The results indicate that survivability percent at 7 and 14 days were higher on littered floor than cage. At 21, 28 and 35 days of age survivability percent was tended to be lower on floor housed than cage rearing. Significant effects of rearing system on survivability were found during the experimental periods. The effect of two rearing systems (cage and littered floor) on survivability percent was significantly higher on floor at 7th and 14th days. But in 3rd, 4th, 5th week, the higher survivability rate was observed in cage than floor. The recent study showed higher survivability percentage in quails reared in cage than those reared on littered floor (69.70% vs. 65.91%) which was consistence with the findings of Roshdy et al. (2010), Padmakumar et al. (2000b) and Akram et al. (2000).

Meat yield characteristics
The results on meat yield characteristics are presented in the Table 4.

The differences in live weight, heart weight and gizzard weight were statistically significant (P<0.01). Live weight was highest in cage housed birds. Significantly (P<0.01) higher heart weight was found in cage housed birds compared to littered floor housed quails. Interaction of sex and rearing system on meat yield characteristics of Japanese quail at 35 days are shown in Table 4. There was no difference in the weight of head, liver, breast weight, thigh, wing drum stick between cages housed and littered floor house quails.

Gizzard weight was higher in cage housed birds. There was no significant interaction of rearing system and sex in case of breast weight, thigh weight, wing weight, heart weight, liver weight, gizzard weight and head weight. In case of sex, the female birds live weight and head weight were higher in cage housed quails birds but not heart weight, liver weight, gizzard weight, wing weight, drum stick, breast weight, thigh weight, wing weight and drum sticks weight.

From the result, non-significant effect on sex of cage and floor housed quails birds for male and female quails were found on breast weight, thigh weight and drum stick of quails. The current study showed significant effect of interaction of sex and two rearing systems (cage and floor) on meat yield characteristics of Japanese quails at 35 day on live weight, heart weight and gizzard weight and non-significant effect on head, breast, thigh, wing and drumsticks weight.

Live weight, head weight have significant result in cage housed quails birds than floor housed birds for main effect of sex and interaction effect of rearing system and sex. Head weight, liver weight, wing weight have non significant effect on two rearing system.

Table 3 Effects of rearing system on survivability

Age (d)	Cage		Floor		X^2 value (chi square)	Level of significance
	Dead (%)	Alive (%)	Dead (%)	Alive (%)		
07	4^b (12.12)	29^a (87.87)	2^b (6.06)	31^a (93.93)	23.26	**
14	9^c (27.27)	24^b (72.72)	5^d (15.15)	28^a (84.84)	1.48	**
21	9^d (27.27)	24^a (72.72)	12^c (36.36)	21^b (63.63)	0.952	**
28	9^c (27.27)	24^a (72.72)	13^{bc} (39.39)	20^{ab} (60.60)	3.60	*
35	9^c (27.27)	24^a (72.72)	13^{bc} (39.39)	20^{ab} (60.60)	3.60	*

The means within the same row with at least one common letter, do not have significant difference (P>0.05) and (P>0.01).
* (P<0.05) and ** (P<0.01).

Table 4 Interaction of sex and rearing system on meat yield characteristics of Japanese quail at 35 days

Variables	Rearing system				LSD (SED) and level of significant*		
	Sex	Cage	Floor	Mean	RS	S	RS × S
Live weight (g)	male	95.80^a	74.62^b	85.21			
	female	108.50^a	82.20^b	95.35	$(0.465)^{**}$	$(0.919)^{**}$	$(0.431)^{**}$
	mean	102.15^a	78.41^b	90.28			
Heart (g)	male	1.87^a	0.77^b	1.32			
	female	1.39^a	0.65^b	1.02	$(0.144)^{**}$	$(0.101)^{**}$	$(0.101)^{**}$
	mean	1.63^a	0.71^b	1.17			
Liver weight (g)	male	3.03^a	2.94^b	2.99			
	female	2.87^a	2.17^b	2.52	$(0.176)^{ns}$	$(0.337)^{**}$	$(0.227)^{**}$
	mean	2.95^a	2.56^b	2.76			
Gizzard weight (g)	male	3.42^a	2.87^b	3.15			
	female	2.95^a	2.76^b	2.86	$(0.144)^{**}$	$(0.101)^{**}$	$(0.176)^{**}$
	mean	3.18^a	2.82^b	3.00			
Head weight (g)	male	5.22	5.27	5.25			
	female	5.32^a	4.40^b	4.86	$(0.144)^{ns}$	$(0.144)^{**}$	$(0.287)^{**}$
	mean	5.28^a	4.83	5.06			
Breast weight (g)	male	22.23	22.11	22.17			
	female	19.61	19.57	19.59	$(0.203)^{ns}$	$(0.919)^{ns}$	$(0.465)^{ns}$
	mean	20.92	20.84	20.88			
Thigh weight (g)	male	8.23	8.22	8.23			
	female	6.51	6.48	6.50	$(0.144)^{ns}$	$(0.101)^{ns}$	$(0.101)^{ns}$
	mean	7.37	7.35	7.36			
Wing weight (g)	male	5.78	5.76	5.77			
	female	5.06^a	5.02^b	5.04	$(0.176)^{ns}$	$(0.032)^{**}$	$(0.101)^{ns}$
	mean	5.42	5.39	5.41			
Drum stick (g)	male	6.12	6.05	6.09			
	female	5.31	5.22	5.27	$(0.203)^{ns}$	$(0.287)^{ns}$	$(0.175)^{ns}$
	mean	5.72	5.63	5.67			

The means within the same row with at least one common letter, do not have significant difference (P>0.05) and (P>0.01).
* (P<0.05) and ** (P<0.01).
NS: non significant.
LSD: least significant difference.

Wing weights of male and female quail have non-significant result in the study but in case of sex and two rearing system showed significant effect of Japanese quail birds.

CONCLUSION

It is concluded from the results of this study that:

1) Rearing system had significant influence on the performances (body weight, feed intake and feed conversion rat-

io) of growing Japanese quails. So, there was a consistent tendency of quails to perform better in cage than floor in Sylhet region.

2) Higher survivability was found in cage housed birds during the 35 days of experiment.

3) There were no significant difference among average weight of breast, thigh and drum stick between two rearing system.

Further study is needed in Sylhet, which would be help to justify the result of this study.

ACKNOWLEDGEMENT

The authors thank to the personnel of Department of Poultry Science, Faculty of Veterinary and Animal Science, Sylhet Agricultural University, Sylhet and Abul kashum quail farm, Khadimnagor, Sylhet for their valuable assist conducting the research.

REFERENCES

Ahuja S.D., Bandyopadhyay U.K. and Kundu A. (1998). Performance of growing quail for meat under different cage densities. *Indian J. Poult. Sci.* **33(1),** 8-14.

Akram M., Shah A.H. and Imran Khan M. (2000). Effect of varying floor space on productive performance of Japanese quail breeders maintained under litter floor and cage housing systems. *Pakistan Agric. Sci. J.* **37(1),** 42-46.

Alam M.A., Howlider M.A.R., Mondal A., Hossain K. and Bostami R. (2008). Pattern of egg production in Japanese quail reared on littered floor and in cage. *Bangladesh Res. Pub. J.* **1(3),** 239-249.

Anca P., Prelipcean A.A. and Teuşan V. (2012). Investigations on the structure, chemical composition and caloricity of the quail eggs, deposited at the plateau phase of the laying period. *Lucrări Ştiin. Seria Zootech.* **57,** 113-120.

Ayasan T. (2013). Effects of dietary inclusion of protexin (probiotic) on hatchability of Japanese quails. *Indian J. Anim. Sci.* **83(1),** 78-81.

Ayorinde K.L. (1994). Evaluation of the growth and carcass characteristic of the Japanese quail (*Coturnix conturnix japonica*) in Nigeria. *Nigerian J. Anim. Prod.* **21(1),** 94-99.

Dogan N. and Tulin A. (2012). Effects of mass selection based on phenotype and early feed restriction on the performance and carcass characteristics in Japanese quails. *Kafkas Univ. Vet. Fak. Derg.* **18(3),** 425-430.

Fouzder S.K., Ali M.L., Howlider M.A.R. and Khan N.R.M.Z. (1999). Performance of growing Japanese quails in cages, on slatted floor and on littered floor. *Indian J. Anim. Sci.* **69(12),** 1059-1062.

Huque Q.M.E., Poul D.C. and Salahuddin M. (1992). Study of the performance of quail under Bangladesh condition. Pp. 34-37 in Proc. 4th Natal. Conf. Bangladesh Anim. Husb. Assoc. Sylhet, Bangladesh.

Jatoi A.S., Khan M.K., Sahota A.W., Akram M., Javed k., Jaspall M.H. and Khan S.H. (2013). Post-peak egg production in local and imported strains of Japanese quails (*Coturnix coturnix japonica*) as influenced by continuous and intermittent light regimens during early growing period. *J. Anim. Plant. Sci.* **23(3),** 727-730.

Karousa M.M., Souad A., Ahmed Elaithy S.M. and E.A. Elgazar. (2015). Effect of housing system and sex ratio of quails on egg production, fertility and hatchability. *Benha Vet. Med. J.* **28(2),** 241-247.

Narahari D., Ramamurthy N., Viswanathan N., Viswanathan S., Thangavel A., Murugnandam B., Sussarasu V. and Majur K.A. (1986). The effect of rearing systems and marketing age on the performance of Japanese quail (*Coturnix coturnix japonica*). *Cheiron.* **15,** 160-163.

Ojedapo L.O. and Amao S.R. (2014). Sexual dimorphism on carcass characteristics of japanese quail (*Coturnix coturnix japonica*) reared in derived Savanna zone of Nigeria. *Int. J. Sci. Environ. Technol.* **3(1),** 250-257.

Padmakumar B., Reghunathan Nair G., Ratnakrishnan A., Unni A.A.K.K. and Ravindrana N. (2000a). Effect of floor space on egg weight and egg quality traits of Japanese quails reared in cages and deep litter. *J. Vet. Anim. Sci.* **31,** 34-36.

Padmakumar B., Reghunanthan Nair G., Ramakrishnan A., Unni A.A.K. and Ravindranathan N. (2000b). Effect of floor density on production performance of Japanese quails reared in cages and deep litter. *J. Vet. Anim. Sci.* **31,** 37-39.

Panda B., Ahuja S.D., Shrivastav A.K., Singh R.P., Agerwal S.K. and Thomas P.C. (1987). Quail Production Technology. Publication CARI, Izatnagar, India.

Roshdy M., Khalil H.A., Hanafy A.M. and Mady M.E. (2010). Productive and reproductive traits of Japanese quail as affected by two housing system, Egypt. *Poult. Sci.* **30(1),** 55-67.

Sharma G.I. and Panda B. (1978). Studies on some productive traits in Japanese quail (*Coturnix coturnix japonica*). *Indian Poult. Gaz.* **62(1),** 24-30.

Singh K.S. and Panda B. (1988). Poultry Nutrition. Kalyani Publishers, New Delhi, India.

26

Walnut Meal as an Excellent Source of Energy and Protein for Growing Japanese Quails

M.A. Arjomandi[1] and M. Salarmoini[2]*

[1] Department of Animal Science, Faculty of Agriculture, Shahid Bahonar University of Kerman, Kerman, Iran

*Correspondence E-mail: salarmoini@uk.ac.ir

ABSTRACT

The present study was designed to study the chemical composition, apparent and true metabolizable energy values of the walnut meal and to evaluate the effects of different levels of walnut meal (0, 10, 20 and 30%) on Japanese quail's growth performance, blood metabolites, relative weight of different organs, malondialdehyde (MDA) concentration in breast meat and egg yolks' cholesterol. This study was conducted as a completely randomize design with 288 unsexed Japanese quails randomly dividing into 4 treatments with 4 replicates of 18 birds each. As a result of this study, no significant differences were found for feed intake and feed conversion ratio (P>0.05), except the birds fed 30% walnut meal showed lower weight gain compared to the control at 7-21 days of age (P<0.05). There were no significant differences in serum glucose, uric acid, serum aspartate aminotranspherase (AST) and alanine aminotranspherase (ALT) activities between different dietary treatments. The serum low density lipoprotein (LDL), cholesterol and triglyceride tended to decrease linearly (P<0.01) as the walnut meal levels were increased. The serum high density lipoprotein (HDL) level in quails fed 10% walnut meal were significantly higher than control group (P<0.05). Consumption of different levels of walnut meal significantly decreased malondialdehyde (MDA) concentration in breast meat of chicks aged 42 d (P<0.01). Different dietary treatments had no effect on the relative weight of different organs and carcass traits. In general, walnut meal is a good source of energy (apparent metabolizable energy corrected for nitrogen (AMEn) 3689 kcal/kg), oil (23%) and crude protein (40%) and could be used up to 20% for young chicks and 30% for older chicks, without any adverse effect on growth performance.

KEY WORDS growth performance, Japanese quail, meat quality, walnut meal.

INTRODUCTION

According to the Food and Agriculture Organization (FAO, 2012), the top 3 walnut (*Juglans regia*) producers in 2012 respectively were China, the United States and Iran. Walnut with 64000 hectares cultivated area is one of the important horticulture products in Iran. The walnut seed (kernel) represents from 40 to 60% of the nut weight, depending mainly on the variety. The seed has high levels of oil (52-70%) in which polyunsaturated fatty acids predominate (Prasad, 2003; Martinez *et al.* 2006). In addition to oil, walnuts provide appreciable amounts of proteins (up to 24% of the walnut seed weight), carbohydrates (12-16%), fibre (1.5-2%) and minerals (1.7-2%) (Savage, 2001; Prasad, 2003). According to USDA (2013), dry matter, ether extract, crude protein, ash and crude fiber content of the nut are 95.3, 49.42, 21.22, 2.99 and 12.2%, respectively. A lot of studies have been undertaken regarding the health effects of walnut in human nutrition. Walnut and almond can reduce serum low density lipoprotein (LDL), increase serum high density lipoprotein (HDL) and thus, can reduce coronary heart disease risk factors (Jenkins *et al.* 2002; Hyson *et al.* 2002; Kalgaonkar *et al.* 2011). In the past decades, different phenolic compounds with antioxidant activities were characterized and identified in walnut seed extract and its skin, shell and hull as walnut by-products (Fukuda *et al.* 2003; Wijerant *et al.* 2006; Labuckas *et al.* 2008; Oliveira *et al.* 2008).

Walnut meal, a by-product of walnut processing, is obtained after oil has been extracted from the walnut kernels, mostly by cold pressing methods which leaves considerable amount of oil in walnut meal. There is no available report regarding the use of walnut meal in poultry diets. This study was planned to evaluate the effects of using different levels of walnut meal on Japanese quail's growth performance, blood metabolites, relative weight of different organs and egg yolks' cholesterol.

MATERIALS AND METHODS

Metabolizable energy assay
The experiment was conducted at the Shahid Bahonar University of Kerman, Kerman, Iran. In the first trial, apparent metabolizable energy (apparent metabolizable energy (AME) and apparent metabolizable energy corrected for nitrogen (AMEn)) content of walnut meal was determined using total collection method (Macleod, 2002). Walnut meal was substituted with a corn-soybean meal basal diet at 40% level and then AME and AMEn of this experimental diet and basal diet were determined. Twelve adult leghorn cockerels (165 d old) with mean body weight of 1645 ± 30 g were housed in individual cages, with 6 pens per treatment. Both feed (as mash) and water were provided *ad libitum*. After 5 days adaptation to experimental diets, excreta were collected and corresponding feed intake recorded for subsequent three days. Feed was removed overnight at the start and termination of the excreta collection period. The collected excreta were dried at 65 °C for 48 hours in a forced-air oven. Then samples were placed in the laboratory environment for 24 hours to equilibrate with ambient humidity and then were ground before dry matter, gross energy, and nitrogen determinations. All samples were analyzed in duplicates. The ME value of walnut was determined according to the formula:

$$ED = (P \times EF) + (1-P)EB$$

Where:
ED: ME of the experimental diet.
P: level of date palm in the experimental diet.
EF: ME of date palm.
(1-P): level of basal diet in experimental diet.
EB: ME of basal diet (Marquardt, 1962).

In the second trial, twelve adult leghorn cockerels with uniform body weights were used to determine the true metabolizable energy using Sibbald method (Sibbald, 1986).

Birds fasted for 48 hours, then 6 birds received 25 g walnut meal using force feeding procedure and their excreta were collected for 48 hours. Also six cockerels were used fasted to determine the endogenous urinary energy and fecal metabolic energy losses. Finally, true metabolizable energy (TME) and true metabolizable energy corrected for nitrogen (TMEn) were calculated. Chemical compositions of walnut meal and excreta samples were measured according to the prevalent methods (AOAC, 2005).

Quail assay
A total of 288 day-old unsexed Japanese quail chicks were randomly allocated to 4 experimental groups with 4 replicates and 18 chicks in each. The diets were fed to quail chicks for 6 weeks. Four experimental diets were formulated to meet the NRC requirements (NRC, 1994) of quails as shown in Table 1.

All experimental diets were formulated and adjusted to be isonitrogenous and isocaloric. Diets contained four levels of walnut meal (0, 10, 20 and 30%). Feed and water were provided on an *ad libitum* basis. Lighting program was adjusted to meet 24 hours of light daily. Body weight gain and feed intake were determined on a weekly basis.

To measure the weight of different organs, two male chicks from each replication were selected randomly and sacrificed at 21 and 42 days of age. At 42 days of age, blood samples from 2 chicks per replicate were also taken from the neck vein and centrifuged at 3000 rpm for 15 minutes. All serum tests including LDL, HDL, serum aspartate aminotranspherase (AST) and alanine aminotranspherase (ALT) activities, uric acid, triglyceride, cholesterol and glucose were analyzed using commercial Pars-Azmoon kits and an auto analyzer.

Malondialdehyde (MDA) concentration in breast meat (fresh and/or frozen at -20 °C for 30 days) was also determined (Tarladgis *et al.* 1960).

Breast meat pH was also determined by blending 10 g sample in 100 mL distilled water for one minute and pH was measured using a pH meter (model AZ 86502) (Ensoy *et al.* 2004). At the beginning of laying, three eggs per replicates were selected and total cholesterol content in the yolks were determined according to the method of Pasin *et al.* (1998).

Statistical analyses
This experiment was performed as a completely randomized design. The data were analyzed using the general linear (GLM) procedure of SAS (SAS, 2003) and Duncan's multiple range test was used to detect differences among treatment means (P<0.05).

Table 1 Diet formulation and calculated chemical composition of diets (as fed)

Ingredients (%)	Level of walnut meal (%)			
	0 (control)	10	20	30
Corn	43.30	41.63	39.97	38.20
Barley grain	0.20	1.47	2.73	4
Soybean meal	47.79	38.13	28.5	18.92
Walnut meal	0	10	20	30
Soybean oil	5.44	4.29	3.15	2
Calcium carbonate	1.35	1.40	1.40	1.50
Dicalcium phosphate	0.8	0.60	0.40	0.20
Common salt	0.37	0.37	0.37	0.37
DL-methionine	0.15	0.19	0.22	0.26
L-lysine hydro chloride	0	0.25	0.50	0.75
Vitamin premix[1]	0.25	0.25	0.25	0.25
Mineral premix[2]	0.25	0.25	0.25	0.25
Grit	0.1	1.17	2.21	3.3
Chemical composition				
ME (kcal/kg)	3000	3001	3002	3000
Crude protein (%)	24.82	24.82	24.82	24.84
Lysine (%)	1.39	1.388	1.377	1.368
Methionine (%)	0.522	0.531	0.541	0.551
Methionine + cysteine (%)	0.915	0.871	0.828	0.784
Calcium (%)	0.839	0.837	0.836	0.834
Available phosphorus (%)	0.315	0.319	0.319	0.318
Sodium (%)	0.158	0.157	0.157	0.156
Crude fiber (%)	4.3	4.36	4.42	4.48

[1] Provided the following per kg of diet: Retinol acetate: 3.1 mg; Thiamine: 1.8 mg; Riboflavin: 6.6 mg; Niacin: 30 mg; Pantothenic acid: 10 mg; Pyridoxine: 3 mg; Folic acid: 1 mg; Cyanocobalamine: 15 μg; Biotin: 0.1 mg; Cholecalciferol: 0.05 mg; Alpha-tocopherol acetate: 18 mg; Menadion: 2 mg and Choline chloride: 0.4 g.
[2] Fe: 50 mg; Mn: 100 mg; Zn: 85 mg; Cu: 10 mg; Se: 0.2 mg and I: 1 mg.

RESULTS AND DISCUSSION

Dry matter, ether extract, crude fibre, crude protein, ash, calcium, total phosphorus and sodium content of walnut meal were 95, 23, 6.3, 40, 5.5, 5.2, 0.67 and 0.2 %, respectively. AME, AMEn, TME and TMEn of walnut meal were 3478.24 ± 77, 3689.42 ± 71, 3997.46 ± 53 and 3673.83 ± 55 (kcal/kg, as fed basis), respectively.

According to USDA (2013), dry matter, ether extract, crude fibre, crude protein, ash and crude fiber content of walnut meal are 95.3, 49.42, 21.22, 2.99 and 12.2%, respectively. So, this nut has high levels of oil in which polyunsaturated fatty acids predominate (Prasad, 2003; Martinez et al. 2006; Ozcan et al. 2010). The chemical composition of walnuts in different reports is not the same. It seems walnuts provide appreciable amounts of proteins (up to 24% of the walnut meal weight), carbohydrates (12-16%) and minerals (1.7-2%) (Savage, 2001; Prasad, 2003).

There is no available report regarding the chemical composition and ME content of walnut meal. Considering walnut meal composition, it should be an appropriate source of energy and protein.

Performance parameters and carcass traits

The effects of different levels of walnut meal on feed intake, weight gain and feed conversion ratio of Japanese quails are shown in Table 2. During the whole experimental period extended from 1-42 days of age the obtained results showed that growth performance of the birds fed diets with different levels of walnut meal were not statistically different compared to control, although the birds fed 30% walnut meal showed lower weight gain as compared with the control and 10% walnut meal at 7-21 days of age (P<0.05). Thus it seems that walnut meal should be limited to 20% for young chicks.

There is no report regarding the effect of walnut meal on poultry growth performance. However, recent results have shown that the encapsulation of intracellular lipids by the cell walls of almond, restricts their digestion in the stomach and small intestine and therefore available for fermentation in the colon by the gut microbiota (Mandalari et al. 2008a; Mandalari et al. 2008b). These results should be explained the poorer performance of the younger chicks fed 30% walnut diet.

The effects of different dietary treatments on the relative weight of different internal organs and carcass traits at 21 d and 42 d of age are shown in Tables 3 and 4 and found no significant effects among the groups.

Blood analysis

The effect of different dietary treatments on some blood serum metabolites at 42 days of age are presented in Table 5. There were no significant differences in glucose, uric acid, AST and ALT levels between different dietary treatments. The serum LDL, cholesterol and triglyceride tended to decrease linearly (P<0.01) as the walnut meal levels were increased. The serum HDL level in quails fed 10% walnut meal were significantly higher than control group (P<0.05). The results presented here are in agreement with the previous reports showing that the application of walnut and almond in human diets can reduce blood total cholesterol and LDL and also can increase blood HDL (Hyson et al. 2002; Jenkins et al. 2002; Kalgaonkar et al. 2011). These effects of walnuts are mediated by components in the oil fraction of this nut (Hyson et al. 2002) or probably in part because of the nonfat (protein and fiber) and monounsaturated fatty acid components of the nut (Jenkins et al. 2002).

Meat quality

The effects of walnut meal on fresh and frozen breast meat quality and egg yolk cholesterol are given in Table 6. There were no significant differences in pH and CP content of fresh breast meat and egg yolk cholesterol concentration between different dietary treatments.

Table 2 The effects of different levels of walnut meal on feed intake (FI, g/bird/day), body weight gain (WG, g/bird/day) and feed conversion ratio (FCR) of Japanese quails

Walnut meal level (%)	7-21 (day)			21-42 (day)			7-42 (day)		
	FI	WG	FCR	FI	WG	FCR	FI	WG	FCR
0 (control)	14.72	6.79[a]	2.16	26.48	5.03	5.25	21.78	5.73	3.79
10	14.63	6.74[a]	2.17	25.75	5.40	4.78	21.30	5.94	3.59
20	14.64	6.61[ab]	2.21	25.98	5.18	5.04	21.45	5.75	3.73
30	14.40	6.28[b]	2.29	25.57	5.14	4.99	21.10	5.60	3.76
SEM	0.34	0.12	0.04	0.60	0.16	0.21	0.42	0.10	0.10
P-values	0.917	0.045	0.176	0.73	0.49	0.51	0.72	0.18	0.55

The means within the same column with at least one common letter, do not have significant difference (P>0.05).
SEM: standard error of the means.

Table 3 The effects of different dietary treatments on the relative weight of different organs and carcass traits in Japanese quails aged 21d (% of live body weight)

Walnut meal level (%)	Heart	Liver	Spleen	Large intestine	Small intestine	Ceca	Bursa of fabricius	Pancreas	Thighs	Breast
0 (control)	0.73	2.42	0.06	0.98	2.22	0.71	0.11	0.28	13.15	21.59
10	0.73	2.48	0.07	0.94	2.03	0.68	0.11	0.20	12.94	23.07
20	0.73	2.36	0.05	0.81	1.53	0.64	0.10	0.28	14.03	22.97
30	0.61	2.51	0.06	0.75	1.56	0.56	0.08	0.29	13.73	22.79
SEM	0.049	0.090	0.007	0.081	0.275	0.113	0.011	0.035	0.395	0.560
P-values	0.27	0.68	0.59	0.16	0.22	0.81	0.25	0.29	0.20	0.23

SEM: standard error of the means.

Table 4 The Effects of dietary treatments on the relative weight of different organs and carcass traits in Japanese quails aged 42 d (% of live weight)

Walnut meal level (%)	Heart	Liver	Spleen	Large intestine	Small intestine	Ceca	Bursa of fabricius	Pancreas	Thighs	Breast
0 (control)	0.82	1.39	0.038	0.58	1.25	0.45	0.085	0.18	13.89	25.31
10	0.78	1.43	0.033	0.62	1.43	0.54	0.052	0.18	13.95	23.63
20	0.74	1.61	0.068	0.64	1.47	0.50	0.052	0.18	12.85	24.05
30	0.80	1.30	0.055	0.53	1.33	0.57	0.09	0.29	14.34	25.35
SEM	0.049	0.104	0.005	0.043	0.101	0.058	0.013	0.025	0.623	0.495
P-values	0.4	0.85	0.21	0.88	0.67	0.55	0.38	0.59	0.83	0.42

SEM: standard error of the means.

Table 5 Effects of dietary treatments on blood serum parameters in Japanese quails aged 42 d

Walnut meal level (%)	LDL (mg/dL)	HDL (mg/dL)	AST (IU/L)	ALT (IU/L)	Uric acid (mg/dL)	Triglyceride (mg/dL)	Cholesterol (mg/dL)	Glucose (mg/dL)
0 (control)	63.6[a]	60.1[b]	7.6	199	10.2	61.2[a]	136[a]	275
10	48.5[ab]	79.0[a]	8.7	230	12.0	65.0[a]	140[a]	249
20	37.1[b]	63.6[b]	9.5	198	11.4	55.7[a]	112[b]	253
30	32.2[c]	69.8[ab]	7.5	206	10.1	41.0[b]	110[b]	275
SEM	5.57	4.06	0.83	11.11	0.90	4.39	5.76	8.69
P-values	0.008	0.03	0.32	0.19	0.39	0.01	0.004	0.1

LDL: low density lipoprotein; HDL: high density lipoprotein; AST: aspartate aminotranspherase and ALT: alanine aminotranspherase.
The means within the same column with at least one common letter, do not have significant difference (P>0.05).
SEM: standard error of the means.

Table 6 Effects of dietary treatments on breast meat quality and egg yolk cholesterol content in Japanese quails

Walnut meal level (%)	CP (%)	pH	MDA-42[1]	MDA-42[2]	MDA-21[3]	Cholesterol (mg/g yolk)
0 (control)	20.5	6.36	0.23	2.64[a]	0.88	13.54
10	21.21	6.36	0.21	0.95[bc]	0.69	14.08
20	21.48	6.41	0.21	0.78[c]	0.80	14.06
30	21.13	6.36	0.23	1.16[b]	0.84	13.47
SEM	0.429	0.018	0.024	0.097	0.046	0.34
P-values	0.45	0.29	0.79	0.0001	0.058	0.53

[1] Malondialdehyde concentration in fresh breast of chicks aged 42 d (mg MDA/kg).
[2] Malondialdehyde concentration in breast meat after 30 d freezing in chicks aged 42 d (mg MDA/kg).
[3] Malondialdehyde concentration in breast meat after 30 d freezing in chicks aged 21 d (mg MDA/kg).
The means within the same column with at least one common letter, do not have significant difference (P>0.05).
SEM: standard error of the means.

The MDA concentration in fresh breast meat of quails aged 42 d and in frozen meat of birds aged 21 d were not significantly affected by the dietary treatments. However, MDA concentration in breast meat after 30 d freezing in chicks aged 42 d was significantly affected by the dietary treatments. Consumption of different levels of walnut meal significantly decreased MDA concentration (P<0.01).

Some studies showed the scavenging activity of superoxide and hydroxyl radicals by different walnut extracts. Wijerant *et al.* (2006) showed brown skin extract at 50 ppm effectively inhibited copper-induced oxidation of human LDL cholesterol compared to whole seed and green shell cover extracts. Torabian *et al.* (2009) reported that the consumption of walnut seed increased serum polyphenol concentrations, increased total antioxidant capacity and reduced serum lipid peroxidation. The antioxidant activity of walnut has been attributed to its phenolic compounds (Fukuda *et al.* 2003; Wijerant *et al.* 2006; Labuckas *et al.* 2008; Oliveira *et al.* 2008).

CONCLUSION

The finding presented here provided new information for using walnut meal in quail diets. In general, walnut meal is a good source of energy (AMEn 3689 kcal/kg), oil (23%) and crud protein (40%) and could be used up to 20% for young chicks and 30% in finisher diets, without any adverse effect on growth performance.

ACKNOWLEDGEMENT

This work was supported by the Shahid Bahonar University of Kerman, which is gratefully acknowledged.

REFERENCES

AOAC. (2005). Official Methods of Analysis. 18th Ed. Association of Official Analytical Chemists, Arlington, VA, USA.

Ensoy U., Cadogan K., Kolsarici N., Karslioglu B. and Cizmeci M. (2004). Influence of acetic acid and lactic acid treatment on lipid changes and color of chicken legs. Pp. 874-874 in Proc. 22nd World's Poult. Congr. Istanbul, Turkey.

FAO. (2012). Food and Agriculture Organization of the United Nations the State of Food Insecurity in the World.

Fukuda T., Ito H. and Yoshida T. (2003). Antioxidative polyphenols from walnuts (*Juglans regia*). *Phytochemistry.* 63, 795-801.

Hyson D., Scheneeman B.O. and Davis P.A. (2002). Almond and almond oil have similar effects on serum lipids and LDL oxidation in healthy men and women. *J. Nutr.* 132, 703-707.

Jenkins D.J., Kendall C.W., Marchie A., Parker T.L., Connelly P.W., Qian W., Haight J.S., Faulkner D., Vidgen E., Lapsley K.G. and Spiller G.A. (2002). Dose response of almond on coronary heart disease risk factors: blood lipids, oxidized low-density lipoproteins, homocysteine and pulmonary nitric ox-

ide: a randomized, controlled, crossover trail. *Circulation.* 106, 1327-1332.

Kalgaonkar S., Almario R.U., Gurusinghe D., Garamendi E.M., Buchan W., Kim K. and Karakas S.E. (2011). Differential effects of walnuts *vs.* almonds on improving metabolic and endocrine parameters in PCOS. *European J. Clin. Nutr.* 65, 386-393.

Labuckas D.O., Maestri D.M., Perello M. and Martinez M.L. (2008). Phenolics from walnut (*Juglans regia*) kernels: antioxidant activity and interactions with proteins. *Food Chem.* 107, 607-612.

Macleod M.G. (2002). Energy utilization: measurement and prediction. Pp. 191-220 in Poultry Feedstuffs: Supply, Composition and Nutritive Value. J. MacNab, Ed. CABI. Publishing, New York.

Mandalari G., Faulks R.M., Rich G.T., LoTurco V., Picout D.R., Lo Curto R.B., Bisignano G., Dugo P., Dugo G., Waldron K.W., Ellis P.R. and Wickham M.S.J. (2008a). Release of protein, lipid and vitamin E from walnut seeds during digestion. *J. Agric. Food Chem.* 56, 3409-3416.

Mandalari G., Nueno-Palop C., Bisignano G., Wickham M.S.J. and Narbad A. (2008b). Potential prebiotic properties of almond (*Amygdalus communis*) seeds. *Appl. Environ. Microb.* 74, 4264-4270.

Marquardt D.W. (1962). An algorithm for least squares estimation of nonlinear parameters. *J. Soc. Ind. Appl. Math.* 11, 431-441.

Martinez M.L., Mattea M. and Maestri D.M. (2006). Varietal and crop year effects on lipid composition of walnut (*Juglans regia*) genotypes. *J. Am .Oil. Chem. Soc.* 83, 791-796.

NRC. (1994). Nutrient Requirements of Poultry, 9th Rev. Ed. National Academy Press, Washington, DC., USA.

Oliveira I., Sousa A., Ferreira I., Bento A., Estevinho L. and Pereira J.A. (2008). Total phenols, antioxidant potential and antimicrobial activity of walnut (*Juglans regia*) green husks. *Food Chem. Toxicol.* 46, 2326-2331.

Ozcan M., Iman C. and Arsaln D. (2010). Physico-chemical properties, fatty acid and mineral content of some walnuts (*Juglans regia*) types. *Agric. Sci.* 1, 62-67.

Pasin G., Smith G.M. and Mahony M.O. (1998). Rapid determination of total cholesterol in egg yolk using commercial diagnostic cholesterol reagent. *Food Chem.* 61, 255-259.

Prasad R.B.N. (2003). Walnuts and pecans. Pp. 6071-6079 in Encyclopedia of Food Science and Nutrition. B. Caballero, L.C. Trugo and P.M. Finglas, Eeds. Academic Press, London, UK.

SAS Institute. (2003). SAS®/STAT Software, Release 9.1. SAS Institute, Inc., Cary, NC. USA.

Savage G.P. (2001). Chemical composition of walnuts (*Juglans regia*) grown in New Zealand. *Plant Food Hum. Nutr.* 56, 75-82.

Sibbald I.R. (1986). The TME System of Feed Evaluation. Animal Research Center, Agriculture Canada, Ottawa, Canada.

Tarladgis B.G., Watts B.M. and Younathan M.T. (1960). A distillation method for the quantitative determination of malondialdehyde in rancid foods. *J. Am .Oil. Chem. Soc.* 37, 44-48.

Torabian S., Haddad E. and Rajaram S. (2009). Acute effect of nut consumption on plasma total polyphenols, antioxidant capacity and lipid peroxidation. *J. Hum. Nutr. Diet.* 22, 64-71.

USDA. (2013). U.S. Department of Agriculture, Agricultural Research Service. USDA National Nutrient Database for Standard Reference, Release 26. Available at: http://www.ars.usda.gov/ba/bhnrc/ndl.

Wijerant S.S., Abou-Zaid M.M. and Shahidi F. (2006). Antioxidant polyphenols in almond and its co products. *J. Agric. Food Chem.* **54,** 312-318.

27

The Effect of a Dietary Innovative Multi-Material on Sex Hormones and Molting Period of Canaries and Laying-Hens

J. Salary[1], H.R. Hemati Matin[2] and H. Hajati[3*]

[1] Department of Animal Science, Faculty of Agriculture, Bu Ali Sina University, Hamedan, Iran
[2] Department of Animal Science, Faculty of Agriculture, Tarbiat Modares University, Tehran, Iran
[3] Department of Animal Science, Faculty of Agriculture, Ferdowsi University of Mashhad, Mashhad, Iran

*Correspondence E-mail: h.hajati2010@gmail.com

ABSTRACT

Two experiments were conducted to determine the effect of offering a multi-material innovative (MMI) feed including: *Vitex agnus-castus*, *Thymus vulgaris*, *Lavandula angustifolia*, Marigold (*Calendula officinalis*) on curtails molting and sex hormone concentrations in canaries and laying hens. In the first study, a total of 120 female molted canaries were allotted in to 12 cages of 10 birds with 4 replicates for 135 d. Treatments were control (drinking water without MMI) and 1.25 g or 2.25 g MMI dissolved in 1 L of drinking water. In the second study, a total of 72 molted laying hens were allotted to 24 cages of 3 birds with 8 replicates fed with similar treatments of the first study for 21 d. Results showed that at 45, 90, and 135 d, a linear and quadratic decrease (L and Q: P≤0.006) in plasma prolactin and a linear increase (L: P≤0.037) in plasma progestron and estrogen concentrations of canaries as MMI levels increased. On d 21, the plasma concentrations of prolactin (L and Q: P=0.001), progesterone (L: P=0.042), and estrogen (Q: P=0.036) increased in laying hens along with the increasing of MMI levels. Also, the feather rejuvenation of canaries and laying hens was decreased (L and Q: P≤0.018) with increasing MMI levels. The egg production increased in the first (L: P=0.045), second, and third (L and Q: P≤0.005) weeks with no pronounced trends in feed intake and drinking water of laying hens during the experiment. The use of MMI in the drinking water of canaries increased sex hormones production and shortened molting duration. Also, the innovative dietary additive has stimulating effects on egg production of laying hens without impressed feed or water consumption.

KEY WORDS canary, herbal additive, layer hens, molting.

INTRODUCTION

Hormones have major metabolic roles in reproductive function (e.g. egg production) and molting progress of poultry (Herremans *et al.* 1988; Hoshino *et al.* 1988; Dicerman and Bahr, 1988; Peebles *et al.* 1994; Renden *et al.* 1994), especially hormones upon the hypothalamo-hypophyseal-ovarian axis. Reproductive function in the female is controlled primarily by the interaction of the ovarian sex steroids progesterone and estradiol. The preovulatory rise of

plasma progesterone is directly correlated to the ovulation of mature follicles (Johnson, 1984) and precedes and stimulates luteinizing hormone rise. A positive feedback mechanism between progesterone and luteinizing hormone for ovulation induction resulted in positive correlations between circulating levels of progesterone and egg production in layers (Tanabe *et al.* 1983). The importance of progesterone in controlling ovulation has been previously reported in chickens (Kappauf and Van Tienhoven, 1972) and turkeys (Mashaly *et al.* 1979). In mallard ducks, the cessation of egg laying was associated with low circulating levels of both progesterone and estradiol (Bluhm *et al.* 1983). In turkey, both non-laying and laying hens with low egg production also have low levels of circulating progesterone (Mashaly *et al.* 1979). Thus, the modulation of endocrine factors may lead to the improvement of egg productivity in poultry. Kang *et al.* (2001) reported that high egg production is associated with higher weights of ovary and follicles and higher progesterone confirmed the positive relationship of progesterone expression with egg production. Several studies were conducted to reduce the secretion of prolactin through active and / or passive immunization in bantam hens to prevent development of broodiness (Sharp *et al.* 1989). These studies were targeted through dopamine system because of dopamine inhibits prolactin secretion via hypothalamus. In contrast of mammals, the biological effect of inhibition of prolactin in avian species is not well established (Reddy *et al.* 2006). Molting in avian species is defined as the periodic shedding and replacement of feathers (Berry, 2003).

Poultry normally go through periodic molts and do not have reproduction during molting. Molting naturally can be problematic for the producer because it is erratic and does not have economic benefits (Oguike *et al.* 2005) as well as results in severe stress, lowered immunity and increased mortality rate in birds (Holt, 2003). The canary (*Serinus canarius*) can be considered as a domesticated species of the order Passeriformes. People like this specious very much. Most canaries will stop singing and loose plumage when they begin to drop their feathers, which lead to economical losses.

On the other hand, natural molting in laying hens generally takes four months (North and Bell, 1990), which raises economic concerns as the hens continue to be fed during non-productive times (McDaniel and Aske, 2000). Innovative molting techniques should be safe, animal friendly, minimize bird stress, and duration of molting. Therefore, this study was designed to investigate the effect of offering a multi-material innovative on molting and sex hormone concentrations in canaries along with egg production of laying hens.

MATERIALS AND METHODS

Experimental design
The Animal Ethics Committee of the Agricultural Research Center of Qom, Qom-Iran approved the experiments. The experiment was carried out in the company of Simorgh in Esfahan (Esfahan-Iran). The MMI was the mixture of plant hydro-alcoholic extracts including; *Vitex agnus-castus*, *Thymus vulgaris*, *Lavandula angustifolia*, Marigold (*Calendula officinalis)*.

Experiment 1
A total of 120 female canaries (Munich-initial body weight=29±1 g) in last egg production season were housed in an environmentally controlled room. Canaries were assigned to 12 wire cages (50×40×40 cm). There were 3 dietary treatments, 4 replicates and 10 birds each replicate. The experiment started when the birds began to drop their feathers, which lead to economic losses. The cages were equipped with 2 water cups and 2 feeding troughs. Canaries were exposed to 16 hours daily photoperiod at 23 ˚C. The diets were formulated according to experienced canaries' producers and were based on cotton seed and millet (75:25%) and last for 135 d. Treatments were including: control [(C) water without MMI] and 1.25 g or 2.25 g MMI dissolved in 1 L of drinking water.

Experiment 2
In total, a total of 72 (30 weeks age) Bowens laying hens (with initial body weight=1195 g) were used in a house equipped with cages (50×45×45 cm). They exposed to 16 hours daily photoperiod at 18 ˚C. The experiment was started when the laying hens were dropped in egg production phase and had low egg production (about 5 to 7% flock). The laying hens with lower than 20% egg production were selected for the experiment. Birds distributed among 24 cages so that 3 hens were allotted to each cage. Treatments were similar to the experiment 1 and the experiment was lasted for 21 d. Laying hens were fed with basal diets formulated according to the Bowens strain guideline.

Sampling and data collection
Two representative canaries from each cage were randomly selected and 0.5 mL of blood from the brachial vein at 3 phases was collected (45, 90, and 135 d after the commence of experiment). In addition, two representative laying hens from each cage were randomly selected on d 21 after the commence of experiment. Blood samples were collected and transferred to vial tubes containing sodium heparin. The tubes were centrifuged (5000×g for 20 min at room temperature) and the supernatant (plasma) was removed. The samples were stored at -20 ˚C. They were assayed for estrogen and progesterone using ELISA by specific kits (DRG®, USA) and prolactin using gamma-counter by specific kits (Kavoushyar® Co. Tehran, Iran). Feed intake (FI), drinking water (DW) and the number of egg production were recorded for laying hens.

Statistical analysis
Data from both trials was analyzed in a completely randomized design using the GLM procedure of the SAS software (SAS, 2008).

All data were analyzed for normal distribution using the NORMAL option of the UNIVARIATE procedure. Pre-planned contrasts were carried out for investigate MMI effects. The P-value of ≤ 0.05 was deemed as significant.

RESULTS AND DISCUSSION

The plasma prolactin concentrations of canaries decreased (L and Q: P≤0.006) with increasing MMI levels at 45, 90, and 135 d of the experiment (Table 1).

Table 1 Ingredinets and chemical composition of the experimental

Ingredients (%)	Amount (%)
Corn grain	55.07
Soybean meal	27.45
Wheat barn	0.2
Oil	3.0
Oyster shell	9.37
Dicalcium phosphate	1.61
Common salt	0.31
DL-methionine	0.22
L-lysine HCL	0.03
L-threonine	0.05
Vitamin and mineral premix[1]	0.50
Enzyme	0.050
Vitamin A	0.1
Vitamin B	0.1
KCO_3	0.05
$NaHCO_3$	0.1
Calculated contents	
Metabolizable energy (kcal/kg)	2780
Crude protein %	16.5
Calcium %	4.02
Available phosphorus %	0.43
Sodium %	0.17
Lysine %	1.12
Methionine %	0.38
Met + Cys %	0.65
Threonine %	0.62

[1] Vitamin and mineral premix supplied per kilogram of diet: vitamin A: 22500 IU; vitamin D_3: 5000 IU; vitamin E: 45 IU; vitamin K_3: 5 mg; B_{12}: 0.04 mg; Choline: 625 mg; Mn: 74400 mg; Fe: 75000 mg; Zn: 64.675 mg; Cu: 6000 mg and Se: 200 mg.

The plasma prolactin concentrations reduced (P<0.001) in canaries fed with MMI than control group at 45, 90 and 135 d of the experiment. The plasma progestron and estrogen concentrations of canaries increased linearly (L: P≤0.037) with increasing MMI levels at 45, 90 and 135 d of the experiment. The canaries that received MMI had greater (P=0.003) plasma progestron concentrations than control group at 135 d of the experiment. In addition, the plasma estrogen concentrations in canaries that received by MMI was greater (P≤0.039) than those of control group at 45, 90 and 135 d of the experiment. On d 21, the plasma concentrations of measured hormones [prolactin (L and Q: P=0.001), progesterone (L: P=0.042) and estrogen (Q: P=0.036)] increased in laying hens with increasing MMI levels.

The laying hen that received MMI had greater (P≤0.03) plasma prolactin, progesterone and estrogen concentrations than the control group. The feather dropped and rejuvena-tion time of canaries and laying hens decreased (L and Q: P≤0.018) with increasing MMI levels (Table 2). Similar responses were observed in canaries and laying hens that received MMI) than those received the control diet (P≤0.001). The number of egg produced increased in the first (L: P=0.045), second and third weeks (L and Q: P≤0.005) after the beginnings of experiment (Table 3). These parameters were greater (P<0.001) in laying hens fed with MMI rather those fed with the control in second and third weeks. No trends were observed in FI and DW of lay-ing hen subjected to treatments during the experiment.

Data of the plasma sex steroid hormones changes during molting in laying hens are so limit and almost there is no information in canaries. Prolactin is the mainly peptide at hypothalamic level. The elevated levels of prolactin in birds play a negative role on reproductive performance. This is further concurred when dopamine stimulates the prolactin secretion from anterior pituitary gland (Youngren et al. 1998). The egg production in birds declines with increasing the concentration of prolactin (Table 4). It seems that the offering MMI containing plant extracts to canaries led to change the reproductive system of the females from low functional sate to a high functional state. Action of plant extracts attributed to lowering prolactin concentration and low levels of prolactin, in turn, increases progesterone lev-els by inhibiting enzymes, which catabolizes progesterone, that observed for canaries and laying hens in the present study. On the other hand, progesterone is the only steroid produced by the ovulating follicle (Decuypere et al. 1993). The effects of progesterone on avian reproductive behavior are to a large extent unknown and often contradictory among studies (Blas et al. 2010). The peak of progesterone levels are attained in females during incubation suggests that progesterone could be involved in the expression of parental behaviors. This is supported by recent studies showing that non-breeding females have lower circulating progesterone levels compared to breeding females during the incubation period, but not at other times of the cycle (Blas and Hiraldo, 2010). The plasma concentrations of progesterone vary with age (Joyner et al. 1987).

The progesterone levels of 4.29, 3.80 and 1.64 ng/mL were recorded for pullets, old layers and old non-laying hens, respectively (Joyner et al. 1987). Increasing levels of the plasma progesterone also was found in the current study with aging, which can imply that ovulating follicle of ca-naries was activated and they are ready for reproduction. However, molt induction by reduction in day length and starvation precipitate significant decrease in sex steroids and gonadotropins.

Table 2 The plasma hormonal concentrations (ng/mL) of canaries (1.5, 3 and 4.5 months after the beginnings of trial) and laying hen at d 21 post-trial period subjected to treatments

Item	C	T1	T2	SEM	P-value		
					Linear	Quadratic	Contrast
Canaries							
Prolactin							
1.5	323.2	256.4	219.3	13.1	< 0.001	0.006	< 0.001
3	406.7	299.4	239.4	21.0	< 0.001	0.001	< 0.001
4.5	353.1	326.3	246.6	13.7	< 0.001	< 0.001	< 0.001
Progesterone							
1.5	2.8	3.1	3.8	0.192	0.031	0.569	0.087
3	3.1	3.3	4.1	0.200	0.037	0.420	0.125
4.5	3.4	4.0	4.9	0.214	0.001	0.610	0.003
Estrogen							
1.5	269.3	283.4	329.8	10.21	0.008	0.326	0.039
3	343.5	363.2	443.8	15.01	0.001	0.112	0.007
4.5	319.7	360.8	449.1	19.11	0.001	0.342	0.006
Laying hens							
Prolactin	2.23	2.55	2.51	0.05	0.001	0.001	< 0.001
Progesterone	1.27	0.20	0.47	0.20	0.042	0.073	0.017
Estrogen	697.33	556.00	435.33	47.71	0.136	0.036	0.030

C: control [drinking water without multi-material innovative] and T1 and T2: 1.25 g or 2.25 g dissolved multi-material innovative in 1 L of drinking water.
SEM: standard error of the means.
Contrast: C *vs.* T1/T2.

Table 3 The feather dropped and complete rejuvenation time in canaries and laying hens (day)

Item	C	T1	T21	SEM	P-value		
					Linear	Quadratic	Contrast
Canaries							
Dropped	40	38	19	2.84	< 0.001	0.003	0.001
Rejuvenation	124	110	77	5.99	< 0.001	0.001	< 0.001
Laying hens							
Dropped	35	21	15	2.59	< 0.001	0.018	< 0.001
Rejuvenation	130	88	81	6.59	< 0.001	< 0.001	< 0.001

C: control [drinking water without multi-material innovative] and T1 and T2: 1.25 g or 2.25 g dissolved multi-material innovative in 1 L of drinking water.
SEM: standard error of the means.
Contrast: C *vs.* T1/T2.

Table 4 The egg production (egg/hen) in the first, second, and third weeks of the trial as well as the average feed intake (g/d) and drinking water (mL/d) of laying hen subjected to treatments

Week	C	T1	T2	SEM	P-value		
					Linear	Quadratic	Contrast
First	2.71	3.50	4.75	0.41	0.045	0.791	0.084
Second	2.86	5.71	5.57	0.34	< 0.001	0.001	< 0.001
Third	3.57	7.43	7.00	0.45	< 0.001	0.005	< 0.001
Average feed intake	88.43	92.50	92.03	124	0.264	0.408	0.177
Water intake	179.00	181.75	180.50	0.55	0.870	0.802	0.790

C: control [drinking water without multi-material innovative] and T1 and T2: 1.25 g or 2.25 g dissolved multi-material innovative in 1 L of drinking water.
SEM: standard error of the means.
Contrast: C *vs.* T1/T2.

Tanabe *et al.* (1981) reported that plasma levels of progesterone and oestradiol in laying hens decreased during starvation with the plasma progesterone levels ranged between 0.51 and 0.72 ng/mL during starvation, in contrast with 1.38 ng/mL when fed *ad libitum*. In addition, it was reported that molting occurs when plasma concentrations of steroids and gonadotropins are low (Etches, 1996) and changes in the reproductive functions during forced-molting were associated with reduced levels of luteinizing hormone and sex steroids in laying hens (Decuypere and Verheyen, 1986; Jacquet *et al.* 1993), Humboltd penguins (*Spheniscus humboltdi*) (Otsuka *et al.* 1998), the emperor (*Aptenodytes forsteri*), adelie (*Pygoscelis adeliae*) penguins (Groscolas and Leloup, 1986), mallard ducks (Bluhm *et al.* 1983) and turkey (Mashaly *et al.* 1979). Declined progesterone during the molt was coincident with the cessation of egg production (Hoshino *et al.* 1988), which be explained by the absence of the large yellow yolk filled follicles in the ovaries. Thus, the rising levels of plasma progesterone by the use of MMI in DW could be attributed to the redevelopment and / or regeneration of the large yellow follicles in the ovary.

The use of MMI in DW for canaries increased sex steroids in the current study which indicated that MMI have the potential for shortening molting. An observation obtained in the present study (almost 21 d for canaries and 47 d for laying hens). On the other hand, *in vitro* studies describe dopaminergic effects of *Vitex agnus-castus*, yielding potent inhibition of prolactin in cultured pituitary cells. The flavonoid apigenin can be isolated from *Vitex agnus-castus* and has selective binding affinity for the β-estrogen receptor subtype (Wuttke *et al.* 2003). Additional *in vitro* studies provide evidence of prolactin inhibition with direct binding to dopamine receptors. These changes are because of low concentration of prolactin by the actions of plant extracts at ovarian level. Low concentration of the steroid hormones in birds fed with control diet seems because of interference of high prolactin concentration (Reddy *et al.* 2006), which can explain the lower egg production in birds fed with control diet. In agreement with the results of Ariana *et al.* (2011), the egg production of laying hens with MMI treatments increased in the present study. The increased egg production with the use of MMI might be related to their effects on stimulating bird digestive systems and feed efficacy. In this regards, the supplementation of broiler chicken diets with plant extracts results in growth promotion, nutrient digestibility enhancement, and improvement of feed efficacy (Papageorgiou *et al.* 2003; Yakhkeshi *et al.* 2012). It was reported that the plant extracts increased the production of digestive enzymes and the improved utilization of digestive products through enhanced liver functions (Langhout, 2000; Hertrampf, 2001; Williams and Losa, 2001).

CONCLUSION

The results indicated that the use of MMI in drinking water of canaries resulted in shortening molting duration (almost 30 d) associated with greater steroids hormones and lower prolactin production. It can provide a condition to return the birds on reproductive phase fast. The innovative plant additives have stimulating effects on egg production without impress feed or water consumption in laying hens.

ACKNOWLEDGEMENT

This project is certified by Qom Agricultural and Natural Resources Research Center. The authors would like to thank Dr Hamid Rreza Sirous, laboratory head and Mrs Qarakhani, laboratory staff, for their kindly cooperation.

REFERENCES

Ariana M., Samie A., Edriss M.A. and Jahanian R. (2011). Effects of powder and extract form of green tea and marigold and α-

tocopheryl acetate on performance, egg quality and egg yolk cholesterol levels of laying hens in late phase of production. *J. Med. Plant. Res.* **5**, 2710-2716.

Berry W.D. (2003). The physiology of induced molting. *Poult. Sci.* **82**, 971-980.

Blas J. and Hiraldo F. (2010). Proximate and ultimate factors explaining floating behavior in long-lived birds. *Horm. Behav.* **57**, 169-176.

Blas J., Lopez L., Tanferna A., Sergio F. and Hiraldo F. (2010). Reproductive endocrinology of wild, long-lived raptors. *Gen. Comp. Endocr.* **1**, 22-28.

Bluhm C.K., Phillips R.E. and Burke W.H. (1983). Serum levels of luteinizing hormone, prolactin, estradiol and progesterone in laying and nonlaying Mallards (*Anas platyrhynchos*). *Biol. Reprod.* **28**, 295-305.

Decuypere E., Leenstra F., Huybrechts L.M., Feng P.Y., Arnonts S., Herremans M. and Nys M. (1993). Selection for weight gain or food conversion in broiler affects the progesterone production capacity of large follicles in reproductive adult breeders. *Br. Poult. Sci.* **34**, 543-552.

Decuypere E. and Verheyen G. (1986). Physiological basis of inducedmoulting and tissue regeneration in fowls. *World's Poult. Sci.* **42**, 56-68.

Dicerman R.W. and Bahr J.M. (1988). Physiology and reproduction. *Poult. Sci.* **68**, 1402-1408.

Etches R.J. (1996). Reproduction in Poultry. CAB International, Wallingford, UK.

Groscolas R. and Leloup J. (1986). The endocrine control of reproduction and molt in emperor (*Aptenodytes forsteri*) and adelie (*Pygoscelis adeliae*) penguins. II. Annual changes in plasma levels of thyroxine and triiodothyronine. *Gen. Compar. Endocrin.* **63**, 264-274.

Herremans M., Decuypere E. and Chiason R.B. (1988). Role of ovarian steroids in the control of moult induction in laying fowls. *Br. Poult. Sci.* **29**, 125-136.

Hertrampf J.W. (2001). Alternative antibacterial performance promoters. *Poult. Int.* **40**, 50-52.

Holt P.S. (2003). Molting and *Salmonella enterica* serovar enteritidis infection: the problem and some solutions. *Poult. Sci.* **82**, 1008-1010.

Hoshino S., Suzuki M., Kakegawa K., Imal M., Wakita M., Kobayashi Y. and Yamada Y. (1988). Changes in plasma thyroid hormones, luteinizing hormone (LH), estradiol, progestrone and corticosterone of laying hens during a forced molt. *Br. Poult. Sci.* **29**, 238-247.

Jacquet J.M., Seigneurin F. and De Rivers M. (1993). Inducedmoulting in cockrels effect on sperm production, plasma concentration of luteinizing hormone, testosterone and thyroxine on pituitary sensitivity to luteinizing hormone-releasing hormone. *Br. Poult. Sci.* **34**, 765-775.

Johnson A.L. (1984). Interactions of progesterone and luteinizing hormone leading to ovulation in the domestic hen. Pp 133-143 in Proc. Rep. Biol. Poult. British Poult. Sci. Edinburgh, UK.

Joyner C.J., Peddie M.J. and Taylor T.G. (1987). The effect of age on egg production in the domestic hen. *Gen. Compar. Endocrinol.* **65**, 331-336.

Kang W.J., Yun J.S., Seo D.S., Hong K.C. and Ko Y. (2001).

Relationship among egg productivity, steroid hormones (progesterone and estradiol) and ovary in korean native ogol chicken. *Asian-Australuan J. Anim. Sci.* **14,** 922-928.

Kappauf B. and Van Tienhoven A. (1972). Progesterone concentrations in peripheral plasma of laying hens in relation to the time of ovulation. *Endocrinology.* **90,** 1350-1355.

Langhout P. (2000). New additives for broiler chickens. *World Poult. Elsevier.* **16,** 22-27.

Mashaly M.M., Proudman J.A. and Wentworth B.C. (1979). Reproductive performance of turkey hens as influenced by hormone implants. *Br. Poult. Sci.* **20,** 19-26.

Mcdaniel B.A. and Aske D.R. (2000). Egg prices, feed costs and the decision to molt. *Poult. Sci.* **79,** 1243-1245.

North M.O. and Bell D.D. (1990). Commercial Chicken Production Management. Chapman and Hall, New York.

Oguike M.A., Igboeli G., Ibe S.N., Iromkwe M.O., Akomas S.C. and Uzoukwu M. (2005). Plasma progesterone profile and ovarian activity of forced-moult layers. *African J. Biotechnol.* **4,** 1005-1009.

Otsuka R., Hori H., Aoki K. and Wada M. (1998). Changes in circulating LH, sex steroid hormones, thyroid hormones and corticosterone in relation to breeding and molting in captive Humboltd penguins (*Spheniscus humboltdi*) kept in an outdoor open display. *Zool. Sci.* **15,** 103-109.

Papageorgiou G., Botsoglou N., Govaris A., Giannenas I., Iliadis S. and Botsoglou E. (2003). Effect of dietary oregano oil and α-tocopheryl acetate supplementation on ironinduced lipid oxidation of turkey breast, thigh, liver and heart tissues. *J. Anim. Physiol. Anim. Nutr.* **87,** 324-335.

Peebles E.D., Miller C.R., Boyle C.R., Brake J.D. and Latour M.A. (1994). Effects of dietary thiouracil on thyroid activity, egg production and eggshell quality in commercial layers. *Poult. Sci.* **73,** 1829-1837.

Reddy I.J., David C.G. and Raju S.S. (2006). Chemical control of prolactin secretion and its effects on pause days, egg production and steroid hormone concentration in girirani birds. *Int. J. Poult. Sci.* **5,** 685-692.

Renden J.A., Lien R.J., Oates S.S. and Bilgili F.S. (1994). Plasma concentrations of corticosterone and thyroid hormones in broilers provided various lighting schedules. *Poult. Sci.* **62,** 1080-1083.

SAS Institute. (2008). SAS®/STAT Software, Release 9.2. SAS Institute, Inc., Cary, NC. USA.

Sharp P.J., Sterling R.J., Talbot R.T. and Huskisson N.S. (1989). The role of hypothalamic vasoactive intestinal polypeptide in the maintenance of prolactin secretion in incubating bantam hens. Observations using passive immunization, radioimmunoassay and immunohistochemistry. *J. Endocrinol.* **122,** 5-13.

Tanabe Y., Ogowa T. and Nakamura T. (1981). The effect of short-term starvation on pituitary and plasma luteinizing hormone, plasma oestradiol and progesterone, and pituitary response to luteinizing hormone-releasing hormone in the laying hen (*Gallus domesticus*). *Gen. Compar. Endocrinol.* **43,** 392-398.

Tanabe Y., Hirose K., Nakamura T., Watanabe K. and Ebisawa S. (1983). Relationship between the egg production rate and plasma estradiol, progesterone and testeron concentration in White Leghorn, Rhode Island Red and their hybrid pullets at various ages. *Japanese Soc. Zootec. Sci.* **2,** 99-100.

Williams P. and Losa R. (2001). The use of essential oils and their compounds in poultry nutrition. *World Poult. Elsevier.* **17,** 14-15.

Wuttke W., Jarry H., Christoffel V., Spengler B. and Seidlova-Wuttke D. (2003). Chaste tree. (*Vitex agnus-castus*)–pharmacology and clinical indications. *Phytomedicine.* **10,** 348-357.

Yakhkeshi S., Rahimi S. and Hemati Matin H.R. (2012). Effects of yarrow (*Achillea millefolium*), antibiotic and probiotic on performance, immune response, serum lipids and microbial population of broilers. *J. Agric. Sci. Technol.* **14,** 799-810.

Youngren O.M., Chaischa Y. and El-Halawani M.E. (1998). Serotogenic stimulation of avian prolactin secretion requires an intact system. *Gen. Compar. Endocrinol.* **112,** 63-68.

PERMISSIONS

LIST OF CONTRIBUTORS

A. Ghahremani, A.A. Sadeghi and M. Chamani
Department of Animal Science, Science and Research Branch, Islamic Azad University, Tehran, Iran

S. Hesaraki
Deptartment of Veterinary Medicine, Science and Research Branch, Islamic Azad University, Tehran, Iran

P. Shawrang
Nuclear Agriculture Research School, Nuclear Science and Technology Research Institute, Atomic Energy Organization of Iran, Karaj, Iran

S. Arab Ameri, F. Samadi, B. Dastar and S. Zarehdaran
Department of Animal Science, University of Agricultural Science and Natural Resources of Gorgan, Gorgan, Iran

N. Jabbari, A. Fattah and F. Shirmohammad
Department of Animal Science, Shahr-e-Qods Branch, Islamic Azad University, Tehran, Iran

A. A. Gheisari and A. Azarbayejani
Department of Animal Science, Isfahan Research Center for Agriculture and Natural Resources, Isfahan, Iran

G. Maghsoudinejad
Animal Science Research Institute, Karaj, Iran

E. Ebdi and A. Nobakht
Department of Animal Science, Maragheh Branch, Islamic Azad University, Maragheh, Iran

F. Asadi, F. Shariatmadari and M.A. Karimi-Torshizi
Department of Poultry Science, Faculty of Agriculture, Tarbiat Modares University, Tehran, Iran

M. Mohiti-Asli and M. Ghanaatparast-Rashti
Department of Animal Science, Faculty of Agricultural Science, University of Guilan, Rasht, Iran

R. Abdulkarimi and M.H. Shahir
Department of Animal Science, Faculty of Agriculture, University of Zanjan, Zanjan, Iran

K. Shirzadegan and H.R. Taheri
Department of Animal Science, Faculty of Agriculture, University of Zanjan, Zanjan, Iran

B. Darabighane, F. Mirzaei Aghjeh Gheshlagh and B. Navidshad
Department of Animal Science, Faculty of Agricultural Science, University of Mohaghegh Ardabili, Ardabil, Iran

A. Mahdavi
Department of Animal Husbandry, Faculty of Veterinary Medicine, Semnan University, Semnan, Iran

A. Zarei
Department of Animal Science, Karaj Branch, Islamic Azad University, Karaj, Iran

S. Nahashon
Department of Agricultural Science, Tennessee State University, Nashville, USA

K. G. S. C. Katukurunda
Department of Food Science and Technology, Faculty of Applied Science, University of Sri Jayewardenepura, Gangodawila, Nugegoda, Sri Lanka

N.S.B.M. Atapattu and P.W.A. Perera
Department of Animal Science, Faculty of Agriculture, University of Ruhuna, Mapalana, Kamburupitiya, Sri Lanka

S. M. Akbari, A. A. Sadeghi, M. Amin Afshar and M. Chamani
Department of Animal Science, Science and Research Branch, Islamic Azad University, Tehran, Iran

M. Faramarzzadeh
Department of Animal Science, Islamic Azad University, Shabestar Branch, Shabestar, Iran

M. Behroozlak
Department of Animal Science, Faculty of Agriculture, Payame Noor University, Tehran, Iran

F. Samadian
Department of Animal Science, College of Agriculture, Yasouj University, Yasouj, Iran

V. Vahedi
Department of Animal Science, Moghan College of Agriculture and Natural Resources, University of Mohaghegh Ardabili, Ardabil, Iran

A. Abedpour, S. M. A. Jalali and F. Kheiri
Department of Animal Science, Faculty of Agriculture and Natural Resources, Shahrekord Branch, Islamic Azad University, Shahrekord, Iran

A. Sylvia John-Jaja and A. R. Abdullah
Department of Animal Science, College of Agriculture, Babcock University, Ilshan Remo, Nigeria

C. Samuel Nwokolo
Department of Physics, Faculty of Science, University of Calabar, Calabar, Nigeria

E. Opoola, S. O. Ogundipe, G. S. Bawa and P. A. Onimisi
Department of Animal Science, Ahmadu Bello University, Zaria, Nigeria

J. Nasr
Department of Animal Science, Saveh Branch, Islamic Azad University, Saveh, Iran

F. Kheiri
Department of Animal Science, Shahrekord Branch, Islamic Azad University, Shahrekord, Iran

S.R. Hashemi and B. Dastar
Department of Animal Science, Gorgan University of Agricultural Science and Natural Resources, Gorgan, Iran

D. Davoodi
Department of Nanotechnology, Agricultural Biotechnology Research Institute of Iran (ABRII), Karaj, Iran

E. Rezvannejad and E. Nasirifar
Department of Biotechnology, Institute Science and High Technology and Environmental Science, Graduate University of Advanced Technology, Kerman, Iran

A.R. Akhavast and M. Daneshyar
Department of Animal Science, Faculty of Agriculture, Urmia University, Urmia, Iran

S. K. Ebrahimzadeh, B. Navidshad and F. Mirzaei Aghjehgheshlagh
Department of Animal Science, Faculty of Agricultural Science, University of Mohaghegh Ardabili, Ardabil, Iran

P. Farhoomand
Department of Animal Science, Faculty of Agriculture, Urmia University, Urmia, Iran

H. Jahanian Najafabadi, A. A. Saki, Z. Bahrami, A. Ahmadi, D. Alipour and M. Abdolmaleki
Department of Animal Science, Faculty of Agriculture, Bu Ali Sina University, Hamedan, Iran

A. O. Adeyina, K. M. Okukpe, A. S. Akanbi, M. D. Ajibade, T. T. Tiamiyu and O. A. Salami
Department of Animal Production, Faculty of Agriculture, University of Ilorin, Ilorin, Nigeria

A. A. Annongu, J. O. Atteh, M.A.Belewu, A. O. Adeyina, A. S. Akanbi, A. T. Yusuff and F. E. Sola - Ojo
Department of Animal Production, Division of Nutritional Biochemistry and Toxicology, P. M. B. 1515 University of Ilorin, Ilorin, Nigeria

J. K. Joseph
Department of Home Economics and Food Science, University of Ilorin, Ilorin, Nigeria

S. O. Ajide and V. O. Chimezie
Landmark University, Km 4 Ipetu, Omu Aran, PMB 1001, Omu Aran, Kwara State of Nigeria

J.H. Edoh
Department of Animal Production, Division of Nutritional Biochemistry and Toxicology, P.M.B. 1515 University of Ilorin, Ilorin, Nigeria
Laboratoire de Recherche Avicole et de Zoo - Economie (LaRAZE), Département de Production Animale, Faculté des Science Agronomiques, Université d' Abomey - Calavi, 01 BP 526 Cotonou, Benin

K. Kheiralipour and Z. Payandeh
Department of Mechanical Engineering Biosystems, Ilam University, Ilam, Iran

B. Khoshnevisan
Department of Agricultural Machinery, University of Tehran, Karaj, Iran

A. Razee, A. S. M. Mahbub, M. Y. Miah, M. R. Hasnath and M. K. Hasan
Department of Poultry Science, Sylhet Agricultural University, Sylhet, Bangladesh

M. N. Uddin
Department of Livestock Production and Management, Sylhet Agricultural University, Sylhet, Bangladesh
Department of Animal Science, Chonbuk National University, Jeonju 561 - 756, Republic of Korea

S. A. Belal
Department of Poultry Science, Sylhet Agricultural University, Sylhet, Bangladesh
Department of Animal Biotechnology, Chonbuk National University, Jeonju 561 - 756, Republic of Korea

M. A. Arjomandi and M. Salarmoini
Department of Animal Science, Faculty of Agriculture, Shahid Bahonar University of Kerman, Kerman, Iran

J. Salary
Department of Animal Science, Faculty of Agriculture,
Bu Ali Sina University, Hamedan, Iran

H. R. Hemati Matin
Department of Animal Science, Faculty of Agriculture,
Tarbiat Modares University, Tehran, Iran

H. Hajati
Department of Animal Science, Faculty of Agriculture,
Ferdowsi University of Mashhad, Mashhad, Iran

Index